AF357341

Photoactive Nanomaterials

Photoactive Nanomaterials

Editor

Nurxat Nuraje

MDPI • Basel • Beijing • Wuhan • Barcelona • Belgrade • Manchester • Tokyo • Cluj • Tianjin

Editor
Nurxat Nuraje
Department of Chemical and
Materials Engineering,
School of Engineering and Digital Science,
Nazarbayev University
Kazakhstan

Editorial Office
MDPI
St. Alban-Anlage 66
4052 Basel, Switzerland

This is a reprint of articles from the Special Issue published online in the open access journal *Nanomaterials* (ISSN 2079-4991) (available at: https://www.mdpi.com/journal/nanomaterials/special_issues/photoactive_nano).

For citation purposes, cite each article independently as indicated on the article page online and as indicated below:

LastName, A.A.; LastName, B.B.; LastName, C.C. Article Title. *Journal Name* **Year**, *Volume Number*, Page Range.

ISBN 978-3-0365-0520-6 (Hbk)
ISBN 978-3-0365-0521-3 (PDF)

Contents

About the Editor . . . . **vii**

Nurxat Nuraje
Photoactive Nanomaterials
Reprinted from: *Nanomaterials* **2021**, *11*, 77, doi:10.3390/nano11010077 . . . . **1**

Yong-sheng Fu, Jun Li and Jianguo Li
Metal/Semiconductor Nanocomposites for Photocatalysis: Fundamentals, Structures, Applications and Properties
Reprinted from: *Nanomaterials* **2019**, *9*, 359, doi:10.3390/nano9030359 . . . . **3**

Yerkin Shabdan, Aiymkul Markhabayeva, Nurlan Bakranov and Nurxat Nuraje
Photoactive Tungsten-Oxide Nanomaterials for Water-Splitting
Reprinted from: *Nanomaterials* **2020**, *10*, 1871, doi:10.3390/nano10091871 . . . . **29**

Baglan Bakbolat, Chingis Daulbayev, Fail Sultanov, Renat Beissenov, Arman Umirzakov, Almaz Mereke, Askhat Bekbaev and Igor Chuprakov
Recent Developments of TiO2-Based Photocatalysis in the Hydrogen Evolution and Photodegradation: A Review
Reprinted from: *Nanomaterials* **2020**, *10*, 1790, doi:10.3390/nano10091790 . . . . **67**

Fail Sultanov, Chingis Daulbayev, Seitkhan Azat, Kairat Kuterbekov, Kenzhebatyr Bekmyrza, Baglan Bakbolat, Magdalena Bigaj and Zulkhair Mansurov
Influence of Metal Oxide Particles on Bandgap of 1D Photocatalysts Based on SrTiO3/PAN Fibers
Reprinted from: *Nanomaterials* **2020**, *10*, 1734, doi:10.3390/nano10091734 . . . . **83**

Gauhar Mussabek, Sergei A. Alekseev, Anton I. Manilov, Sergii Tutashkonko, Tetyana Nychyporuk, Yerkin Shabdan, Gulshat Amirkhanova, Sergei V. Litvinenko, Valeriy A. Skryshevsky and Vladimir Lysenko
Kinetics of Hydrogen Generation from Oxidation of Hydrogenated Silicon Nanocrystals in Aqueous Solutions
Reprinted from: *Nanomaterials* **2020**, *10*, 1413, doi:10.3390/nano10071413 . . . . **93**

Askar A. Maxim, Shynggys N. Sadyk, Damir Aidarkhanov, Charles Surya, Annie Ng, Yoon-Hwae Hwang, Timur Sh. Atabaev and Askhat N. Jumabekov
PMMA Thin Film with Embedded Carbon Quantum Dots for Post-Fabrication Improvement of Light Harvesting in Perovskite Solar Cells
Reprinted from: *Nanomaterials* **2020**, *10*, 291, doi:10.3390/nano10020291 . . . . **107**

Rebeca Sola-Llano, Ainhoa Oliden-Sánchez, Almudena Alfayate, Luis Gómez-Hortigüela, Joaquín Pérez-Pariente, Teresa Arbeloa, Johan Hofkens, Eduard Fron and Virginia Martínez-Martínez
White Light Emission by Simultaneous One Pot Encapsulation of Dyes into One-Dimensional Channelled Aluminophosphate
Reprinted from: *Nanomaterials* **2020**, *10*, 1173, doi:10.3390/nano10061173 . . . . **115**

Qingrui Zeng, Suyue Guo, Yuanbo Sun, Zhuojuan Li and Wei Feng
Protonation-Induced Enhanced Optical-Light Photochromic Properties of an Inorganic-Organic Phosphomolybdic Acid/Polyaniline Hybrid Thin Film
Reprinted from: *Nanomaterials* **2020**, *10*, 1839, doi:10.3390/nano10091839 . . . . **131**

Shifeng Wang, Yong Li, Annie Ng, Qing Hu, Qianyu Zhou, Xin Li and Hao Liu
2D Bi_2Se_3 van der Waals Epitaxy on Mica for Optoelectronics Applications
Reprinted from: *Nanomaterials* **2020**, *10*, 1653, doi:10.3390/nano10091871 . . . . **149**

About the Editor

Nurxat Nuraje is an Associate Professor in the Department of Chemical and Materials Engineering at Nazarbayev University (NU). He is also the Head of the Advanced Functional Materials and Devices cluster. Before joining NU, he held an Assistant Professor position at Texas Tech University, USA, and was a research scientist at the Department of Materials Science and Engineering at Massachusetts Institute of Technology (MIT), USA. Dr. Nuraje has published more than 70 journal articles and an RSC book, *Green Photoactive Nanomaterials*, as a primary editor. In 2008, the Materials Research Society (USA) awarded him the "Graduate Student Silver Medal" for his excellence in the Ph.D. thesis. For his outstanding research in nanoscience, he was selected as the winner of the "Joseph Wang Award" by Cognizure in 2015 and named as one of the 2018 RSC emerging investigators by *Journal of Materials Chemistry A*. In 2017, for his excellent performance in teaching, Dr. Nuraje was recognized as the most influential faculty member by the College of Engineering of Texas Tech University. He was awarded the US Air Force Summer Faculty Fellowship for the years 2017, 2018, and 2019. He has served as a chair/co-chair for four technical sessions in the AICHE meeting (USA) and as a grant reviewer for US NSF, US DOE, and ACS PRF.

 *nanomaterials*

Editorial

Photoactive Nanomaterials

Nurxat Nuraje

Department of Chemical and Materials Engineering, School of Engineering and Digital Science,
Nazarbayev University, 53 Kabanbay Batyr Avenue, Nur-Sultan 010000, Kazakhstan; nurxat.nuraje@nu.edu.kz

Received: 25 December 2020; Accepted: 29 December 2020; Published: 1 January 2021

With the depletion of carbon-based energy resources and the consideration of global warming, renewable energy is considered a promising energy source for future energy. Solar energy has shown the greatest potential to meet global energy needs with its ample source and free availability. Although primitive photovoltaic work gets credited to the silicon-based works done in the bell lab, recently, progress in the photovoltaic cell is significant. For example, many different types of solar cells have appeared, which include dye-sensitized solar cells, quantum-based solar cells, polymer solar cells, perovskite solar cells, etc. Here, we can point out the perovskite solar cell which reaches 25% efficiency within a short period. According to the literature search done on the Web of science, the number of publications on perovskite solar cells has increased exponentially within the last 10 years.

In addition to photovoltaic application, many research areas in utilization of solar energy have been extensively exploited to form distinct disciplines. One of them, photocatalytic water splitting, has molded a promising research pathway with exponentially increasing publications. Other fields include fuel generation from carbon dioxide, nitrogen conversion, photodegradation of organic and inorganic pollutants, and optoelectronic devices. Besides the fields mentioned, photoactive nanomaterials have been used in other fields. Experimental and theoretical investigations of solar energy conversion using various simulation software, from material design to reaction kinetics, lay the groundwork to improve the efficiency of the above applications. Furthermore, the fundamental study of photoactive nanomaterials in solar energy conversion, which is done through designing advanced functional materials and advanced experimental techniques, focuses on light harnessing, charge separation/recombination, and catalytic reaction kinetics.

Therefore, it is very important to deliver the recent research progress in the field of photoactive nanomaterials to the scientific community. Thus, the scope of this Special Issue not only covers a basic study of the solar energy conversion process with simulations but also targets certain applications including environmental remediation, optoelectronic devices, and solar fuel generation. This Special Issue includes six research articles and three review articles to focus on photoactive nanomaterials from fundamental research on their applications. Three research works [1–3] reported recent emerging materials, including 2D Bi_2Se_3 in optoelectronic applications, and discussed fundamental behaviors of dyes in confined media. Another research work reported by Maxim et al. [4] explores the enhancement of light harnessing in perovskite solar cells with embedded carbon quantum dots. Most of the articles are focused on photocatalytic solar fuel generation. Hydrogen generation kinetics is studied via silicon nanocrystals in aqueous media [5]. Nanocomposite-based photocatalysts including TiO_2, $SrTiO_3$, and Polyaniline (PANI) are also reviewed to give quick updates regarding photocatalytic water splitting [6–8]. Shabdan et al. updated with recent progress in a promising oxygen evolution of WO_3-based photo-catalysts [9].

Therefore, I think it is essential to present a Special Issue which provide recent research works on photoactive nanomaterials, from fundamental studies to recent applications including optoelectronic devices, perovskite solar cells, and photocatalytic water splitting.

Funding: This research was funded by Nazarbayev University FGRG grant number SEDS2020 016.

Institutional Review Board Statement: Not applicable.

Informed Consent Statement: Not applicable.

Conflicts of Interest: The authors declare no conflict of interest.

References

1. Sola-Llano, R.; Oliden-Sánchez, A.; Alfayate, A.; Gómez-Hortigüela, L.; Pérez-Pariente, J.; Arbeloa, T.; Hofkens, J.; Fron, E.; Martínez-Martínez, V. White Light Emission by Simultaneous One Pot Encapsulation of Dyes into One-Dimensional Channelled Aluminophosphate. *Nanomaterials* **2020**, *10*, 1173. [CrossRef] [PubMed]

2. Wang, S.; Li, Y.; Ng, A.; Hu, Q.; Zhou, Q.; Li, X.; Liu, H. 2D Bi_2Se_3 van der Waals Epitaxy on Mica for Optoelectronics Applications. *Nanomaterials* **2020**, *10*, 1653. [CrossRef] [PubMed]

3. Zeng, Q.; Guo, S.; Sun, Y.; Li, Z.; Feng, W. Protonation-Induced Enhanced Optical-Light Photochromic Properties of an Inorganic-Organic Phosphomolybdic Acid/Polyaniline Hybrid Thin Film. *Nanomaterials* **2020**, *10*, 1839. [CrossRef] [PubMed]

4. Maxim, A.A.; Sadyk, S.N.; Aidarkhanov, D.; Surya, C.; Ng, A.; Hwang, Y.-H.; Atabaev, T.S.; Jumabekov, A.N. PMMA Thin Film with Embedded Carbon Quantum Dots for Post-Fabrication Improvement of Light Harvesting in Perovskite Solar Cells. *Nanomaterials* **2020**, *10*, 291. [CrossRef] [PubMed]

5. Mussabek, G.; Alekseev, S.A.; Manilov, A.I.; Tutashkonko, S.; Nychyporuk, T.; Shabdan, Y.; Amirkhanova, G.; Litvinenko, S.V.; Skryshevsky, V.A.; Lysenko, V. Kinetics of Hydrogen Generation from Oxidation of Hydrogenated Silicon Nanocrystals in Aqueous Solutions. *Nanomaterials* **2020**, *10*, 1413. [CrossRef] [PubMed]

6. Fu, Y.-S.; Li, J.; Li, J. Metal/Semiconductor Nanocomposites for Photocatalysis: Fundamentals, Structures, Applications and Properties. *Nanomaterials* **2019**, *9*, 359. [CrossRef] [PubMed]

7. Bakbolat, B.; Daulbayev, C.; Sultanov, F.; Beissenov, R.; Umirzakov, A.; Mereke, A.; Bekbaev, A.; Chuprakov, I. Recent Developments of TiO_2-Based Photocatalysis in the Hydrogen Evolution and Photodegradation: A Review. *Nanomaterials* **2020**, *10*, 1790. [CrossRef] [PubMed]

8. Sultanov, F.; Daulbayev, C.; Azat, S.; Kuterbekov, K.; Bekmyrza, K.; Bakbolat, B.; Bigaj, M.; Mansurov, Z. Influence of Metal Oxide Particles on Bandgap of 1D Photocatalysts Based on $SrTiO_3$/PAN Fibers. *Nanomaterials* **2020**, *10*, 1734. [CrossRef] [PubMed]

9. Shabdan, Y.; Markhabayeva, A.; Bakranov, N.; Nuraje, N. Photoactive Tungsten-Oxide Nanomaterials for Water-Splitting. *Nanomaterials* **2020**, *10*, 1871. [CrossRef] [PubMed]

Publisher's Note: MDPI stays neutral with regard to jurisdictional claims in published maps and institutional affiliations.

 *nanomaterials*

Review

Metal/Semiconductor Nanocomposites for Photocatalysis: Fundamentals, Structures, Applications and Properties

Yong-sheng Fu, Jun Li * and Jianguo Li

School of Materials Science and Engineering, Shanghai Jiao Tong University, Shanghai 200240, China;
fuys648@163.com (Y.-s.F.); lijg@sjtu.edu.cn (J.L.)
* Correspondence: li.jun@sjtu.edu.cn; Tel.: +86-156-0189-7272

Received: 1 February 2019; Accepted: 19 February 2019; Published: 4 March 2019

Abstract: Due to the capability of utilizing light energy to drive chemical reactions, photocatalysis has been widely accepted as a green technology to help us address the increasingly severe environment and energy issues facing human society. To date, a large amount of research has been devoted to enhancing the properties of photocatalysts. As reported, coupling semiconductors with metals is one of the most effective methods to achieve high-performance photocatalysts. The excellent properties of metal/semiconductor (M/S) nanocomposite photocatalysts originate in two aspects: (a) improved charge separation at the metal-semiconductor interface; and (b) increased absorption of visible light due to the surface plasmon resonance of metals. So far, many M/S nanocomposite photocatalysts with different structures have been developed for the application in environmental remediation, selective organic transformation, hydrogen evolution, and disinfection. Herein, we will give a review on the M/S nanocomposite photocatalysts, regarding their fundamentals, structures (as well as their typical synthetic approaches), applications and properties. Finally, we will also present our perspective on the future development of M/S nanocomposite photocatalysts.

Keywords: nanocomposite photocatalyst; environmental remediation; selective organic transformation; hydrogen evolution; disinfection

---

1. Introduction

Since the discovery of water splitting on a TiO_2 electrode under irradiation of ultraviolet (UV) light by Fujishima and Honda in 1972, many semiconductors, such as CdS, ZnO, $SrTiO_3$ and g-C_3N_4, have been demonstrated to exhibit photocatalytic properties and, through the efforts of researchers, the application areas of semiconductor photocatalysts have been greatly extended [1–14]. Unfortunately, due to fast charge separation and limited light absorption, the properties of semiconductor photocatalysts are relatively unsatisfactory for practical application [15–17]. To improve the properties of semiconductor photocatalysts, several strategies have been proposed by researchers, such as doping, dye-sensitization, coupling, etc. [18–27]. However, each of these strategies has its own pros and cons. Toward the strategy of doping, the band structures of semiconductor photocatalysts could be modulated by the doping atoms to exhibit better properties for light absorption, but the doped semiconductor photocatalysts could be more susceptible to photo-corrosion and the charge recombination of the doped semiconductor photocatalysts could be aggravated at the defects introduced by the doping atoms [19,21,27]. As for the strategy of dye-sensitization, although the light-sensitive dyes can absorb light more efficiently and transfer the photoexcited electrons to the semiconductor photocatalysts, the light-sensitive dyes are susceptible to chemical corrosion, resulting in the poor stability of dye-sensitized semiconductor photocatalysts [24–26]. Coupling, as a strategy with a relatively short history, has aroused great interest among researchers since its very beginning.

When coupled with metals, especially the noble metals, the properties of semiconductor photocatalysts can be considerably improved due to the enhanced charge separation at the metal–semiconductor interface and the enhanced visible light absorption caused by the surface plasmon resonance (SPR) of metals [28–30]. Except for the high cost of noble metals, the main drawback of the coupling strategy used to be the poor control of the process of coupling semiconductors with metals. However, with the development of nanotechnology, the coupling process now could be delicately controlled and several new structures have been synthesized to improve the properties of metal/semiconductor (M/S) nanocomposite photocatalysts [29,31].

Herein, we would like to review the M/S nanocomposite photocatalysts regarding their fundamentals, structures (as well as their typical synthetic approaches), applications and properties. Finally, we will also present our perspective on the future development of M/S nanocomposite photocatalysts.

2. Fundamentals

2.1. Principles of Photocatalysis

As defined by the International Union of Pure and Applied Chemistry (IUPAC), photocatalysis is "A catalytic reaction involving light absorption by substrate" [32]. When semiconductor photocatalysts are irradiated by light with the photon energy larger than their band gap (BG) energy, the electrons in the valence band (VB) will be excited into the conduction band (CB), leaving positive holes in the VB. Because photocatalytic reactions are reactions happening at the surface of photocatalysts, the photo-induced free charge carriers need first to diffuse into the active sites on the surface of photocatalysts before they can induce photocatalytic reactions [33–35]. The photocatalytic process is schematically presented in Figure 1. However, for a particular substrate, whether it can undergo chemical reactions on the semiconductor photocatalysts depends on the relative positions between its redox potentials and the band edges of semiconductor photocatalysts [36–38]. There are four possibilities as follows:

(1) If the redox potential of the substrate is lower than the CB edge of the semiconductor photocatalyst, then the substrate can undergo reductive reactions.
(2) If the redox potential of the substrate is higher than the VB edge of the semiconductor photocatalyst, then the substrate can undergo oxidative reactions.
(3) If the redox potential of the substrate is higher than the CB edge or lower than the VB of the semiconductor photocatalyst, then the substrate can undergo neither reductive nor oxidative reactions.
(4) If the redox potential of the substrate is lower than the CB edge and higher than the VB of the semiconductor photocatalyst, then the substrate can undergo either reductive or oxidative reactions.

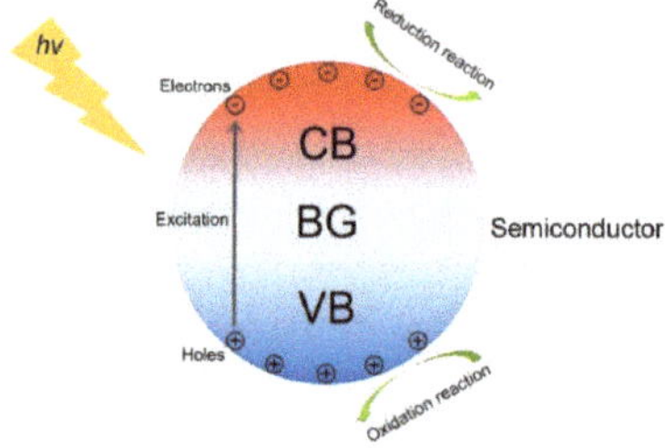

Figure 1. Schematic diagram for the photocatalytic process.

2.2. Mechanisms for the Enhanced Properties of Metal/Semiconductor (M/S) Nanocomposite Photocatalysts

As mentioned above, by coupling with different metals, the mechanisms for the enhanced properties of M/S nanocomposite photocatalysts can be divided into two categories, i.e., the enhanced charge separation at the metal-semiconductor interface and the enhanced visible light absorption due to the SPR of metals.

2.2.1. Enhanced Charge Separation

The enhanced charge separation of M/S nanocomposite photocatalysts originates from the electron transfer across the metal–semiconductor interface [31,39–41]. When a semiconductor is coupled with a metal, which possesses a higher work function than that of the semiconductor (it is the most common case for M/S nanocomposite photocatalysts), then the electrons will flow from the semiconductor to the metal until their Fermi levels (i.e., $E_{F,m}$ and $E_{F,s}$) are aligned, leading to the upward bending of the band edges in the semiconductor as revealed in Figure 2. Accompanied with the band bending, a Schottky barrier (i.e., ϕ_{SB} in Figure 2) forms at the metal–semiconductor interface [39]. As has been shown, the efficiency of electron transfer across the metal–semiconductor interface increases with the height of the Schottky barrier [42]. In addition, the metals in M/S nanocomposite photocatalysts not only act as the electron reservoir to enhance the charge separation of the semiconductors, but also provide active sites for reductive reactions, hence dramatically improving the properties of M/S nanocomposite photocatalysts [43].

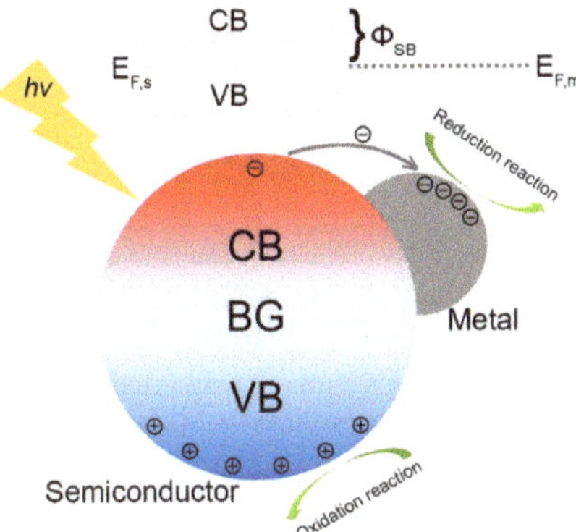

Figure 2. Enhanced charge separation at the metal-semiconductor interface.

2.2.2. Enhanced Visible Light Absorption

When coupled with metals such as Ag and Au, the SPR of metals could endow the M/S nanocomposite photocatalysts with enhanced absorption toward visible light [44,45]. However, the mechanism governing the SPR-enhanced visible light absorption of M/S nanocomposite photocatalysts is still unclear [28,29,46,47]. So far, there exist three non-mutually exclusive mechanisms for the explanation of the SPR-enhanced properties of M/S nanocomposite photocatalysts under visible light irradiation, i.e., (a) SPR-induced electron injection from metals to semiconductors; (b) charge separation induced by near-field electric field (NFEF); (c) scattering-enhanced light absorption.

Surface Plasmon Resonance (SPR)-Induced Electron Injection

Due to the SPR of metals such as Au and Ag, the energy of excited electrons in the metals can be excited to the range between 1.0 and 4.0 eV above their Fermi levels. Then, the energetic electrons will overcome the Schottky barrier at the metal–semiconductor interface and transfer to the CB of semiconductors, leaving energetic holes in the metals, as schematically shown in Figure 3a. Thus, the hot electrons in the semiconductors will drive reduction reactions and the hot holes in the metals will drive oxidation reactions.

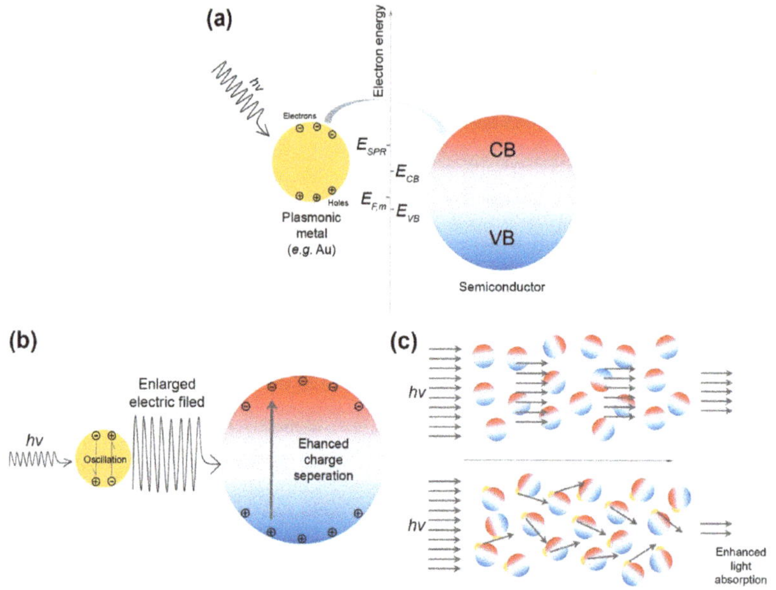

Figure 3. (a) SPR-induced electron injection; (b) Charge separation induced by NFEF; (c) Scattering-enhanced light absorption.

Charge Separation Induced by Near-Field Electric Field (NFEF)

Under visible light irradiation, the SPR-excited hot electrons in the metal particles of M/S nanocomposite photocatalysts will generate intense electric field in their proximity (i.e., NFEF), which could be up to 100–10000 times larger than the electric field of incident light. Since the formation rate of electron-hole pairs in the semiconductors of M/S nanocomposite photocatalysts is proportional to the intensity of the NFEF (more specifically, $|E|^2$), the generation of electron-hole pairs in M/S nanocomposite photocatalysts can be significantly enhanced. The process of NFEF-induced charge separation is schematically shown in Figure 3b.

Scattering-Enhanced Light Absorption

If the size of the metal particles in M/S nanocomposite photocatalysts is larger than 50 nm, the SPR-excited metal particles can efficiently scatter the incident light, leading to the increase of the path length of light through the M/S nanocomposite photocatalysts. Thus, the light absorption of M/S nanocomposite photocatalysts will be improved due to the increased path length of light, resulting in the enhancement of the properties of M/S nanocomposite photocatalysts. Figure 3c schematically reveals the scattering-enhanced photon absorption mechanism.

3. Structures of M/S Nanocomposite Photocatalysts

Based on the morphologies and synthetic approaches, the structures of M/S nanocomposite photocatalysts could be divided into six categories, i.e., the conventional structure, core-shell structure, yolk-shell structure, Janus structure, array structure and multi-junction structure, which are schematically shown in Figure 4.

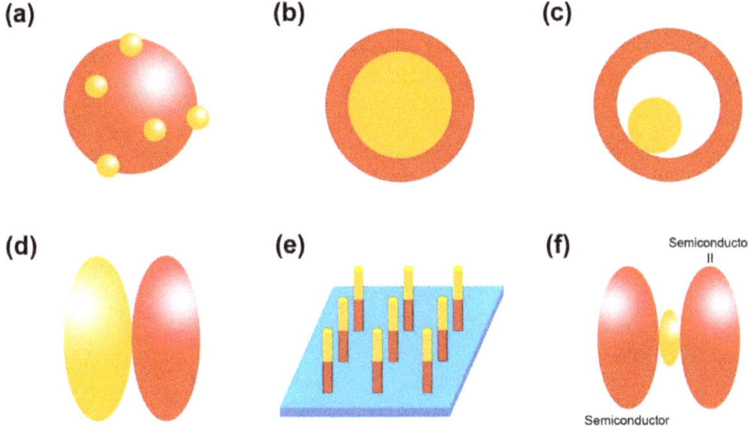

Figure 4. (**a**) Conventional structure; (**b**) Core-shell structure; (**c**) Yolk-shell structure; (**d**) Janus structure; (**e**) Array structure; (**f**) Multi-junction structure.

3.1. Conventional Structure

The conventional structure of M/S nanocomposite photocatalysts refers to the structure, in which the combination of semiconductor nanoparticles and metal nanoparticles is not delicately controlled. So far, many approaches have been developed to synthesize M/S nanocomposite photocatalysts with conventional structure, which include photoreduction, impregnation, deposition-precipitation, chemical vapor deposition (CVD) etc. [48–51].

3.1.1. Photoreduction

Since some chemical compounds containing metal element can be reduced to metal under irradiation, photoreduction is regarded as a facile approach to decorate semiconductor nanostructures with noble metal. Typically, in the work of Yu et al., hydrothermally-synthesized TiO_2 nanosheets were added to an aqueous solution of H_2PtCl_6. Under stirring and the irradiation of ultraviolet (UV) light, H_2PtCl_6 was reduced to Pt and deposited on TiO_2 nanosheets, leading to the formation of Pt-decorated TiO_2 nanosheets [48].

3.1.2. Impregnation

During the impregnation process, the combination of noble metal and semiconductor nanostructures is achieved through the attachment of noble metal precursor onto semiconductor substrate due to the van der Waals force, coulomb force or some other interactions between them. Typically, in the work of Lin et al., H_2PtCl_6 was added into a suspension of TiO_2 (B) nanofibers. Then the solvent was evaporated under stirring. Then, the as-obtained powders were heated in a reducing atmosphere to reduce H_2PtCl_6 absorbed on TiO_2 (B) nanofibers, leading to the formation of Pt-decorated TiO_2 (B) nanofibers [49].

3.1.3. Deposition-Precipitation

Deposition-precipitation is also a facile approach to synthesize M/S nanocomposites. Typically, in the work of Wu et al., TiO_2 nanoparticles were dispersed in an aqueous solution of chloroauric acid. Then NaOH solution was added into the suspension to adjust the pH value to the desired level. After stirring, the precipitates were filtered, washed and dried. Then, the as-obtained products were calcined to form the Au-decorated TiO_2 nanoparticles [50].

3.1.4. Chemical Vapor Deposition (CVD)

As is well known, CVD is a powerful approach in synthesizing nanostructure and is often utilized to prepare metal-decorated nanostructures. Typically, in the work of Shi et al., $HAuCl_4$ was vaporized at a high temperature in a tube furnace. Then, the vapor was blown at the TiO_2 nanorod arrays at a relatively low temperature with the carrier gas of nitrogen. Thus, the gold would deposit on the TiO_2 nanorods, leading to the formation of Au-decorated TiO_2 nanorod arrays [51].

3.2. Core-Shell Structure

The core-shell structure was first developed to improve the quantum yield of quantum dots in the 1990s [52]. After that, the core-shell structure attracted great interest from researchers and its application areas were greatly extended. In photocatalysis, M/S nanocomposite photocatalysts with core-shell structure occupy an important position due to their outstanding photocatalytic properties [53]. Besides, the enhanced charge transfer between metal and semiconductor, the core-shell structure can also hinder the aggregation of particles and protect the metal core from undesired corrosion or dissolution during the photocatalytic process.

The synthesis of core-shell M/S nanoparticles usually involves the coating of a semiconductor layer on metal nanoparticles. For example, in the work of Sudeep, photocatalytic $Ag@TiO_2$ core-shell nanoparticles were synthesized through controlled hydrolysis of titanium-(triethanolaminato) isopropoxide (TTEAIP) on the surface of Ag nanoparticles. Briefly, TTEAIP and silver nitrate were added into 2-propanol followed by stirring. Then dimethyl formamide (DMF) was added to the solution. Next, the solution was heated and refluxed. During this process, silver nitrate was first reduced to Ag nanoparticles by DMF. Then, due to the interaction between Ag nanoparticles and triethanolamine ligands, TTEAIP hydrolyzed on the surface of Ag nanoparticles, leading to the formation of $Ag@TiO_2$ core-shell nanoparticles [54].

3.3. Yolk-Shell Structure

The yolk-shell structure originates from the pioneer work of Xia's group as a variation of the core-shell structure [55]. Under the efforts of researchers, the yolk-shell structure has shown promising application in nanoreactors, drug delivery and lithium ion batteries [56]. Recently, some researchers have also investigated the application of yolk-shell structures in photocatalysis [57–59].

The synthesis of yolk-shell structures usually needs a sacrificial template. In the work of Li et al., gold nanoparticles were first coated with a layer of SiO_2 by the hydrolysis of tetraethyl orthosilicate (TEOS). Next, the SiO_2-coated gold nanoparticles were coated with a layer of TiO_2 by the hydrolysis of tetrabutyl titanate (TBOT). Then, the as-prepared gold nanoparticles were coated with a layer of SiO_2 again to protect the TiO_2 layer during the calcination process. After calcination, the two SiO_2 layers were removed by the etch of NaOH solution. Thus, the $Au@TiO_2$ yolk-shell nanoparticles were prepared. The SiO_2 layers served as the sacrificial template during the synthetic process [57].

3.4. Janus Structure

Janus particles, as first described by de Gennes in 1991, refer to the particles with anisotropic structure, which are composed of two distinct parts [60]. The dual nature of Janus particles endows themselves with fascinating properties, such as unique surface properties, controlled self-assembly behavior and response to multiple stimuli, etc. [61,62]. When applied in photocatalysis, Janus nanoparticles composed of metal and semiconductor can also exhibit remarkable photocatalytic properties.

Different from conventional structures, the synthesis of Janus structures needs delicate control over the combination of two distinct parts. As demonstrated by the work of Seh et al., Au/TiO_2 Janus nanoparticles were synthesized through controlled hydrolysis of titanium diisopropoxide bis(acetylacetonate) (TAA) on the surface of gold nanoparticles. The reason for choosing TAA was that the hydrolysis rate of TAA was rather slow, which had a significant influence on the structure of the products. Thus, during the slow hydrolysis of TAA in the alkaline suspension of gold nanoparticles in isopropanol, TiO_2 combined with gold nanoparticles to form the Au/TiO_2 Janus nanoparticles [63].

3.5. Array Structure

Due to the promising application in nanodevices, the array structure has been widely researched during the past few decades [64–66]. In photocatalysis, the array structure has also been utilized to improve the photocatalytic properties of M/S nanocomposites [67].

The synthesis of array structures is mainly achieved through a template-assisted deposition process. In the work of Wang et al., an anodic aluminum oxide (AAO) template with one side deposited with a layer of gold and connected to a piece of aluminum foil was utilized as the working electrode in an electrochemical cell. For the deposition of CdS, the working electrode was biased to -2.5 V vs. standard calomel electrode (SCE) in the electrolyte of sulfur and cadmium chloride dissolved in dimethyl sulfoxide (DMSO). For the deposition of Au, the working electrode was biased to -0.95 V vs. SCE in commercial Au Orotemp 24. After alternate deposition of CdS and Au, the AAO template was removed through chemical etching. Thus, photocatalytic Au/CdS nanorod arrays were obtained [67].

3.6. Multi-Junction Structure

M/S nanocomposite photocatalysts with multi-junction structures are formed by sandwiching metal nanoparticles between two semiconductors. Thus, the electron transfer process in multi-junction M/S nanocomposite photocatalysts is very similar to that of Z-scheme photocatalytic systems in nature, which could endow them excellent photocatalytic properties [68–70].

Take the pioneer work of Tada et al. for example. The $CdS/Au/TiO_2$ multi-junction nanoparticles as the all-solid-state Z-scheme photocatalytic system were synthesized through a two-step deposition process. In the first step, gold was deposited on the TiO_2 nanoparticles through an impregnation process with chloroauric acid as the precursor. In the second step, cadmium sulfide was deposited on the Au/TiO_2 nanoparticles through the UV-induced reaction of sulfur and cadmium perchlorate. Thus, $CdS/Au/TiO_2$ multi-junction nanoparticles were prepared [71].

4. Applications and Properties of M/S Nanocomposite Photocatalysts

To date, the application areas of photocatalysis are mainly focused on environmental remediation, selective organic transformation, hydrogen evolution and disinfection.

4.1. Environmental Remediation

The photo-excited electrons and holes in semiconductor photocatalysts as well as the associated radicals possess high chemical activity, which can degrade pollutants to low- or non-hazardous substances [72–75]. As reported, most of the researches in photocatalytic environmental remediation were focused on degradation of organic pollutants such as Rhodamine B (RhB), Methylene blue (MB), methyl orange (MO) and so on. To the best of our knowledge, the only inorganic pollutant that has been involved in the research of photocatalytic environmental remediation is nitric oxide (NO). Table 1 lists some of the most representative M/S nanocomposite photocatalysts applied in environmental remediation [51,59,71,76–100].

Table 1. Metal/semiconductor (M/S) nanocomposite photocatalysts for environmental remediation.

Photocatalyst	Structure	Pollutant	Light source	Reference
Au/TiO$_2$	Conventional structure	RhB	Visible light	[51]
Au/ZnO	Conventional structure	RhB, phenol red, procion red	Ultraviolet (UV) light	[76]
Au/ZnO	Conventional structure	RhB	UV light	[77]
Ag/ZnO	Conventional structure	MB	UV light	[78]
Ag/ZnO	Conventional structure	RhB	UV light	[79]
Ag/ZnO	Conventional structure	RhB	Simulated sunlight	[80]
Au/TiO$_2$	Conventional structure	MB, MO, *p*-nitrophenol	Visible light	[81]
Ag/ZnO	Conventional structure	RhB	Visible light	[82]
Au/TiO$_2$, Pt/TiO$_2$, Ir/TiO$_2$	Conventional structure	Azo dye	UV light	[83]
Ag/TiO$_2$	Conventional structure	RhB, ciprofloxacin	Visible light	[84]
Cu/Cu$_2$O	Conventional structure	RhB, MO, MB	Visible light	[85]
Bi/amorphous bismuth oxide (A-BO)	Conventional structure	NO	Visible light	[86]
Bi/(BiO)$_2$CO$_3$	Conventional structure	NO	Visible light	[87]
Bi/g-C$_3$N$_4$	Conventional structure	NO	Visible light	[88]
Bi/ZnWO$_4$	Conventional structure	NO	Visible light	[89]
Bi/BiOCl	Conventional structure	NO	Visible light	[90]
M@TiO$_2$ (M = Au, Pd, Pt)	Core–shell structure	RhB	UV light, visible light	[91]
Au@Cu$_2$O	Core–shell structure	MO	Visible light	[92]
Au@Cu$_2$O	Core–shell structure	MO	Visible light	[93]
Au@CdS	Core–shell structure	RhB	Visible light	[94]
Au@TiO$_2$	Core–shell structure	Acetaldehyde	UV light, visible light	[95]
Bi@Bi$_2$O$_3$	Core–shell structure	NO	Visible light	[96]
Ag@Cu$_2$O	Core–shell structure	MO	Visible light	[97]
Au@TiO$_2$	Yolk–shell structure	RhB	Visible light, simulated sunlight	[59]
Cu@CuS	Yolk–shell structure	MB	Simulated sunlight	[98]
Au/ZnO	Janus structure	RhB	UV light	[99]
Au/ZnO	Janus structure	MO	UV light	[100]
CdS/Au/TiO$_2$	Multi-junction structure	Methyl viologen (MV)	UV light	[71]

Due to the facile synthesis, more than half of the M/S nanocomposite photocatalysts listed in Table 1 possess the conventional structure. In the work of Lin et al., through coupling ZnO nanofibers with Ag, the as-synthesized Ag/ZnO photocatalysts exhibited much better properties for photocatalytic degradation of RhB than pristine ZnO photocatalysts, which was revealed in Figure 5a [79]. As revealed in Figure 5b, the properties of Ag/ZnO photocatalysts did not increase linearly with the Ag loading content and reached a peak at the Ag loading content of 7.5 at%. In addition, the repeatability test results in Figure 5c manifested the attenuation of the properties of Ag/ZnO photocatalysts was negligible after three cycles. The excellent properties of Ag/ZnO photocatalysts originated from the enhanced charge separation at the Ag-ZnO interface, which was evidenced by the photoluminescence (PL) spectra in Figure 5d. Toward removal of NO, Li et al. synthesized Bi/A-BO photocatalysts which could remove NO effectively under visible light illumination [86]. The remarkable properties of Bi/A-BO photocatalysts originated from SPR of Bi which could absorb visible light effectively and enhance the charge separation in A-BO. Li et al. also utilized electron spin resonance (ESR) spectroscopy for in-situ investigation of the reactive species during the photocatalytic process. (For detailed information about utilizing ESR spectroscopy to detect radicals, these two reviews, [101,102], are recommended.) The results revealed superoxide radicals (O$_2^{-\bullet}$) radicals were the major active species for photocatalytic NO oxidation.

For the M/S nanocomposite photocatalysts possessing the core-shell structure, their properties in photocatalytic environmental remediation were in strong relationship with the metal core and shell thickness. In the work of Zhang et al., the effect of Au, Pd and Pt cores on the photocatalytic activity of M@TiO$_2$ (M = Au, Pd and Pt) core-shell nanoparticles for degradation of RhB was investigated [91]. As revealed in Figure 6, core-shell M@TiO$_2$ (M = Au, Pd and Pt) exhibited different photocatalytic properties under the irradiation of UV light and visible light. Under irradiation of UV light, the photocatalytic properties of core-shell M@TiO$_2$ (M = Au, Pd and Pt) followed the order P25 TiO$_2$ > Pt@TiO$_2$ > Pd@TiO$_2$ > Au@TiO$_2$, while under irradiation of visible light the photocatalytic properties of core-shell M@TiO$_2$ (M = Au, Pd and Pt) followed the order Pd@TiO$_2$ > Pt@TiO$_2$ > Au@TiO$_2$ > P25 TiO$_2$. The results indicated the metal core had two impacts on the photocatalytic properties of core-shell M@TiO$_2$ (M = Au, Pd and Pt), i.e., improving charge separation at the metal-TiO$_2$ interface and modulating light absorption of TiO_2. In addition, Zhang et al. also demonstrated that the hydroxyl radicals ($^\bullet OH$) generated on core-shell $M@TiO_2$ played a more important role than superoxide radicals and holes in degradation of RhB under irradiation of UV light, while under the irradiation of visible light holes generated on core-shell $M@TiO_2$ played a predominant role in degradation of RhB. The effect of shell thickness on core-shell M/S nanocomposite photocatalysts was revealed by Kong et al. [93]. In their research, the properties of core-shell $Au@Cu_2O$ photocatalysts for degradation of MO varied with the Cu_2O shell thickness in the order 127 nm > 96 nm > 197 nm > 250 nm > 158 nm. Kong et al. thought the nonlinear relationship between the properties of core-shell $Au@Cu_2O$ photocatalysts and the shell thickness could be attributed to several factors, i.e., surface area, light absorption and scattering, charge-carrier dynamics and core-shell interactions, which competed with each other.

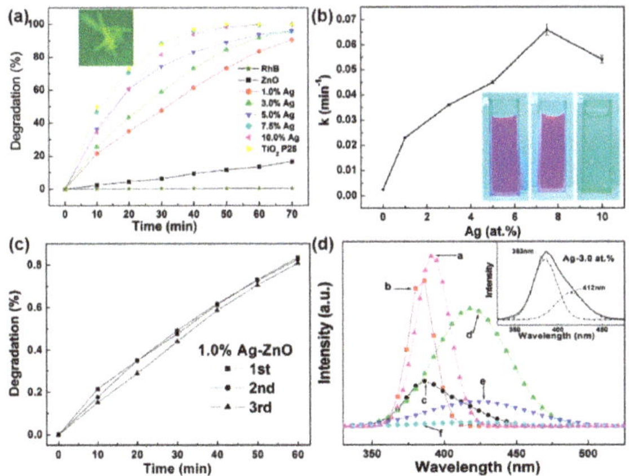

Figure 5. (**a**) Kinetics of the photodegradation of RhB by Ag/ZnO nanoparticles; (**b**) The degradation rate constant versus the Ag loading content; (**c**) Photodegradation of RhB by Ag/ZnO nanoparticles for three cycles; (**d**) PL spectra of the Ag/ZnO nanofibers with various Ag loading contents ((a) pure ZnO, (b) 1 atom% Ag, (c) 3 atom% Ag, (d) 5 atom% Ag, (e) 7.5 atom% Ag, (f) 10 atom% Ag). Reproduced from [79], with copyright permission from American Chemical Society, 2009.

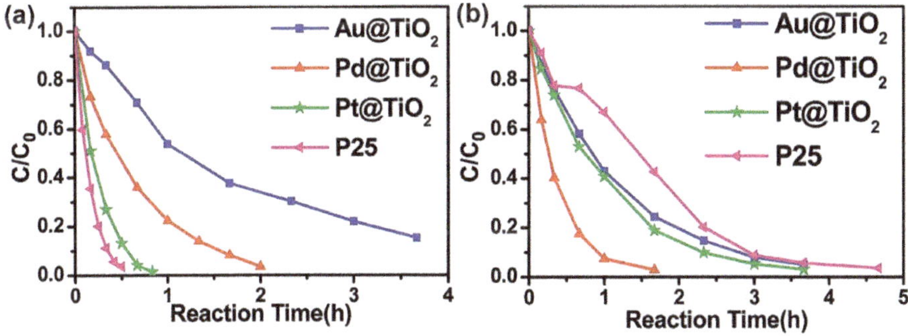

Figure 6. (**a**) Degradation of RhB by P25 and $M@TiO_2$ (M = Au, Pd and Pt) under UV irradiation; (**b**) Degradation of RhB by P25 and $M@TiO_2$ (M = Au, Pd and Pt) under visible light irradiation. Reproduced from [90], with copyright permission from American Chemical Society, 2011.

Compared with M/S nanocomposite photocatalysts possessing the conventional and core-shell structures, there is less research on M/S nanocomposite photocatalysts with the yolk-shell, Janus and multi-junction structures for environmental remediation. Wang et al. investigated the properties of yolk-shell $Au@TiO_2$ photocatalysts for degradation of RhB [59]. In their research, they also introduced reduced graphene oxide (r-GO) into the TiO_2 shell. The results demonstrated the properties of yolk-shell Au@r-GO/TiO_2 photocatalysts were better than yolk-shell Au@TiO_2 photocatalysts, which indicated the charge separation in the TiO_2 shell increased with the electron conductivity of the TiO_2 shell. In the work of Yao et al., the properties of Janus Au/ZnO photocatalysts were investigated for degradation of MO [100]. The results demonstrated the properties of Janus Au/ZnO photocatalysts for degradation of MO were higher than that of pristine ZnO photocatalysts and increased with the size of Au nanoparticles under UV light illumination. In the pioneer work of Tada et al., multi-junction CdS/Au/TiO_2 photocatalysts were synthesized as the all-solid-state Z-scheme photocatalysts for the degradation of MV [71]. As revealed in Figure 7a, multi-junction CdS/Au/TiO_2 photocatalysts exhibited much higher properties than CdS/TiO_2, Au/TiO_2 and TiO_2 photocatalysts, which revealed the charge separation caused by $TiO_2 \rightarrow$ Au $\rightarrow$ CdS Z-scheme electron transfer was more efficient than that of the single- or double-component systems. The Z-scheme electron transfer was illustrated in Figure 7b. Besides, Tada et al. pointed out that the properties of all-solid-state Z-scheme photocatalysts could be further improved by modifying the energy band structures of the semiconductor components.

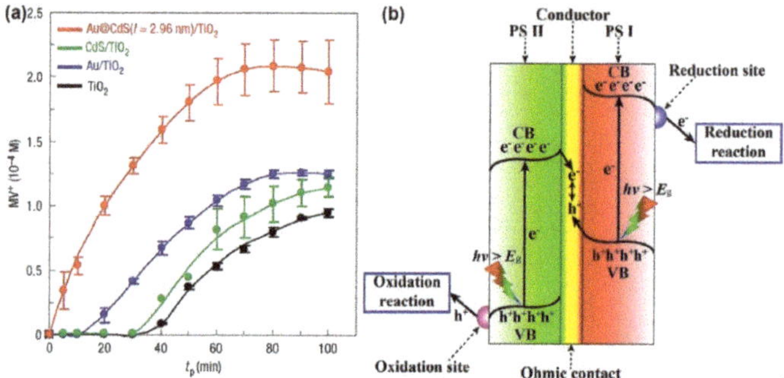

Figure 7. (**a**) Photocatalytic reduction of MV by TiO_2, Au/TiO_2, CdS/TiO_2 and Au@CdS/TiO_2 (CdS/Au/TiO_2). Reproduced from [71], with copyright permission from Springer Nature, 2006; (**b**) Schematic diagram of Z-scheme electron transfer. Reproduced from [68], with copyright permission from John Wiley & Sons, 2014.

4.2. Selective Organic Transformation

Unlike the non-selective mineralization process in environmental remediation, photocatalysis can also drive organic transformation processes to selectively synthesize valuable chemicals [103–105]. Compared with thermochemical synthetic process, the photocatalytic selective organic transformation process often requires milder conditions and shorter reaction sequences and can exclude some undesirable side reactions. Based on the reported research, the photocatalytic selective organic transformation reactions can be divided into three categories, i.e., the oxidation reaction, reduction reaction and coupling reaction [103]. In Table 2, some of the most representative M/S nanocomposite photocatalysts applied in selective organic transformation are listed [106–120].

Table 2. M/S nanocomposite photocatalysts for selective organic transformation.

Photocatalyst	Structure	Reaction Type	Reactant	Light Source	Reference
Pt/TiO$_2$, Pd/TiO$_2$	Conventional structure	Oxidation reaction	Benzene	UV light	[106]
Au/TiO$_2$	Conventional structure	Oxidation reaction	Benzene	Simulated sunlight	[107]
Pt/TiO$_2$	Conventional structure	Oxidation reaction	Benzene	UV light	[108]
Pt/TiO$_2$	Conventional structure	Oxidation reaction	(substituted) Benzene	UV light	[109]
Au/CeO$_2$	Conventional structure	Oxidation reaction	Aromatic alcohols	Visible light	[110]
Pt@CeO$_2$	Core-shell, yolk-shell structure	Oxidation reaction	Benzyl alcohol	Visible light	[111]
Au/TiO$_2$	Janus structure	Oxidation reaction	Methanol	UV light	[112]
Ag/TiO$_2$	Conventional structure	Reduction reaction	Nitrobenzene	UV light	[113]
M/TiO$_2$ (M = Pt, Pd, Rh, Ag and Au)	Conventional structure	Reduction reaction	Carbon dioxide (CO$_2$)	UV light	[114]
Pt-Cu/TiO$_2$	Conventional structure	Reduction reaction	CO$_2$	Visible light	[115]
Ag/TiO$_2$	Conventional structure	Reduction reaction	CO$_2$	Simulated sunlight	[116]
Ag/La$_2$Ti$_2$O$_7$	Conventional structure	Reduction reaction	CO$_2$	UV light	[117]
Pt/TiO$_2$	Conventional structure	Reduction reaction	CO$_2$	Simulated sunlight	[118]
Au-Pd/ZrO$_2$	Conventional structure	Oxidation, coupling reaction	Benzylamine, benzyl alcohol etc.	Visible light	[119]
Pd/SiC	Conventional structure	Coupling reaction	Iodobenzene, phenylboronic acid	Visible light	[120]

Towards the photocatalytic selective oxidation reactions, Yuzawa et al. investigated the effect of Pt loading on the properties of Pt/TiO$_2$ photocatalysts for amination of benzene to aniline and unfolded the reaction mechanism governing the photocatalytic amination process [108]. As revealed in Figure 8, the conversion yield of Pt/TiO$_2$ did not increase linearly with the Pt loading content and reached a peak at about 0.1 wt% Pt loading content, while the aniline selectivity of Pt/TiO$_2$ photocatalysts was hardly influenced by the Pt loading content and remained about 97%. The mechanism governing the photocatalytic amination process was clarified through ESR spectroscopy. First, the holes on the TiO$_2$ surface oxidized an ammonia molecule to form a neutral amide radical, which then reacted with the aromatic ring to form an intermediate, and afterwards the hydrogen of the intermediate was abstracted by the active sites on the Pt surface, leading to the formation of aniline. Thus, it could be deduced that the charge separation at the Pt-TiO$_2$ interface could significantly enhance the properties of Pt/TiO$_2$ photocatalysts for the amination of benzene. Zhang et al. investigated the properties of core-shell and yolk-shell Pt@CeO$_2$ photocatalysts for oxidation of benzyl alcohol to benzaldehyde [111]. As revealed in Figure 9, core-shell Pt@CeO$_2$ photocatalysts exhibited much higher benzaldehyde yield and selectivity than yolk-shell Pt@CeO$_2$ photocatalysts. Zhang et al. attributed the unsatisfactory properties of yolk-shell Pt@CeO$_2$ photocatalysts to the loose contact between the Pt core and CeO$_2$ shell, which deteriorated the charge separation at the Pt-CeO$_2$ interface.

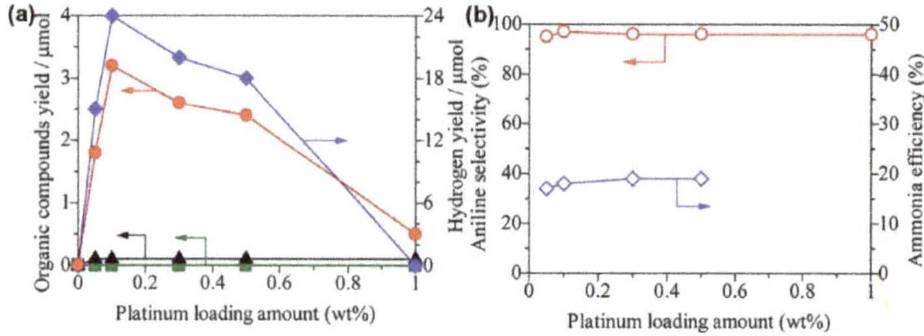

Figure 8. (**a**) Conversion yield of Pt/TiO$_2$ versus the Pt loading amount; (**b**) Aniline selectivity of Pt/TiO$_2$ versus the Pt loading amount. Reproduced from [107], with copyright permission from American Chemical Society, 2013.

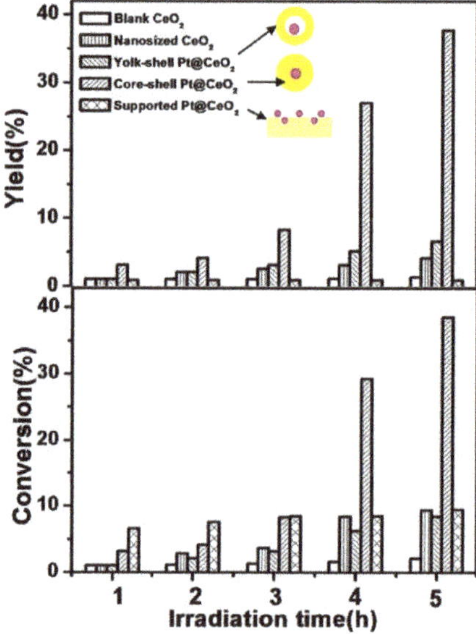

Figure 9. Results of photocatalytic selective oxidation of benzyl alcohol to benzyl aldehyde over the core-shell Pt@CeO$_2$, yolk-shell Pt@CeO$_2$, supported Pt/CeO$_2$, CeO$_2$ nanoparticles and blank CeO$_2$. Reproduced from [110], with copyright permission from The Royal Society of Chemistry, 2011.

As listed in Table 2, the photocatalytic selective reduction reactions mainly involve hydrogenation of nitroaromatics and reduction of CO$_2$. The investigation of Tada et al. on hydrogenation of nitroaromatics demonstrated that the life time of electrons could also be prolonged by the charge separation at the Ag-TiO$_2$ interface, leading to the enhanced properties of Ag/TiO$_2$ photocatalysts for hydrogenation of nitrobenzene to aniline [113]. For better properties of photocatalytic reduction of CO$_2$ to CH$_4$, Xie et al. combined P25 TiO$_2$ with five noble metals, i.e., Pt, Pd, Rh, Ag and Au [114]. During the photocatalytic test, the properties of M/TiO$_2$ photocatalysts increased in the sequence of Ag/TiO$_2$ < Rh/TiO$_2$ < Au/TiO$_2$ < Pd/TiO$_2$ < Pt/TiO$_2$, which indicated the charge separation at the metal-TiO$_2$ interface improved with the Schottky barrier height.

Towards the photocatalytic coupling reactions, Jiao et al. investigated the properties of Pd/SiC photocatalysts for Suzuki coupling of iodobenzene and phenylboronic acid [120]. Under visible light illumination, the conversion of iodobenzene and selectivity for the main product of Pd/SiC both reached nearly 100%. Jiao et al. also evaluated the effect of photogenerated electrons and holes in Pd/SiC on the Suzuki coupling process. After adding the electron-capturing agent (or hole-capturing agent) into the reaction system, the photocatalytic properties of Pd/SiC decreased dramatically, which indicated both the photogenerated electrons and holes contributed to the Suzuki coupling process.

4.3. Hydrogen Evolution

As a significant process to convert solar energy into chemical energy, photocatalytic hydrogen evolution has always been the hot spot in the research field of photocatalysis [47,121–123]. Table 3 lists some of the most representative M/S photocatalysts for hydrogen evolution [48,63,67,124–144].

Table 3. M/S nanocomposite photocatalysts for hydrogen evolution.

Photocatalyst	Structure	Light Souce	Reference
Pt/TiO_2	Conventional structure	UV light	[27]
$PtNi_x/g$-C_3N_4	Conventional structure	UV light	[124]
Au/TiO_2, Ag/TiO_2	Conventional structure	Visible light	[125]
Cu/TiO_2, Ni/TiO_2, $CuNi/TiO_2$	Conventional structure	UV light	[126]
Pt_3Co/CdS, Pt_3Co/TiO_2	Conventional structure	UV light	[127]
Au/ZnO	Conventional structure	Simulated sunlight	[128]
Au/MoS_2	Conventional structure	Visible light	[129]
Au/ZnO	Conventional structure	Visible light	[130]
Au/CdS	Conventional structure	Simulated sunlight	[131]
Ni/CdS	Conventional structure	UV light	[132]
Pt/TiO_2, Au/TiO_2	Conventional structure	Simulated sunlight	[133]
$SnRu_x/TiO_2$	Conventional structure	UV light	[134]
Au/Cu_2O	Conventional, core-shell structure	Visible light	[135]
$Au@CdS$	Core-shell structure	Visible light	[136]
$Au@TiO_2$	Yolk-shell structure	UV light	[137]
$Au@TiO_2$	Yolk-shell structure	UV light	[138]
Au/TiO_2	Janus structure	Visible light	[63]
Au/CdS	Array structure	Visible light	[67]
$CdS/Au/ZnO$	Multi-junction structure	UV light	[139]
$CdS/Au/MoS_2$	Multi-junction structure	Visible light	[140]
$CdS/Pt/ZnO$	Multi-junction structure	Simulated sunlight	[141]
$CdS/M/TiO_2$ (M = Au, Ag, Pt, Pd)	Multi-junction structure	Simulated sunlight	[142]
$CoO_x/Ir/Ta_3N_5$	Multi-junction structure	Simulated sunlight	[143]
$Cr_2O_3/Rh/(Ga_{1-x}Zn_x)(N_{1-x}O_x)$	Multi-junction structure	Visible light	[144]

For the M/S nanocomposite photocatalysts listed in Table 3, their properties for photocatalytic hydrogen evolution achieved remarkable enhancement from the charge separation at the metal-semiconductor interface and the SPR of the metal component. In the work of Bi et al., they utilized Pt-Ni alloy nanoparticles to decorate g-C_3N_4 for saving the usage of Pt [124]. The results revealed the $PtNi_x/g$-C_3N_4 photocatalysts could exhibit comparable properties to that of Pt/g-C_3N_4 photocatalysts, which provided a possible approach for developing M/S nanocomposite photocatalysts with excellent properties as well as low cost. In the work of Ingram et al., the difference in the properties of Ag/TiO_2 and Au/TiO_2 (TiO_2 was doped by nitrogen atoms, i.e., N-TiO_2) photocatalysts for visible light-driven hydrogen evolution was investigated, which shed light on more effective utilization of SPR [125]. As revealed in Figure 10a, the SPR of Ag and Au was excited at the wavelengths of about 400 and 500 nm, respectively, while the absorption edge of N-TiO_2 was around 400 nm. Therefore, the SPR of Ag was more efficient for exciting electron-hole pairs in N-TiO_2 than Au, which was supported by the photocatalytic hydrogen evolution test results in Figure 10b.

Towards the core-shell and yolk-shell structure, Ma et al. investigated the effect of Au cores on the properties of core-shell Au@CdS photocatalysts for hydrogen evolution [136]. As revealed in Figure 11, under irradiation of visible light with wavelength $\geq$ 420 nm, core-shell Au@CdS exhibited apparently higher photocatalytic properties than pristine CdS, while under irradiation of visible light with wavelength $\geq$500 nm, the photocatalytic properties of core-shell Au@CdS were only slightly higher than pristine CdS. Thus, Ma et al. thought the radiative energy transfer from SPR-excited Au to CdS was the main contribution to the enhanced photocatalytic properties of core-shell Au@CdS, rather than the electron transfer from SPR-excited Au to CdS. In the work of Ngaw et al., the effect of Au content on the photocatalytic properties of yolk-shell Au@TiO_2 for water splitting was investigated [137]. The photocatalytic properties of yolk-shell Au@TiO$_2$ also did not improve linearly with the Au content under the irradiation of both visible light and UV light, and reached a peak at 2 wt% Au content. In addition, Ngaw et al. ascribed the enhanced photocatalytic properties of yolk-shell Au@TiO$_2$ to the void space and highly porous shell in yolk-shell Au@TiO$_2$, which provided more active sites for H$^+$ ions to be reduced and more channels for reactants to diffuse into and out of the photocatalytic particles.

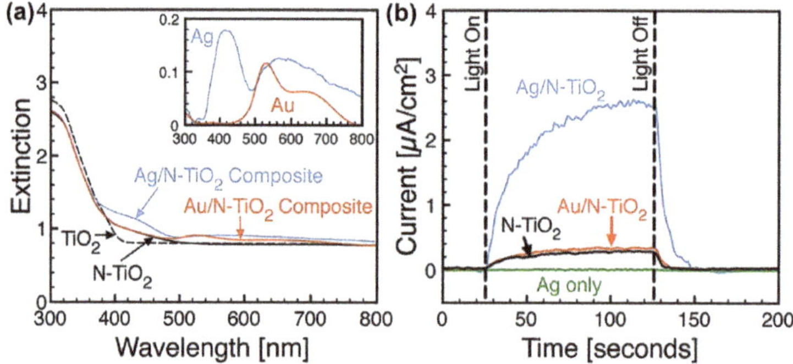

Figure 10. (**a**) Ultraviolet-visible extinction spectra of TiO$_2$, N-TiO$_2$, Ag/N-TiO$_2$ and Au/N-TiO$_2$ samples. The inset shows difference spectra for Ag and Au (i.e., the Ag/N-TiO$_2$ or Au/N-TiO$_2$ spectrum minus the N-TiO$_2$ spectrum); (**b**) Photocurrent responses (per macroscopic electrode area) under visible light illumination. Reproduced from [124], with copyright permission from American Chemical Society, 2011.

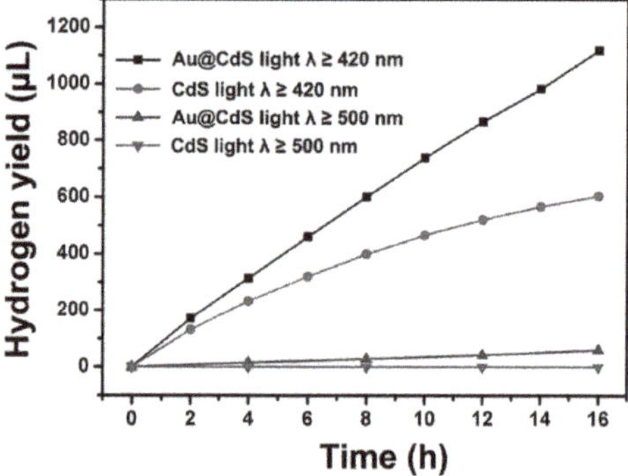

Figure 11. Photocatalytic water splitting of core-shell Au@CdS and CdS nanoparticles under irradiation of visible light with wavelength more than 420 nm and 500 nm, respectively. Reproduced from [133], with copyright permission from John Wiley & Sons, 2014.

In the work of Seh et al., the properties of Janus Au/TiO$_2$ and core-shell Au@TiO$_2$ photocatalysts for hydrogen evolution were compared and investigated [63]. As revealed in Figure 12a,b, Janus Au/TiO$_2$ photocatalysts exhibited higher properties than core-shell Au@TiO$_2$ photocatalysts and the properties of Janus Au/TiO$_2$ photocatalysts increased with the size of Au nanoparticles. Moreover, Seh et al. attributed the enhanced charge separation in Janus Au/TiO$_2$ photocatalysts to the strong plasmonic near-fields localized closely to the Au-TiO$_2$ interface, which was supported by the discrete-dipole approximation simulation results.

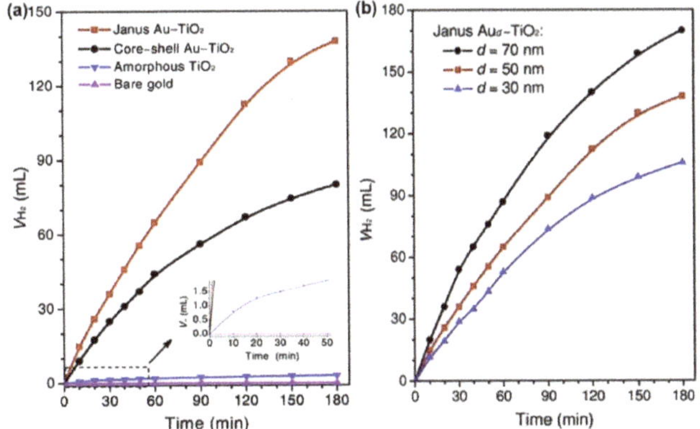

Figure 12. (**a**) Hydrogen evolution through photocatalytic water splitting on Janus Au/TiO$_2$, core-shell Au@TiO$_2$, amorphous TiO$_2$ and pristine Au nanoparticles; (**b**) Hydrogen evolution through photocatalytic water splitting on Janus Au/TiO$_2$ with different Au nanoparticle sizes. Reproduced from [63], with copyright permission from John Wiley & Sons, 2012.

As for the array structure, Wang et al. demonstrated the excellent properties of multi-segmented Au/CdS nanorod arrays (NRAs) for photocatalytic hydrogen evolution [67]. As revealed in Figure 13a, the activities of Au/CdS NRAs for hydrogen evolution increased with the number of Au-CdS segments under irradiation of simulated sunlight. In addition, compared with pristine CdS NRAs, the activities of Au/CdS NRAs for hydrogen evolution gained significant enhancement under irradiation of simulated sunlight, which was presented in Figure 13b. Wang et al. attributed the excellent properties of Au/CdS NRAs for hydrogen evolution to the charge separation at the Au-CdS interface under irradiation of simulated sunlight.

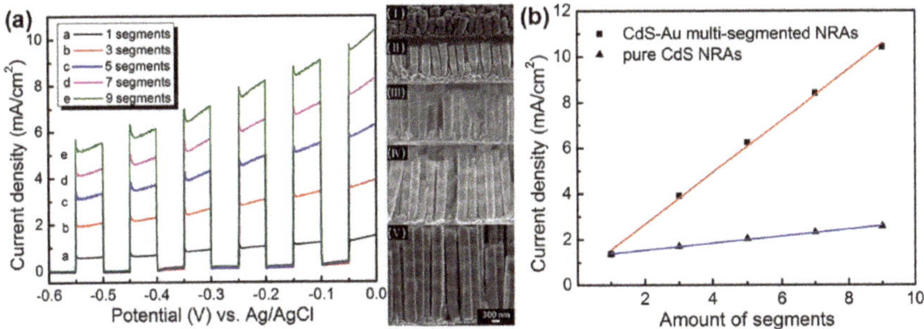

Figure 13. (**a**) Hydrogen evolution on the multi-segmented CdS/Au NRAs in the photoelectrochemical cell under irradiation of simulated sun light. (I-V) SEM images of the side view of the CdS-Au NRAs. (**b**) Comparison of the activities for hydrogen evolution on the CdS/Au NRAs and pristine CdS NRAs under irradiation of simulated sunlight. Reproduced from [67], with copyright permission from John Wiley & Sons, 2014.

Toward the multi-junction structure, Yu et al. demonstrated the excellent properties of multi-junction CdS/Au/ZnO photocatalysts for hydrogen evolution [139]. To reveal the high charge separation efficiency bought by the Z-scheme electron transfer, Yu el al. also synthesized Au/CdS/ZnO photocatalysts by depositing Au on CdS/ZnO photocatalysts. As shown in Figure 14a, the properties of multi-junction CdS/Au/ZnO photocatalysts was 4.5 times higher than that of CdS/ZnO photocatalysts, while the properties of Au/CdS/ZnO photocatalysts achieved only a small enhancement compared with CdS/ZnO photocatalysts. The PL spectra in Figure 14b further confirmed the high charge separation efficiency due to the Z-scheme electron transfer in multi-junction CdS/Au/ZnO photocatalysts.

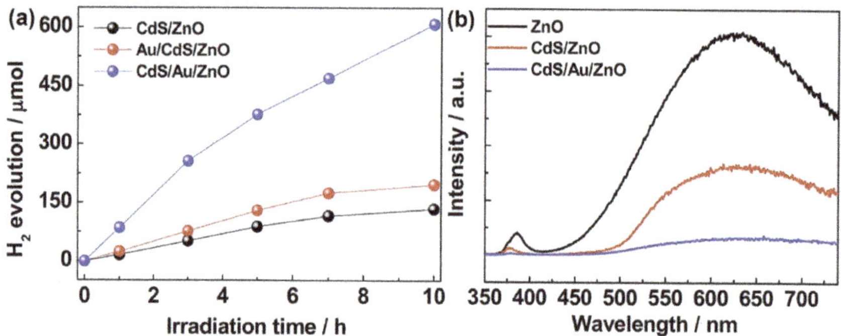

Figure 14. (a) Hydrogen evolution through photocatalytic water splitting on CdS/ZnO, Au/CdS/ZnO and CdS/Au/ZnO, respectively. (b) PL spectra of ZnO, CdS/ZnO and CdS/Au/ZnO at a 270 nm excitation wavelength. Reproduced from [136], with copyright permission from The Royal Society of Chemistry, 2013.

4.4. Disinfection

In 1985, Matsunaga et al. first reported the photocatalytic inactivation of bacteria on the surface of TiO_2 [145]. Since then, a lot of research has been devoted to the photocatalytic inactivation of microorganisms such as bacteria, viruses, protozoa and so on [18,146–150]. Furthermore, due to the nonexistence of secondary pollution, photocatalytic disinfection is a promising alternative approach for water purification. Table 4 lists some of the most representative M/S nanocomposite photocatalysts for disinfection [151–159].

Table 4. M/S nanocomposite photocatalysts for disinfection.

Photocatalyst	Structure	Bacteria	Light Source	Reference
$Ag/g\text{-}C_3N_4$	Conventional structure	*Escherichia coli*	UV and visible light	[151]
Ag/TiO_2	Conventional structure	*Escherichia coli*	UV light	[152]
Ag/TiO_2	Conventional structure	*Escherichia coli*	Sunlight	[153]
Ag/TiO_2	Conventional structure	*Escherichia coli*	Simulated sunlight	[154]
Cu/TiO_2	Conventional structure	*Escherichia coli*	UV and visible light	[155]
Ag/BiOI	Conventional structure	*Escherichia coli*	Visible light	[156]
Ag/AgX (X = Cl, Br, I)	Conventional structure	*Escherichia coli*	Visible light	[157]
Ag/ZnO	Conventional structure	*Escherichia coli*	Visible light	[158]
Ag@ZnO	Core–shell structure	*Vibrio cholerae*	Sunlight	[159]

Taking into account the practical application of photocatalytic disinfection for water purification, most of the light sources involved in the research are visible light as revealed in Table 4. Therefore, the enhanced properties of M/S nanocomposite photocatalysts for disinfection are mainly attributed to the SPR of the metal component. In the work of Shi et al., due to the SPR of Ag, the Ag/AgX (X = Cl, Br, I) photocatalysts exhibited remarkable properties for the inactivation of *Escherichia coli* under visible light illumination [157]. Shi el al. also evaluated the contribution of different photo-generated reactive species to the disinfection process by adding scavengers into the reaction system. The results indicated holes are the dominant reactive species over other reactive species such as electrons, $^{\bullet}OH$, H_2O_2 and so on.

5. Conclusions and Perspectives

In this review, we demonstrate the properties of M/S nanocomposite photocatalysts in relation to their structures for application in environmental remediation, selective organic transformation, hydrogen evolution and disinfection. Due to the enhanced charge separation at the metal-semiconductor interface and increased absorption of visible light induced by the SPR of metals, M/S nanocomposite photocatalysts usually exhibit much better properties than pristine semiconductor photocatalysts.

For future development of M/S nanocomposite photocatalysts, our perspectives can be summarized as the following four points:

(1) To date, most of metals utilized to combine with semiconductor photocatalysts are noble metals, which are scarce in nature and expensive. To save the use of noble metals, coupling semiconductors with alloys composed of noble and non-noble metals is highly recommended. In addition, due to the non-linear relationship between the properties of M/S nanocomposite photocatalysts and metal loading, precise control over the metal loading in the M/S nanocomposite photocatalysts deserve further research.

(2) For more efficient utilization of SPR to enhance the properties of M/S nanocomposite photocatalysts, the SPR excitation wavelength of the metal nanoparticles should overlap the absorption edge of the semiconductor nanoparticles, which could be achieved by changing the shape and particle size of the metal nanoparticles and modulating the band structure of the semiconductor nanoparticles.

(3) Synergistic utilization of enhanced charge separation at the metal-semiconductor interface and SPR of metals might endow the M/S nanocomposite photocatalysts with even better properties under visible light illumination, because the SPR-induced charge separation in M/S nanocomposite photocatalysts could be further enhanced by introducing another metal co-catalyst with large work function to the M/S nanocomposite photocatalysts.

(4) Due to their intrinsic ability to prohibit particle agglomeration, the core-shell, yolk-shell and array structures (especially the array structure) might be the ideal structures for M/S nanocomposite photocatalysts. Therefore, there exists a strong demand for more facile synthesis of these structures.

Author Contributions: Y.-s.F., J.L. and J.G.L. conceptualized this review; Y.-s.F. collected and analyzed the data; Y.-s.F. and J.L. wrote the paper.

Funding: This research was funded by National Key Research & Development Program of China through the grant number 2017YFB0305300 and Joint Funds of the National Natural Science Foundation of China through the grant number U1660203.

Acknowledgments: The authors would like to thank Tao Deng and Jianbo Wu for their valuable suggestions to this review.

References

1. Fujishima, A.; Honda, K. Electrochemical Photolysis of Water at a Semiconductor Electrode. *Nature* **1972**, *238*, 37. [CrossRef] [PubMed]
2. Huang, Y.; Sun, F.; Wang, H.; He, Y.; Li, L.; Huang, Z.; Wu, Q.; Yu, J.C. Photochemical growth of cadmium-rich CdS nanotubes at the air–water interface and their use in photocatalysis. *J. Mater. Chem.* **2009**, *19*, 6901–6906. [CrossRef]
3. Wang, S.; Wang, X. Photocatalytic CO_2 reduction by CdS promoted with a zeolitic imidazolate framework. *Appl. Catal. B Environ.* **2015**, *162*, 494–500. [CrossRef]
4. An, X.; Yu, X.; Yu, J.C.; Zhang, G. CdS nanorods/reduced graphene oxide nanocomposites for photocatalysis and electrochemical sensing. *J. Mater. Chem. A* **2013**, *1*, 5158–5164. [CrossRef]
5. Wang, Y.; Shi, R.; Lin, J.; Zhu, Y. Enhancement of photocurrent and photocatalytic activity of ZnO hybridized with graphite-like C3N4. *Energy Environ. Sci.* **2011**, *4*, 2922–2929. [CrossRef]
6. McLaren, A.; Valdes-Solis, T.; Li, G.; Tsang, S.C. Shape and Size Effects of ZnO Nanocrystals on Photocatalytic Activity. *J. Am. Chem. Soc.* **2009**, *131*, 12540–12541. [CrossRef] [PubMed]

7.   Elmolla, E.S.; Chaudhuri, M. Degradation of amoxicillin, ampicillin and cloxacillin antibiotics in aqueous solution by the UV/ZnO photocatalytic process. *J. Hazard. Mater.* **2010**, *173*, 445–449. [CrossRef] [PubMed]

8.   Miyauchi, M.; Takashio, M.; Tobimatsu, H. Photocatalytic Activity of $SrTiO_3$ Codoped with Nitrogen and Lanthanum under Visible Light Illumination. *Langmuir* **2004**, *20*, 232–236. [CrossRef] [PubMed]

9.   Konta, R.; Ishii, T.; Kato, H.; Kudo, A. Photocatalytic Activities of Noble Metal Ion Doped $SrTiO_3$ under Visible Light Irradiation. *J. Phys. Chem. B* **2004**, *108*, 8992–8995. [CrossRef]

10.   Ye, S.; Wang, R.; Wu, M.-Z.; Yuan, Y.-P. A review on g-C3N4 for photocatalytic water splitting and CO_2 reduction. *Appl. Surf. Sci.* **2015**, *358*, 15–27. [CrossRef]

11.   Lam, S.-M.; Sin, J.-C.; Mohamed, A.R. A review on photocatalytic application of g-C3N4/semiconductor (CNS) nanocomposites towards the erasure of dyeing wastewater. *Mater. Sci. Semicond. Process.* **2016**, *47*, 62–84. [CrossRef]

12.   Li, J.; Wu, N. Semiconductor-based photocatalysts and photoelectrochemical cells for solar fuel generation: A review. *Catal. Sci. Technol.* **2015**, *5*, 1360–1384. [CrossRef]

13.   Wang, Y.; Wang, Q.; Zhan, X.; Wang, F.; Safdar, M.; He, J. Visible light driven type II heterostructures and their enhanced photocatalysis properties: A review. *Nanoscale* **2013**, *5*, 8326–8339. [CrossRef] [PubMed]

14.   Montini, T.; Melchionna, M.; Monai, M.; Fornasiero, P. Fundamentals and Catalytic Applications of CeO_2-Based Materials. *Chem. Rev.* **2016**, *116*, 5987–6041. [CrossRef] [PubMed]

15.   Rao, P.M.; Cai, L.; Liu, C.; Cho, I.S.; Lee, C.H.; Weisse, J.M.; Yang, P.; Zheng, X. Simultaneously Efficient Light Absorption and Charge Separation in $WO_3/BiVO_4$ Core/Shell Nanowire Photoanode for Photoelectrochemical Water Oxidation. *Nano Lett.* **2014**, *14*, 1099–1105. [CrossRef] [PubMed]

16.   Wang, H.; Zhang, L.; Chen, Z.; Hu, J.; Li, S.; Wang, Z.; Liu, J.; Wang, X. Semiconductor heterojunction photocatalysts: Design, construction, and photocatalytic performances. *Chem. Soc. Rev.* **2014**, *43*, 5234–5244. [CrossRef] [PubMed]

17.   Daghrir, R.; Drogui, P.; Robert, D. Modified TiO2 For Environmental Photocatalytic Applications: A Review. *Ind. Eng. Chem. Res.* **2013**, *52*, 3581–3599. [CrossRef]

18.   Dong, S.; Feng, J.; Fan, M.; Pi, Y.; Hu, L.; Han, X.; Liu, M.; Sun, J.; Sun, J. Recent developments in heterogeneous photocatalytic water treatment using visible light-responsive photocatalysts: A review. *RSC Adv.* **2015**, *5*, 14610–14630. [CrossRef]

19.   Burda, C.; Lou, Y.; Chen, X.; Samia, A.C.S.; Stout, J.; Gole, J.L. Enhanced Nitrogen Doping in TiO_2 Nanoparticles. *Nano Lett.* **2003**, *3*, 1049–1051. [CrossRef]

20.   Yu, J.C.; Yu, J.; Ho, W.; Zhang, J. Effects of F- Doping on the Photocatalytic Activity and Microstructures of Nanocrystalline TiO_2 Powders. *Chem. Mater.* **2002**, *14*, 3808–3816. [CrossRef]

21.   Liu, L.; Ouyang, S.; Ye, J. Gold-Nanorod-Photosensitized Titanium Dioxide with Wide-Range Visible-Light Harvesting Based on Localized Surface Plasmon Resonance. *Angew. Chem.* **2013**, *125*, 6821–6825. [CrossRef]

22.   Zhang, M.; Chen, C.; Ma, W.; Zhao, J. Visible-Light-Induced Aerobic Oxidation of Alcohols in a Coupled Photocatalytic System of Dye-Sensitized TiO_2 and TEMPO. *Angew. Chem. Int. Ed.* **2008**, *47*, 9730–9733. [CrossRef] [PubMed]

23.   Asahi, R.; Morikawa, T.; Ohwaki, T.; Aoki, K.; Taga, Y. Visible-Light Photocatalysis in Nitrogen-Doped Titanium Oxides. *Science* **2001**, *293*, 269. [CrossRef] [PubMed]

24.   Dhanalakshmi, K.B.; Latha, S.; Anandan, S.; Maruthamuthu, P. Dye sensitized hydrogen evolution from water. *Int. J. Hydrogen Energy* **2001**, *26*, 669–674. [CrossRef]

25.   Youngblood, W.J.; Lee, S.-H.A.; Maeda, K.; Mallouk, T.E. Visible Light Water Splitting Using Dye-Sensitized Oxide Semiconductors. *Acc. Chem. Res.* **2009**, *42*, 1966–1973. [CrossRef] [PubMed]

26.   Zhang, X.; Peng, T.; Song, S. Recent advances in dye-sensitized semiconductor systems for photocatalytic hydrogen production. *J. Mater. Chem. A* **2016**, *4*, 2365–2402. [CrossRef]

27.   Khaki, M.R.D.; Shafeeyan, M.S.; Raman, A.A.A.; Daud, W.M.A.W. Application of doped photocatalysts for organic pollutant degradation—A review. *J. Environ. Manag.* **2017**, *198*, 78–94. [CrossRef] [PubMed]

28.   Wu, N. Plasmonic metal–semiconductor photocatalysts and photoelectrochemical cells: A review. *Nanoscale* **2018**, *10*, 2679–2696. [CrossRef] [PubMed]

29.   Linic, S.; Christopher, P.; Ingram, D.B. Plasmonic-metal nanostructures for efficient conversion of solar to chemical energy. *Nat. Mater.* **2011**, *10*, 911. [CrossRef] [PubMed]

30. Marschall, R. Semiconductor Composites: Strategies for Enhancing Charge Carrier Separation to Improve Photocatalytic Activity. *Adv. Funct. Mater.* **2014**, *24*, 2421–2440. [CrossRef]

31. Qu, Y.; Duan, X. Progress, challenge and perspective of heterogeneous photocatalysts. *Chem. Soc. Rev.* **2013**, *42*, 2568–2580. [CrossRef] [PubMed]

32. Schneider, J.; Bahnemann, D.; Ye, J.; Li, G. *Photocatalysis: Fundamentals and Perspectives*; Royal Society of Chemistry: London, UK, 2016. [CrossRef]

33. Castellote, M.; Bengtsson, N. Principles of TiO$_2$ Photocatalysis. In *Applications of Titanium Dioxide Photocatalysis to Construction Materials: State-of-the-Art Report of the RILEM Technical Committee 194-TDP*; Ohama, Y., Van Gemert, D., Eds.; Springer: Dordrecht, The Netherlands, 2011; pp. 5–10. [CrossRef]

34. Peter, L.M. CHAPTER 1 Photoelectrochemistry: From Basic Principles to Photocatalysis. In *Photocatalysis: Fundamentals and Perspectives*; The Royal Society of Chemistry: London, UK, 2016; pp. 1–28. [CrossRef]

35. Mills, A.; Le Hunte, S. An overview of semiconductor photocatalysis. *J. Photochem. Photobiol. A Chem.* **1997**, *108*, 1–35. [CrossRef]

36. Chen, S.; Wang, L.-W. Thermodynamic Oxidation and Reduction Potentials of Photocatalytic Semiconductors in Aqueous Solution. *Chem. Mater.* **2012**, *24*, 3659–3666. [CrossRef]

37. Ohtani, B. Revisiting the fundamental physical chemistry in heterogeneous photocatalysis: Its thermodynamics and kinetics. *Phys. Chem. Chem. Phys.* **2014**, *16*, 1788–1797. [CrossRef] [PubMed]

38. Liu, B.; Zhao, X.; Terashima, C.; Fujishima, A.; Nakata, K. Thermodynamic and kinetic analysis of heterogeneous photocatalysis for semiconductor systems. *Phys. Chem. Chem. Phys.* **2014**, *16*, 8751–8760. [CrossRef] [PubMed]

39. Zhang, Z.; Yates, J.T. Band Bending in Semiconductors: Chemical and Physical Consequences at Surfaces and Interfaces. *Chem. Rev.* **2012**, *112*, 5520–5551. [CrossRef] [PubMed]

40. Kamat, P.V. Manipulation of Charge Transfer Across Semiconductor Interface. A Criterion That Cannot Be Ignored in Photocatalyst Design. *J. Phys. Chem. Lett.* **2012**, *3*, 663–672. [CrossRef] [PubMed]

41. Cai, Y.-Y.; Li, X.-H.; Zhang, Y.-N.; Wei, X.; Wang, K.-X.; Chen, J.-S. Highly Efficient Dehydrogenation of Formic Acid over a Palladium-Nanoparticle-Based Mott–Schottky Photocatalyst. *Angew. Chem.* **2013**, *125*, 12038–12041. [CrossRef]

42. Liu, Y.; Gu, X.; Qi, W.; Zhu, H.; Shan, H.; Chen, W.L.; Tao, P.; Song, C.Y.; Shang, W.; Deng, T.; et al. Enhancing the Photocatalytic Hydrogen Evolution Performance of a Metal/Semiconductor Catalyst through Modulation of the Schottky Barrier Height by Controlling the Orientation of the Interface. *ACS Appl. Mater. Interfaces* **2017**, *9*, 12494–12500. [CrossRef] [PubMed]

43. Yang, J.H.; Wang, D.G.; Han, H.X.; Li, C. Roles of Cocatalysts in Photocatalysis and Photoelectrocatalysis. *Accounts Chem. Res.* **2013**, *46*, 1900–1909. [CrossRef] [PubMed]

44. Kubacka, A.; Fernández-García, M.; Colón, G. Advanced Nanoarchitectures for Solar Photocatalytic Applications. *Chem. Rev.* **2012**, *112*, 1555–1614. [CrossRef] [PubMed]

45. Teoh, W.Y.; Scott, J.A.; Amal, R. Progress in Heterogeneous Photocatalysis: From Classical Radical Chemistry to Engineering Nanomaterials and Solar Reactors. *J. Phys. Chem. Lett.* **2012**, *3*, 629–639. [CrossRef] [PubMed]

46. Hou, W.; Cronin, S.B. A Review of Surface Plasmon Resonance-Enhanced Photocatalysis. *Adv. Funct. Mater.* **2013**, *23*, 1612–1619. [CrossRef]

47. Warren, S.C.; Thimsen, E. Plasmonic solar water splitting. *Energy Environ. Sci.* **2012**, *5*, 5133–5146. [CrossRef]

48. Yu, J.; Qi, L.; Jaroniec, M. Hydrogen Production by Photocatalytic Water Splitting over Pt/TiO$_2$ Nanosheets with Exposed (001) Facets. *J. Phys. Chem. C* **2010**, *114*, 13118–13125. [CrossRef]

49. Lin, C.-H.; Chao, J.-H.; Liu, C.-H.; Chang, J.-C.; Wang, F.-C. Effect of Calcination Temperature on the Structure of a Pt/TiO2 (B) Nanofiber and Its Photocatalytic Activity in Generating H2. *Langmuir* **2008**, *24*, 9907–9915. [CrossRef] [PubMed]

50. Wu, G.; Chen, T.; Su, W.; Zhou, G.; Zong, X.; Lei, Z.; Li, C. H-2 production with ultra-low CO selectivity via photocatalytic reforming of methanol on Au/TiO$_2$ catalyst. *Int. J. Hydrogen Energy* **2008**, *33*, 1243–1251. [CrossRef]

51. Shi, H.; Wang, X.; Zheng, M.; Wu, X.; Chen, Y.; Yang, Z.; Zhang, G.; Duan, H. Hot-Electrons Mediated Efficient Visible-Light Photocatalysis of Hierarchical Black Au-TiO$_2$ Nanorod Arrays on Flexible Substrate. *Adv. Mater. Interfaces* **2016**, *3*, 1600588. [CrossRef]

52. Ghosh Chaudhuri, R.; Paria, S. Core/Shell Nanoparticles: Classes, Properties, Synthesis Mechanisms, Characterization, and Applications. *Chem. Rev.* **2012**, *112*, 2373–2433. [CrossRef] [PubMed]

53. Zhang, N.; Liu, S.Q.; Xu, Y.J. Recent progress on metal core@semiconductor shell nanocomposites as a promising type of photocatalyst. *Nanoscale* **2012**, *4*, 2227–2238. [CrossRef] [PubMed]

54. Sudeep, P.K.; Takechi, K.; Kamat, P.V. Harvesting photons in the infrared. Electron injection from excited tricarbocyanine dye (IR-125) into TiO_2 and Ag@TiO_2 core-shell nanoparticles. *J. Phys. Chem. C* **2007**, *111*, 488–494. [CrossRef]

55. Kamata, K.; Lu, Y.; Xia, Y.N. Synthesis and characterization of monodispersed core-shell spherical colloids with movable cores. *J. Am. Chem. Soc.* **2003**, *125*, 2384–2385. [CrossRef] [PubMed]

56. Liu, J.; Qiao, S.Z.; Chen, J.S.; Lou, X.W.; Xing, X.R.; Lu, G.Q. Yolk/shell nanoparticles: New platforms for nanoreactors, drug delivery and lithium-ion batteries. *Chem. Commun.* **2011**, *47*, 12578–12591. [CrossRef] [PubMed]

57. Li, A.; Zhang, P.; Chang, X.; Cai, W.; Wang, T.; Gong, J. Gold Nanorod@TiO_2 Yolk-Shell Nanostructures for Visible-Light-Driven Photocatalytic Oxidation of Benzyl Alcohol. *Small* **2015**, *11*, 1892–1899. [CrossRef] [PubMed]

58. Tu, W.; Zhou, Y.; Li, H.; Li, P.; Zou, Z. Au@TiO2 yolk-shell hollow spheres for plasmon-induced photocatalytic reduction of CO_2 to solar fuel via a local electromagnetic field. *Nanoscale* **2015**, *7*, 14232–14236. [CrossRef] [PubMed]

59. Wang, M.; Han, J.; Xiong, H.; Guo, R. Yolk@Shell Nanoarchitecture of Au@r-GO/TiO_2 Hybrids as Powerful Visible Light Photocatalysts. *Langmuir* **2015**, *31*, 6220–6228. [CrossRef] [PubMed]

60. de Gennes, P.-G. Soft Matter (Nobel Lecture). *Angew. Chem. Int. Ed.* **1992**, *31*, 842–845. [CrossRef]

61. Lattuada, M.; Hatton, T.A. Synthesis, properties and applications of Janus nanoparticles. *Nano Today* **2011**, *6*, 286–308. [CrossRef]

62. Walther, A.; Müller, A.H.E. Janus Particles: Synthesis, Self-Assembly, Physical Properties, and Applications. *Chem. Rev.* **2013**, *113*, 5194–5261. [CrossRef] [PubMed]

63. Seh, Z.W.; Liu, S.; Low, M.; Zhang, S.-Y.; Liu, Z.; Mlayah, A.; Han, M.-Y. Janus Au-TiO_2 Photocatalysts with Strong Localization of Plasmonic Near-Fields for Efficient Visible-Light Hydrogen Generation. *Adv. Mater.* **2012**, *24*, 2310–2314. [CrossRef] [PubMed]

64. Wang, Z.L.; Song, J. Piezoelectric Nanogenerators Based on Zinc Oxide Nanowire Arrays. *Science* **2006**, *312*, 242–246. [CrossRef] [PubMed]

65. Zhu, K.; Neale, N.R.; Miedaner, A.; Frank, A.J. Enhanced Charge-Collection Efficiencies and Light Scattering in Dye-Sensitized Solar Cells Using Oriented TiO_2 Nanotubes Arrays. *Nano Lett.* **2007**, *7*, 69–74. [CrossRef] [PubMed]

66. Lin, H.-W.; Lu, Y.-J.; Chen, H.-Y.; Lee, H.-M.; Gwo, S. InGaN/GaN nanorod array white light-emitting diode. *Appl. Phys. Lett.* **2010**, *97*, 073101.

67. Wang, X.; Liow, C.; Qi, D.; Zhu, B.; Leow, W.R.; Wang, H.; Xue, C.; Chen, X.; Li, S. Programmable Photo-Electrochemical Hydrogen Evolution Based on Multi-Segmented CdS-Au Nanorod Arrays. *Adv. Mater.* **2014**, *26*, 3506–3512. [CrossRef] [PubMed]

68. Zhou, P.; Yu, J.G.; Jaroniec, M. All-Solid-State Z-Scheme Photocatalytic Systems. *Adv. Mater.* **2014**, *26*, 4920–4935. [CrossRef] [PubMed]

69. Low, J.; Jiang, C.; Cheng, B.; Wageh, S.; Al-Ghamdi, A.A.; Yu, J. A Review of Direct Z-Scheme Photocatalysts. *Small Methods* **2017**, *1*, 1700080. [CrossRef]

70. Qi, K.; Cheng, B.; Yu, J.; Ho, W. A review on TiO2-based Z-scheme photocatalysts. *Chin. J. Catal.* **2017**, *38*, 1936–1955. [CrossRef]

71. Tada, H.; Mitsui, T.; Kiyonaga, T.; Akita, T.; Tanaka, K. All-solid-state Z-scheme in CdS–Au–TiO_2 three-component nanojunction system. *Nat. Mater.* **2006**, *5*, 782. [CrossRef] [PubMed]

72. McCullagh, C.; Skillen, N.; Adams, M.; Robertson, P.K.J. Photocatalytic reactors for environmental remediation: A review. *J. Chem. Technol. Biotechnol.* **2011**, *86*, 1002–1017. [CrossRef]

73. Khin, M.M.; Nair, A.S.; Babu, V.J.; Murugan, R.; Ramakrishna, S. A review on nanomaterials for environmental remediation. *Energy Environ. Sci.* **2012**, *5*, 8075–8109. [CrossRef]

74. Singh, S.; Mahalingam, H.; Singh, P.K. Polymer-supported titanium dioxide photocatalysts for environmental remediation: A review. *Appl. Catal. A Gen.* **2013**, *462–463*, 178–195. [CrossRef]

75. Hoffmann, M.R.; Martin, S.T.; Choi, W.; Bahnemann, D.W. Environmental applications of semiconductor photocatalysis. *Chem. Rev.* **1995**, *95*, 69–96. [CrossRef]

76. Tahir, M.N.; Natalio, F.; Cambaz, M.A.; Panthöfer, M.; Branscheid, R.; Kolb, U.; Tremel, W. Controlled synthesis of linear and branched Au@ZnO hybrid nanocrystals and their photocatalytic properties. *Nanoscale* **2013**, *5*, 9944–9949. [CrossRef] [PubMed]

77. Chen, Y.; Zeng, D.; Zhang, K.; Lu, A.; Wang, L.; Peng, D.-L. Au–ZnO hybrid nanoflowers, nanomultipods and nanopyramids: One-pot reaction synthesis and photocatalytic properties. *Nanoscale* **2014**, *6*, 874–881. [CrossRef] [PubMed]

78. Ren, C.; Yang, B.; Wu, M.; Xu, J.; Fu, Z.; Lv, Y.; Guo, T.; Zhao, Y.; Zhu, C. Synthesis of Ag/ZnO nanorods array with enhanced photocatalytic performance. *J. Hazard. Mater.* **2010**, *182*, 123–129. [CrossRef] [PubMed]

79. Lin, D.; Wu, H.; Zhang, R.; Pan, W. Enhanced Photocatalysis of Electrospun Ag−ZnO Heterostructured Nanofibers. *Chem. Mater.* **2009**, *21*, 3479–3484. [CrossRef]

80. Georgekutty, R.; Seery, M.K.; Pillai, S.C. A highly efficient Ag-ZnO photocatalyst: Synthesis, properties, and mechanism. *J. Phys. Chem. C* **2008**, *112*, 13563–13570. [CrossRef]

81. Kochuveedu, S.T.; Kim, D.-P.; Kim, D.H. Surface-Plasmon-Induced Visible Light Photocatalytic Activity of TiO_2 Nanospheres Decorated by Au Nanoparticles with Controlled Configuration. *J. Phys. Chem. C* **2012**, *116*, 2500–2506. [CrossRef]

82. Liu, Y.; Wei, S.; Gao, W. Ag/ZnO heterostructures and their photocatalytic activity under visible light: Effect of reducing medium. *J. Hazard. Mater.* **2015**, *287*, 59–68. [CrossRef] [PubMed]

83. Subramanian, V.; Wolf, E.; Kamat, P.V. Semiconductor−Metal Composite Nanostructures. To What Extent Do Metal Nanoparticles Improve the Photocatalytic Activity of TiO_2 Films? *J. Phys. Chem. B* **2001**, *105*, 11439–11446. [CrossRef]

84. Jiang, Z.; Wei, W.; Mao, D.; Chen, C.; Shi, Y.; Lv, X.; Xie, J. Silver-loaded nitrogen-doped yolk-shell mesoporous TiO_2 hollow microspheres with enhanced visible light photocatalytic activity. *Nanoscale* **2015**, *7*, 784–797. [CrossRef] [PubMed]

85. Chen, H.; Tu, T.; Wen, M.; Wu, Q. Assembly synthesis of Cu_2O-on-Cu nanowires with visible-light-enhanced photocatalytic activity. *Dalton Trans.* **2015**, *44*, 15645–15652. [CrossRef] [PubMed]

86. Li, X.; Sun, Y.; Xiong, T.; Jiang, G.; Zhang, Y.; Wu, Z.; Dong, F. Activation of amorphous bismuth oxide via plasmonic Bi metal for efficient visible-light photocatalysis. *J. Catal.* **2017**, *352*, 102–112. [CrossRef]

87. Dong, F.; Li, Q.; Sun, Y.; Ho, W.-K. Noble Metal-Like Behavior of Plasmonic Bi Particles as a Cocatalyst Deposited on $(BiO)_2CO_3$ Microspheres for Efficient Visible Light Photocatalysis. *Acs Catal.* **2014**, *4*, 4341–4350. [CrossRef]

88. Dong, F.; Zhao, Z.; Sun, Y.; Zhang, Y.; Yan, S.; Wu, Z. An Advanced Semimetal–Organic Bi Spheres–g-C3N4 Nanohybrid with SPR-Enhanced Visible-Light Photocatalytic Performance for NO Purification. *Environ. Sci. Technol.* **2015**, *49*, 12432–12440. [CrossRef] [PubMed]

89. Gao, Y.; Huang, Y.; Li, Y.; Zhang, Q.; Cao, J.-j.; Ho, W.; Lee, S.C. Plasmonic Bi/ZnWO4 Microspheres with Improved Photocatalytic Activity on NO Removal under Visible Light. *ACS Sustain. Chem. Eng.* **2016**, *4*, 6912–6920. [CrossRef]

90. Dong, F.; Xiong, T.; Yan, S.; Wang, H.; Sun, Y.; Zhang, Y.; Huang, H.; Wu, Z. Facets and defects cooperatively promote visible light plasmonic photocatalysis with Bi nanowires@BiOCl nanosheets. *J. Catal.* **2016**, *344*, 401–410. [CrossRef]

91. Zhang, N.; Liu, S.; Fu, X.; Xu, Y.-J. Synthesis of M@TiO_2 (M = Au, Pd, Pt) Core–Shell Nanocomposites with Tunable Photoreactivity. *J. Phys. Chem. C* **2011**, *115*, 9136–9145. [CrossRef]

92. Cushing, S.K.; Li, J.; Meng, F.; Senty, T.R.; Suri, S.; Zhi, M.; Li, M.; Bristow, A.D.; Wu, N. Photocatalytic Activity Enhanced by Plasmonic Resonant Energy Transfer from Metal to Semiconductor. *J. Am. Chem. Soc.* **2012**, *134*, 15033–15041. [CrossRef] [PubMed]

93. Kong, L.; Chen, W.; Ma, D.; Yang, Y.; Liu, S.; Huang, S. Size control of Au@Cu_2O octahedra for excellent photocatalytic performance. *J. Mater. Chem.* **2012**, *22*, 719–724. [CrossRef]

94. Yang, T.-T.; Chen, W.-T.; Hsu, Y.-J.; Wei, K.-H.; Lin, T.-Y.; Lin, T.-W. Interfacial Charge Carrier Dynamics in Core-Shell Au-CdS Nanocrystals. *J. Phys. Chem. C* **2010**, *114*, 11414–11420. [CrossRef]

95. Wu, X.-F.; Song, H.-Y.; Yoon, J.-M.; Yu, Y.-T.; Chen, Y.-F. Synthesis of Core-Shell Au@TiO_2 Nanoparticles with Truncated Wedge-Shaped Morphology and Their Photocatalytic Properties. *Langmuir* **2009**, *25*, 6438–6447. [CrossRef] [PubMed]

96. Chen, M.; Li, Y.; Wang, Z.; Gao, Y.; Huang, Y.; Cao, J.; Ho, W.; Lee, S. Controllable Synthesis of Core–Shell Bi@Amorphous Bi_2O_3 Nanospheres with Tunable Optical and Photocatalytic Activity for NO Removal. *Ind. Eng. Chem. Res.* **2017**, *56*, 10251–10258. [CrossRef]

97. Li, J.T.; Cushing, S.K.; Bright, J.; Meng, F.K.; Senty, T.R.; Zheng, P.; Bristow, A.D.; Wu, N.Q. Ag@Cu_2O Core-Shell Nanoparticles as Visible-Light Plasmonic Photocatalysts. *ACS Catal.* **2013**, *3*, 47–51. [CrossRef]

98. Li, Q.; Wang, F.; Sun, L.; Jiang, Z.; Ye, T.; Chen, M.; Bai, Q.; Wang, C.; Han, X. Design and Synthesis of Cu@CuS Yolk–Shell Structures with Enhanced Photocatalytic Activity. *Nano-Micro Lett.* **2017**, *9*, 35. [CrossRef] [PubMed]

99. Li, P.; Wei, Z.; Wu, T.; Peng, Q.; Li, Y. Au−ZnO Hybrid Nanopyramids and Their Photocatalytic Properties. *J. Am. Chem. Soc.* **2011**, *133*, 5660–5663. [CrossRef] [PubMed]

100. Yao, K.X.; Liu, X.; Zhao, L.; Zeng, H.C.; Han, Y. Site-specific growth of Au particles on ZnO nanopyramids under ultraviolet illumination. *Nanoscale* **2011**, *3*, 4195–4200. [CrossRef] [PubMed]

101. Muñoz-Batista, M.J.; Ballari, M.M.; Kubacka, A.; Alfano, O.M.; Fernández-García, M. Braiding kinetics and spectroscopy in photo-catalysis: The spectro-kinetic approach. *Chem. Soc. Rev.* **2019**, *48*, 637–682. [CrossRef] [PubMed]

102. Davies, M.J. Detection and characterisation of radicals using electron paramagnetic resonance (EPR) spin trapping and related methods. *Methods* **2016**, *109*, 21–30. [CrossRef] [PubMed]

103. Shiraishi, Y.; Hirai, T. Selective organic transformations on titanium oxide-based photocatalysts. *J. Photochem. Photobiol. C Photochem. Rev.* **2008**, *9*, 157–170. [CrossRef]

104. Lang, X.; Chen, X.; Zhao, J. Heterogeneous visible light photocatalysis for selective organic transformations. *Chem. Soc. Rev.* **2014**, *43*, 473–486. [CrossRef] [PubMed]

105. Kisch, H. Semiconductor photocatalysis for organic synthesis. *Adv. Photochem.* **2001**, *26*, 93–143.

106. Park, H.; Choi, W. Photocatalytic conversion of benzene to phenol using modified TiO_2 and polyoxometalates. *Catal. Today* **2005**, *101*, 291–297. [CrossRef]

107. Ide, Y.; Nakamura, N.; Hattori, H.; Ogino, R.; Ogawa, M.; Sadakane, M.; Sano, T. Sunlight-induced efficient and selective photocatalytic benzene oxidation on TiO_2-supported gold nanoparticles under CO_2 atmosphere. *Chem. Commun.* **2011**, *47*, 11531–11533. [CrossRef] [PubMed]

108. Yuzawa, H.; Kumagai, J.; Yoshida, H. Reaction Mechanism of Aromatic Ring Amination of Benzene and Substituted Benzenes by Aqueous Ammonia over Platinum-Loaded Titanium Oxide Photocatalyst. *J. Phys. Chem. C* **2013**, *117*, 11047–11058. [CrossRef]

109. Yuzawa, H.; Yoshida, H. Direct aromatic-ring amination by aqueous ammonia with a platinum loaded titanium oxide photocatalyst. *Chem. Commun.* **2010**, *46*, 8854–8856. [CrossRef] [PubMed]

110. Tanaka, A.; Hashimoto, K.; Kominami, H. Preparation of Au/CeO_2 Exhibiting Strong Surface Plasmon Resonance Effective for Selective or Chemoselective Oxidation of Alcohols to Aldehydes or Ketones in Aqueous Suspensions under Irradiation by Green Light. *J. Am. Chem. Soc.* **2012**, *134*, 14526–14533. [CrossRef] [PubMed]

111. Zhang, N.; Fu, X.; Xu, Y.-J. A facile and green approach to synthesize Pt@CeO_2 nanocomposite with tunable core-shell and yolk-shell structure and its application as a visible light photocatalyst. *J. Mater. Chem.* **2011**, *21*, 8152–8158. [CrossRef]

112. Pradhan, S.; Ghosh, D.; Chen, S. Janus Nanostructures Based on Au−TiO_2 Heterodimers and Their Photocatalytic Activity in the Oxidation of Methanol. *ACS Appl. Mater. Interfaces* **2009**, *1*, 2060–2065. [CrossRef] [PubMed]

113. Tada, H.; Ishida, T.; Takao, A.; Ito, S. Drastic Enhancement of TiO_2-Photocatalyzed Reduction of Nitrobenzene by Loading Ag Clusters. *Langmuir* **2004**, *20*, 7898–7900. [CrossRef] [PubMed]

114. Xie, S.; Wang, Y.; Zhang, Q.; Deng, W.; Wang, Y. MgO- and Pt-Promoted TiO_2 as an Efficient Photocatalyst for the Preferential Reduction of Carbon Dioxide in the Presence of Water. *ACS Catal.* **2014**, *4*, 3644–3653. [CrossRef]

115. Zhai, Q.; Xie, S.; Fan, W.; Zhang, Q.; Wang, Y.; Deng, W.; Wang, Y. Photocatalytic Conversion of Carbon Dioxide with Water into Methane: Platinum and Copper(I) Oxide Co-catalysts with a Core-Shell Structure. *Angew. Chem. Int. Ed.* **2013**, *52*, 5776–5779. [CrossRef] [PubMed]

116. Liu, L.; Pitts, D.T.; Zhao, H.; Zhao, C.; Li, Y. Silver-incorporated bicrystalline (anatase/brookite) TiO_2 microspheres for CO_2 photoreduction with water in the presence of methanol. *Appl. Catal. A Gen.* **2013**, *467*, 474–482. [CrossRef]

117. Wang, Z.; Teramura, K.; Hosokawa, S.; Tanaka, T. Photocatalytic conversion of CO_2 in water over Ag-modified $La_2Ti_2O_7$. *Appl. Catal. B Environ.* **2015**, *163*, 241–247. [CrossRef]

118. Feng, X.; Sloppy, J.D.; LaTemp, T.J.; Paulose, M.; Komarneni, S.; Bao, N.; Grimes, C.A. Synthesis and deposition of ultrafine Pt nanoparticles within high aspect ratio TiO_2 nanotube arrays: Application to the photocatalytic reduction of carbon dioxide. *J. Mater. Chem.* **2011**, *21*, 13429–13433. [CrossRef]

119. Sarina, S.; Zhu, H.; Jaatinen, E.; Xiao, Q.; Liu, H.; Jia, J.; Chen, C.; Zhao, J. Enhancing Catalytic Performance of Palladium in Gold and Palladium Alloy Nanoparticles for Organic Synthesis Reactions through Visible Light Irradiation at Ambient Temperatures. *J. Am. Chem. Soc.* **2013**, *135*, 5793–5801. [CrossRef] [PubMed]

120. Jiao, Z.; Zhai, Z.; Guo, X.; Guo, X.-Y. Visible-Light-Driven Photocatalytic Suzuki–Miyaura Coupling Reaction on Mott–Schottky-type Pd/SiC Catalyst. *J. Phys. Chem. C* **2015**, *119*, 3238–3243. [CrossRef]

121. Kudo, A.; Miseki, Y. Heterogeneous photocatalyst materials for water splitting. *Chem. Soc. Rev.* **2009**, *38*, 253–278. [CrossRef] [PubMed]

122. Maeda, K.; Domen, K. Photocatalytic Water Splitting: Recent Progress and Future Challenges. *J. Phys. Chem. Lett.* **2010**, *1*, 2655–2661. [CrossRef]

123. Ahmad, H.; Kamarudin, S.K.; Minggu, L.J.; Kassim, M. Hydrogen from photo-catalytic water splitting process: A review. *Renew. Sustain. Energy Rev.* **2015**, *43*, 599–610. [CrossRef]

124. Bi, L.; Gao, X.; Ma, Z.; Zhang, L.; Wang, D.; Xie, T. Enhanced Separation Efficiency of PtNix/g-C3N4 for Photocatalytic Hydrogen Production. *ChemCatChem* **2017**, *9*, 3779–3785. [CrossRef]

125. Ingram, D.B.; Linic, S. Water Splitting on Composite Plasmonic-Metal/Semiconductor Photoelectrodes: Evidence for Selective Plasmon-Induced Formation of Charge Carriers near the Semiconductor Surface. *J. Am. Chem. Soc.* **2011**, *133*, 5202–5205. [CrossRef] [PubMed]

126. Munoz-Batista, M.J.; Meira, D.M.; Colon, G.; Kubacka, A.; Fernandez-Garcia, M. Phase-Contact Engineering in Mono- and Bimetallic Cu-Ni Co-catalysts for Hydrogen Photocatalytic Materials. *Angew. Chem. -Int. Ed.* **2018**, *57*, 1199–1203. [CrossRef] [PubMed]

127. Hu, Z.; Yu, J.C. Pt3Co-loaded CdS and TiO_2 for photocatalytic hydrogen evolution from water. *J. Mater. Chem. A* **2013**, *1*, 12221–12228. [CrossRef]

128. Zhang, X.; Liu, Y.; Kang, Z. 3D Branched ZnO Nanowire Arrays Decorated with Plasmonic Au Nanoparticles for High-Performance Photoelectrochemical Water Splitting. *ACS Appl. Mater. Interfaces* **2014**, *6*, 4480–4489. [CrossRef] [PubMed]

129. Yin, Z.; Chen, B.; Bosman, M.; Cao, X.; Chen, J.; Zheng, B.; Zhang, H. Au Nanoparticle-Modified MoS2 Nanosheet-Based Photoelectrochemical Cells for Water Splitting. *Small* **2014**, *10*, 3537–3543. [CrossRef] [PubMed]

130. Zhang, C.; Shao, M.; Ning, F.; Xu, S.; Li, Z.; Wei, M.; Evans, D.G.; Duan, X. Au nanoparticles sensitized ZnO nanorod@nanoplatelet core-shell arrays for enhanced photoelectrochemical water splitting. *Nano Energy* **2015**, *12*, 231–239. [CrossRef]

131. Sheeney-Haj-Ichia, L.; Pogorelova, S.; Gofer, Y.; Willner, I. Enhanced Photoelectrochemistry in CdS/Au Nanoparticle Bilayers. *Adv. Funct. Mater.* **2004**, *14*, 416–424. [CrossRef]

132. Zhukovskyi, M.; Tongying, P.; Yashan, H.; Wang, Y.; Kuno, M. Efficient Photocatalytic Hydrogen Generation from Ni Nanoparticle Decorated CdS Nanosheets. *ACS Catal.* **2015**, *5*, 6615–6623. [CrossRef]

133. Rosseler, O.; Shankar, M.V.; Du, M.K.-L.; Schmidlin, L.; Keller, N.; Keller, V. Solar light photocatalytic hydrogen production from water over Pt and Au/TiO_2 (anatase/rutile) photocatalysts: Influence of noble metal and porogen promotion. *J. Catal.* **2010**, *269*, 179–190. [CrossRef]

134. Gu, Q.; Long, J.; Fan, L.; Chen, L.; Zhao, L.; Lin, H.; Wang, X. Single-site Sn-grafted Ru/TiO_2 photocatalysts for biomass reforming: Synergistic effect of dual co-catalysts and molecular mechanism. *J. Catal.* **2013**, *303*, 141–155. [CrossRef]

135. Hong, J.W.; Wi, D.H.; Lee, S.-U.; Han, S.W. Metal–Semiconductor Heteronanocrystals with Desired Configurations for Plasmonic Photocatalysis. *J. Am. Chem. Soc.* **2016**, *138*, 15766–15773. [CrossRef] [PubMed]

136. Ma, X.; Zhao, K.; Tang, H.; Chen, Y.; Lu, C.; Liu, W.; Gao, Y.; Zhao, H.; Tang, Z. New Insight into the Role of Gold Nanoparticles in Au@CdS Core–Shell Nanostructures for Hydrogen Evolution. *Small* **2014**, *10*, 4664–4670. [CrossRef] [PubMed]

137. Ngaw, C.K.; Xu, Q.; Tan, T.T.Y.; Hu, P.; Cao, S.; Loo, J.S.C. A strategy for in-situ synthesis of well-defined core–shell Au@TiO$_2$ hollow spheres for enhanced photocatalytic hydrogen evolution. *Chem. Eng. J.* **2014**, *257*, 112–121. [CrossRef]

138. Lee, Y.J.; Joo, J.B.; Yin, Y.; Zaera, F. Evaluation of the Effective Photoexcitation Distances in the Photocatalytic Production of H2 from Water using Au@Void@TiO$_2$ Yolk–Shell Nanostructures. *ACS Energy Lett.* **2016**, *1*, 52–56. [CrossRef]

139. Yu, Z.B.; Xie, Y.P.; Liu, G.; Lu, G.Q.; Ma, X.L.; Cheng, H.-M. Self-assembled CdS/Au/ZnO heterostructure induced by surface polar charges for efficient photocatalytic hydrogen evolution. *J. Mater. Chem. A* **2013**, *1*, 2773–2776. [CrossRef]

140. Chava, R.K.; Do, J.Y.; Kang, M. Smart Hybridization of Au Coupled CdS Nanorods with Few Layered MoS$_2$ Nanosheets for High Performance Photocatalytic Hydrogen Evolution Reaction. *ACS Sustain. Chem. Eng.* **2018**, *6*, 6445–6457. [CrossRef]

141. Kim, Y.G.; Jo, W.-K. Photodeposited-metal/CdS/ZnO heterostructures for solar photocatalytic hydrogen production under different conditions. *Int. J. Hydrogen Energy* **2017**, *42*, 11356–11363. [CrossRef]

142. Zhou, H.; Pan, J.; Ding, L.; Tang, Y.; Ding, J.; Guo, Q.; Fan, T.; Zhang, D. Biomass-derived hierarchical porous CdS/M/TiO$_2$ (M = Au, Ag, pt, pd) ternary heterojunctions for photocatalytic hydrogen evolution. *Int. J. Hydrogen Energy* **2014**, *39*, 16293–16301. [CrossRef]

143. Wang, Q.; Hisatomi, T.; Ma, S.S.K.; Li, Y.; Domen, K. Core/Shell Structured La- and Rh-Codoped SrTiO$_3$ as a Hydrogen Evolution Photocatalyst in Z-Scheme Overall Water Splitting under Visible Light Irradiation. *Chem. Mater.* **2014**, *26*, 4144–4150. [CrossRef]

144. Maeda, K.; Teramura, K.; Lu, D.; Saito, N.; Inoue, Y.; Domen, K. Noble-metal/Cr$_2$O$_3$ core/shell nanoparticles as a cocatalyst for photocatalytic overall water splitting. *Angew. Chem. Int. Ed.* **2006**, *45*, 7806–7809. [CrossRef] [PubMed]

145. Matsunaga, T.; Tomoda, R.; Nakajima, T.; Wake, H. Photoelectrochemical sterilization of microbial cells by semiconductor powders. *FEMS Microbiol. Lett.* **1985**, *29*, 211–214. [CrossRef]

146. Gamage, J.; Zhang, Z. Applications of Photocatalytic Disinfection. *Int. J. Photoenergy* **2010**, *2010*, 764870. [CrossRef]

147. Malato, S.; Fernández-Ibáñez, P.; Maldonado, M.I.; Blanco, J.; Gernjak, W. Decontamination and disinfection of water by solar photocatalysis: Recent overview and trends. *Catal. Today* **2009**, *147*, 1–59. [CrossRef]

148. Byrne, J.A.; Fernandez-Ibañez, P.A.; Dunlop, P.S.M.; Alrousan, D.M.A.; Hamilton, J.W.J. Photocatalytic Enhancement for Solar Disinfection of Water: A Review. *Int. J. Photoenergy* **2011**, *2011*, 798051. [CrossRef]

149. Sarria, V.; Parra, S.; Adler, N.; Péringer, P.; Benitez, N.; Pulgarin, C. Recent developments in the coupling of photoassisted and aerobic biological processes for the treatment of biorecalcitrant compounds. *Catal. Today* **2002**, *76*, 301–315. [CrossRef]

150. Dalrymple, O.K.; Stefanakos, E.; Trotz, M.A.; Goswami, D.Y. A review of the mechanisms and modeling of photocatalytic disinfection. *Appl. Catal. B Environ.* **2010**, *98*, 27–38. [CrossRef]

151. Muñoz-Batista, M.J.; Fontelles-Carceller, O.; Ferrer, M.; Fernández-García, M.; Kubacka, A. Disinfection capability of Ag/g-C3N4 composite photocatalysts under UV and visible light illumination. *Appl. Catal. B Environ.* **2016**, *183*, 86–95. [CrossRef]

152. Pratap Reddy, M.; Venugopal, A.; Subrahmanyam, M. Hydroxyapatite-supported Ag–TiO$_2$ as Escherichia coli disinfection photocatalyst. *Water Res.* **2007**, *41*, 379–386. [CrossRef] [PubMed]

153. Kumar, R.V.; Raza, G. Photocatalytic disinfection of water with Ag–TiO$_2$ nanocrystalline composite. *Ionics* **2009**, *15*, 579–587. [CrossRef]

154. Liu, L.; Liu, Z.; Bai, H.; Sun, D.D. Concurrent filtration and solar photocatalytic disinfection/degradation using high-performance Ag/TiO$_2$ nanofiber membrane. *Water Res.* **2012**, *46*, 1101–1112. [CrossRef] [PubMed]

155. Rtimi, S.; Giannakis, S.; Sanjines, R.; Pulgarin, C.; Bensimon, M.; Kiwi, J. Insight on the photocatalytic bacterial inactivation by co-sputtered TiO$_2$–Cu in aerobic and anaerobic conditions. *Appl. Catal. B Environ.* **2016**, *182*, 277–285. [CrossRef]

156. Zhu, L.; He, C.; Huang, Y.; Chen, Z.; Xia, D.; Su, M.; Xiong, Y.; Li, S.; Shu, D. Enhanced photocatalytic disinfection of E. coli 8099 using Ag/BiOI composite under visible light irradiation. *Sep. Purif. Technol.* **2012**, *91*, 59–66. [CrossRef]

157. Shi, H.; Li, G.; Sun, H.; An, T.; Zhao, H.; Wong, P.-K. Visible-light-driven photocatalytic inactivation of E. coli by Ag/AgX-CNTs (X=Cl, Br, I) plasmonic photocatalysts: Bacterial performance and deactivation mechanism. *Appl. Catal. B Environ.* **2014**, *158–159*, 301–307. [CrossRef]
158. Gao, P.; Ng, K.; Sun, D.D. Sulfonated graphene oxide–ZnO–Ag photocatalyst for fast photodegradation and disinfection under visible light. *J. Hazard. Mater.* **2013**, *262*, 826–835. [CrossRef] [PubMed]
159. Das, S.; Sinha, S.; Suar, M.; Yun, S.-I.; Mishra, A.; Tripathy, S.K. Solar-photocatalytic disinfection of Vibrio cholerae by using Ag@ZnO core–shell structure nanocomposites. *J. Photochem. Photobiol. B Biol.* **2015**, *142*, 68–76. [CrossRef] [PubMed]

 *nanomaterials*

Review

Photoactive Tungsten-Oxide Nanomaterials for Water-Splitting

Yerkin Shabdan [1,2,†], Aiymkul Markhabayeva [2,†], Nurlan Bakranov [3,4,*] and Nurxat Nuraje [5,*]

1   National Laboratory Astana, Nazarbayev University, Nursultan 010000, Kazakhstan;
    yerkin.shabdan@nu.edu.kz
2   Faculty of Physics and Technology, AI-Farabi Kazakh National University, Almaty 050040, Kazakhstan;
    aiko_marx@mail.ru
3   Faculty of General Education, Kazakh-British Technical University, Almaty 050000, Kazakhstan
4   Laboratory of Engineering Profile, Satbayev University, Almaty 050000, Kazakhstan
5   Department of Chemical and Materials Engineering, Nazarbayev University, Nursultan 010000, Kazakhstan
*   Correspondence: bakranov@mail.ru (N.B.); nurxat.nuraje@nu.edu.kz (N.N.)
†   These authors contributed equally to this work.

Received: 16 July 2020; Accepted: 29 August 2020; Published: 18 September 2020

Abstract: This review focuses on tungsten oxide (WO_3) and its nanocomposites as photoactive nanomaterials for photoelectrochemical cell (PEC) applications since it possesses exceptional properties such as photostability, high electron mobility ($\sim$12 cm^2 V^{-1} s^{-1}) and a long hole-diffusion length ($\sim$150 nm). Although WO_3 has demonstrated oxygen-evolution capability in PEC, further increase of its PEC efficiency is limited by high recombination rate of photogenerated electron/hole carriers and slow charge transfer at the liquid–solid interface. To further increase the PEC efficiency of the WO_3 photocatalyst, designing WO_3 nanocomposites via surface–interface engineering and doping would be a great strategy to enhance the PEC performance via improving charge separation. This review starts with the basic principle of water-splitting and physical chemistry properties of WO_3, that extends to various strategies to produce binary/ternary nanocomposites for PEC, particulate photocatalysts, Z-schemes and tandem-cell applications. The effect of PEC crystalline structure and nanomorphologies on efficiency are included. For both binary and ternary WO_3 nanocomposite systems, the PEC performance under different conditions—including synthesis approaches, various electrolytes, morphologies and applied bias—are summarized. At the end of the review, a conclusion and outlook section concluded the WO_3 photocatalyst-based system with an overview of WO_3 and their nanocomposites for photocatalytic applications and provided the readers with potential research directions.

Keywords: WO_3; nanocomposites; heterostructures; water-splitting; oxygen evolution

---

1. Introduction

The conversion of solar-emitted electromagnetic waves to useful forms of energy is a very promising research area in the field of renewable energy production. Although roughly 32×10^{24} J of solar energy reaches the Earth's surface per year, only 0.001% of the incoming solar energy is used for human needs [1]. The conversion of solar light to useful forms of energy is still challenging at the scientific and engineering level in terms of energy production for the needs of human beings. Even though there are many technologies for renewable energy [2], including solar cells, solar collectors and solar fuel reactors (water-splitting), the major challenges we face are to improve efficiency and stability in the conversion of solar energy to other energy forms. Currently, one of the popular research technologies to tackle solar energy conversion is trying to convert photons into chemical energy [3] by using artificial photoelectrochemical (PEC) processes.

Metal-oxide nanomaterials have been thoroughly studied for the conversion of solar energy to hydrogen molecules due to their chemical and physical stability, optical and electronic properties, easy fabrication and low cost. They have shown good properties for use in photoelectrochemical devices such as TiO_2 [4–6], α-Fe_2O_3 [7–9], $BiVO_4$ [10–12], ZnO [13–15] and WO_3 [16–18]. On the above properties of semiconductor materials, both suitable bandgap positions to generate hydrogen and oxygen gases and the ability to absorb a reasonable portion of the solar light spectrum are critical for water-splitting. However, a single metal-oxide photocatalyst cannot simultaneously satisfy all the requirements for solar-to-hydrogen-driven systems since it encounters many problems (including fast recombination of charge carriers, photo corrosion, instability in aggressive electrolytes, short lifetime of charge carriers, improper bandgap or diffusion length of photogenerated electrons and holes). As a result, most of the metal oxide semiconductors are not suitable to split water at the visible light irradiation, which occupies 54% of whole solar spectrum since they either do not have proper bandgaps or only absorb the UV light region. However, the problems stated above have been successfully addressed by introducing heterojunctions, composite nanomaterials, coupling wide band and narrow band materials, doping, surface–interface engineering, dye sensitization, etc.

Among the metal oxides, WO_3 is a promising semiconductor for PEC water-splitting with favorable properties. (These properties include: suitable bandgap (~2.6 eV), good chemical stability under strong solar exposure, oxygen-evolution capability, long minority carrier diffusion length (~500 nm–6 μm [19,20]), absorption of visible light (~12%) and low cost.) The conduction band energy position of WO_3 is 0.25 eV, which is not suitable for reorientation of bonds of hydrogen atoms from the aqueous phase to the gaseous (0 V vs. NHE). On the other hand, the valence band, located at 2.7 eV, is more positive than the oxidation potential of oxygen (1.23 V vs. NHE) and is suitable for oxygen evolution. Although the WO_3 photocatalyst suffers from some limitations such as sluggish charge transfer [21], boosting charge separation can be achieved by modifying WO_3 photoanode with numerous materials including Ag nanoparticles [22] and Au plasmonic particles [23]. Many papers have reported on using WO_3 photoanodes for O_2 evolution study [24,25]. During the study of hydrogen evolution from aqueous phase, various photoelectrochemical systems and configurations integrated with WO_3 and its composites have been developed.

Among the published materials in this prospect, numerous amounts of work can be distinguished: Ji et al. reported a triple layer heterojunction $BiVO_4/WO_3/SnO_2$ material with a perovskite solar cell [26], Liu. et al. prepared a WO_3 photoanode with a tandem cell [27] and Lee used dye-sensitized solar cells to produce hydrogen with bare WO_3 photoanodes [28]. Zhang fabricated the $WO_3@a$–$Fe_2O_3/FeOOH$ photoanode, which exhibits a 120 mV negative shift in onset potential and yields a photocurrent density of 1.12 mA/cm^2 at 1.23 V vs. reversible hydrogen electrode (RHE) [29]. Moreover, some systems use free-particle WO_3 heterostructures based on photochemical cell reactions. Despite the fact that WO_3 cannot generate hydrogen, there are some publications where scientists show high photocatalytic activity for CdS-WO_3 [30] and non-stoichiometric WO_{3-x}/CdS heterostructures for efficient hydrogen generation [31].

Thus, we have conducted a literature survey on the WO_3-based photocatalytic system and found a dramatic increase of publications recently (Figure 1). This indicates that the WO_3 is a very important material for designing efficient photocatalytic systems. The analysis of scientific articles, reviews and conference materials found in the authoritative database revealed few review papers in the use of tungsten trioxide photocatalyst for water-splitting. As shown in Figure 1, the trend of the published papers in the WO_3 photocatalytic research is increasing exponentially. Therefore, in our opinion, it is essential to present a review article to our scientific community with recent research progress of WO_3 in the photocatalytic water-splitting. Although there are some review papers that included the WO_3 and their water-splitting applications, from the best of our knowledge, few papers have been specifically focused on sole WO_3/nanocomposites and their recent photocatalytic application.

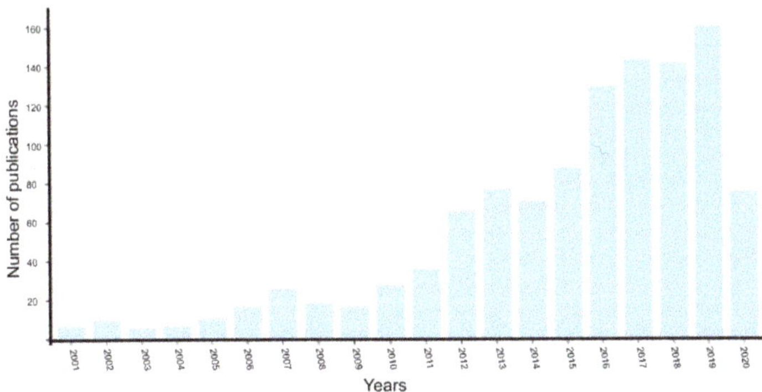

Figure 1. Statistics analysis from Web of Science indicates increase of recent publications in WO$_3$ photocatalytic areas.

The primary focus of our review article is to deliver the recent progress of tungsten-based photocatalytic systems that have been developed. More specifically, it discusses the morphology, crystal, doping, surface–interface engineering effect of WO$_3$ on the heterostructured photocatalytic system, and all of the results in different conditions including electrolyte, power, applied bias, morphology, and synthesis approaches were tabulated for the researchers to check. Therefore, in this review, we try to give comprehensive information on WO$_3$ including the physical chemistry property, crystal structure, and nanomorphology along with their composites including binary and ternary structures used in the particulate, PEC, Z-scheme and tandem configuration for effective water-splitting applications.

2. Basic Principles of the Water-Splitting Reaction

2.1. Thermodynamics of Water-Splitting

In the reaction of water-splitting, solar energy can be directly converted into chemical energy form, hydrogen gas [32–34]. The hydrogen acts as a green energy carrier since it possesses high energy density. When used in a fuel cell, water is the only byproduct.

As early as 1923, J. B. S. Haldane, a British scientist, proposed a concept of photocatalytic hydrogen production. Seeing that there is no naturally produced pure hydrogen on Earth, its resource is highly abundant throughout the universe. Like fossil fuels, water or biomass can be utilized to produce hydrogen or other chemical fuels. Hydrogen gas can be further used in hydrogen fuel cell-powered vehicles, which are much more environmentally friendly than the commonly used nonrenewable fuel options. Increasing the efficiency of water-splitting devices for hydrogen fuel production has a potential to decrease its dependence on using fossil fuels and importation.

There are some other approaches for hydrogen production, however, the most environmentally sustainable, "green" method is photocatalytic or photoelectrochemical water-splitting. The PEC water-splitting works similarly to a solar cell. The main difference is that it converts solar energy to a chemical bond instead of converting directly to electric power, which is beneficial to store energy for later use. PEC consists of three main components: an anode, a cathode and an electrolyte (aqueous media). At the anode, water is oxidized to generate oxygen via the oxygen evolution reaction (OER), whereas at the cathode hydrogen ions are reduced into hydrogen gas via a hydrogen evolution reaction (HER). Based on the configuration of the PEC cell, either the cathode or anode, or both, can be photoactive semiconductors which absorb light. Furthermore, water can also split via connecting a p–n junction solar cell in parallel with a photoelectrochemical cell. This process not only avoids the complicated manufacturing process, but it also reduces the system's cost [35]. Although extensive research has been conducted using many semiconductor configurations, there is still so much that needs to be done to reach the targeted efficiency and stability goals. For a particle-based photo catalytical system, the ideal solar to hydrogen (STH) efficiency should be 10% [36,37]. This efficiency brings cheaper H_2 production.

In 1972, Japanese scientists, Fujishima and Honda first studied TiO_2 as a photonic material and proved that water can be decomposed under UV-light exposure [38]. Since then, scientists have been studying a variety of light-sensitive material, including all inorganic and organic dyes [39–41].

Decomposing water into H_2 and O_2 is an endothermic reaction thermodynamically (+237.2 kJ/mol). This means that additional energy is required to perform the decomposition reaction (E-1):

$$\Delta G^0 = -nF\Delta E^0 = +237.2 \text{ kJ/mol } H_2 \tag{1}$$

where:

F—Faraday's constant (F = 96,485 C/mol),

n—Number of transferred electrons (n = 2)

ΔE^0—standard potential of the electrochemical cell (ΔE^0 = 1.229 V).

The amount of Gibbs free energy required to split a molecule of water into hydrogen and oxygen is ΔG = 237.2 kJ/mol, which is corresponded to $\Delta E^0 \approx 1.23$ eV per electron, transforming the Nernst equation under standard conditions. This means a minimum energy of 1.23 eV per electron should be supplied by the photocatalyst. This process can be written in the following two half-reactions (E-2; E-3; E-4):

$$\text{Water oxidation: } H_2O + 2h^+ \rightarrow \frac{1}{2} O_2 + 2H^+ \text{ (HER)} \tag{2}$$

$$\text{Water reduction: } 2H^+ + 2e^- \rightarrow H_2 \text{ (ORE)} \tag{3}$$

$$H_2O \rightarrow \frac{1}{2} O_2 + H_2 \; \Delta G = +237.2 \text{ kJ/mol} \tag{4}$$

The bandgap (Eg) is the main parameter that defines the light-harvesting ability of an absorber. Photons alone with energies higher than the bandgap can excite electrons in the valence band to the conduction band. The excess energy or the difference in the energy of the absorbed photon and the band gap energy ($E-E_g$), is lost as phonons. The absorption coefficient of the semiconducting materials is another parameter which shows how efficiently a photocatalyst can harness the solar spectrum. One crucial point that needs to be taken into consideration as quantifying the optimal minimum band gap value is the intrinsic loss (E_{loss}), associated with the solar energy conversion process. These losses are connected with the fundamental loss caused by thermodynamics because of non-ideality (kinetic losses) in the conversion process [35,38]. The former loss results from the second law of thermodynamics. In fact, the following equation shows how bandgap energy (Eg) corresponds to the change in internal energy, which is related to the change in Gibbs energy (E-5):

$$\Delta G = \Delta U + P\Delta V - T\Delta S \tag{5}$$

where U, P, V, T and S indicate the internal energy, pressure, volume, temperature and entropy, respectively. When the semiconductor absorbs photons, increasing excited states can be created in addition to ground states, increasing the entropy of the ensemble. The change in entropy or ΔS_{mix}, occurs because there are existing excited states along with the ground states. A volume change (ΔV_{mix}) is also caused by the mixture of excited and ground states. However, this is not true for the ideal chemical system (ΔV_{mix} = 0). Thus, the band gap energy should be greater than the available work under ideal conditions (Gibbs energy change per electron), at least $E_{loss} = T\Delta S_{mix}$ with a minimum of 0.3–0.5 eV. In reality, E_{loss} reaches higher values (roughly 0.8 eV) as a result of kinetic losses and due to non-ideality (overpotential at the anode and cathode, reduction in resistance at the electrolyte, electron–hole pair recombination). Therefore, in order to maximize the chemical conversion efficiency, materials commonly used as photoelectrodes in PEC cells require a band gap of 2.0 to 2.25 eV [35,38].

When UV and/or visible sunlight shine onto a semiconductor photocatalyst, the semiconductor absorbs photons and excites electrons from the semiconductor's valence band to its conduction band, leaving a hole in the valence band, i.e., electron–hole pairs (Figure 2a). This is the so-called, "photo-excited" semiconductor phase. The bandgap is the difference between the maximum valence band energy and the minimum conduction band energy. Ideally, semiconductors have a bandwidth greater than 1.23 V as well as a more negative conduction band relative to the water reduction potential and a more positive valence band relative to the water oxidation potential.

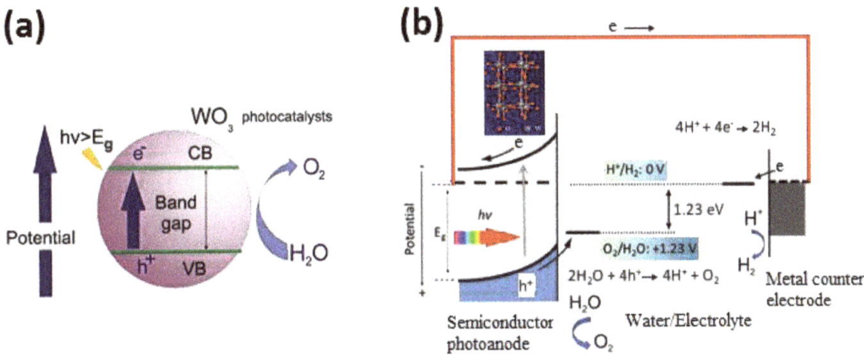

Figure 2. (**a**) Schematic illustration for WO$_3$-particle-based photocatalyst system; (**b**) principle of photoelectrochemical water-splitting.

A typical one-step PEC configuration for water decomposition consists of either a photoanode or a photocathode. N-type tungsten trioxide is mostly represented as a photoanode and the basic principles of such cell can be depicted in Figure 2b. Process of PEC water decomposition is initiated via accepting light photons by photoactive materials. Then, this step is accompanied by generating excitons (electron–hole pairs) inside semiconductors. Photogenerated holes on the surface of WO$_3$ can oxidize water while electrons flow to the Pt electrode to produce hydrogen (Figure 2b). Due to the improper positioning of the conduction and valence bands with respect to the potentials of water reduction and oxidation, external bias voltage is used to separate excitons.

Another thermodynamic precondition is the position of the band edges. For the oxidation reaction to occur, holes move from the photoelectrode to the interface between the semiconductor and the solution freely. The top edge of the valence band must be more positive than the oxidation potential of O$_2$/H$_2$O as seen in Figure 2b. Likewise, the reduction reaction happens if the bottom edge of the conduction band is more negative than the reduction potential of H$^+$/H$_2$.

Figure 3 shows the band structure and bandgap values of some semiconductors [42]. While wide band gap (Eg > 3 eV) photocatalyst can harvest only UV light (a small portion of the solar spectrum, less than 4%), its band gap can be easily engineered to absorb the visible light range via metal and nonmetal doping. Furthermore, narrow bandgap materials (e.g., WO$_3$, Fe$_2$O$_3$) are not able to drive the water reduction and oxidation reactions at the same time since their bandgap energy positions are not properly positioned to the water redox potentials. Therefore, they are commonly used to construct tandem cell structures for the water-splitting reaction.

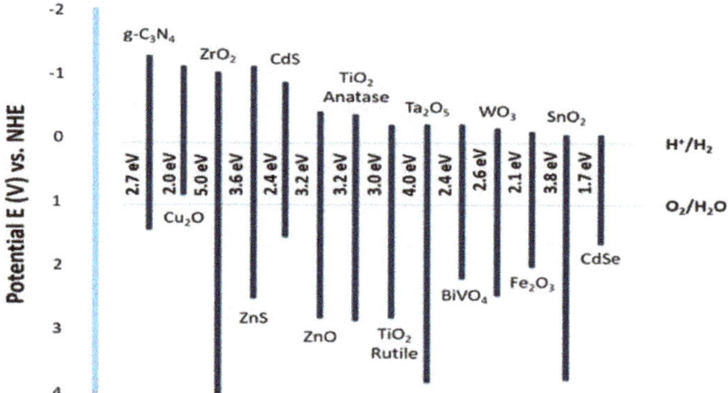

Figure 3. Bandgap positions of typical semiconductors for water-splitting. Reproduced from [42], with permission from MDPI, 2016.

2.2. Device Requirements and Calculation of Their Efficiency

Various types of overall water-splitting techniques include: particulate systems [43], Z-schemes [44–46] and photoelectrochemical cells [46]. The photoelectrodes of PEC water decomposition such as a photoanode and a cathode electrode are made up of photocatalysts. However, fabrication of stable photoelectrodes under the influence of strong sunlight is still a challenging one in PECs. Right now, it is preferred to fabricate the film via direct growth on the photoelectrode, which provides a relatively stable photoactive films.

In order to design an efficient and affordable solar hydrogen production PEC system, the electrode requires low cost materials with the capability of efficient light harnessing and long term stability.

To date, large band gap semiconductors (UV-active specifically) and metal oxides have been extensively investigated for the photocatalytic water-splitting studies due to their robustness and suitable band gap energy positions. One challenge to using these materials is the limitation of solar light harnessing to a small portion of the solar spectrum.

The following formula helps us calculate the theoretical maximum photocurrent J_{max}, which is proportional to the solar–hydrogen conversion efficiency (STH):

$$J_{max} = q \int \Phi_\lambda \left[1 - \exp(-\alpha_\lambda d)\right] d\lambda \tag{6}$$

where λ, q, d and α_λ represent wavelength, electron charge, sample thickness and absorption coefficient under the photon flux of the AM 1.5 G solar spectrum, respectively. Considering the conversion, reflection and other losses, obtaining the goal of 10% STH conversion efficiency is very challenging.

The following efficiencies are usually reported for PEC cells. They are STH conversion efficiency, applied bias photon-current efficiency (ABPE), external quantum efficiency and internal quantum efficiency (IQE) or absorbed photons to current efficiency (APCE).

STH efficiency is commonly used to evaluate PEC device performance and is expressed in the following way:

$$STH = [(H_2 \text{ production rate}) \times (\text{Gibbs free energy per } H_2)]/([\text{incident energy}]) \tag{7}$$

The E-8 formula can be applied to calculate ABPE:

$$ABPE = [J_{ph} \times (1.23 - V_b)]/P_{total} \tag{8}$$

where J_{ph} is the photocurrent density as a bias V_b is applied and P_{total} is the total incident solar light power.

External quantum efficiency defines the photocurrent generation per incident photon flux under a certain irradiation wavelength. Solar-to-hydrogen conversion efficiency can be evaluated via applying the external quantum efficiency data over the total solar spectrum in a two-electrode system. However, applying external quantum efficiency data obtained in a three-electrode system under a bias to estimate solar-to-hydrogen conversion efficiency is not considered to be a valid method. However, it is still considered to be a useful approach for finding PEC cell material properties. The external quantum efficiency (EQE) is expressed by equation (E-9):

$$EQE = (J_{ph} \times hc)/(P_{mono} \times \lambda) \tag{9}$$

where J_{ph} is the photocurrent density, h is Planck's constant, c is the light speed, P_{mono} is the power of calibration and monochromatic illumination, and λ is the wavelength of monochromatic light.

3. WO$_3$ and Its Nanocomposites for Particle-Based Photocatalytic Systems

3.1. Half Reaction Systems

Nowadays, the pursuit for highly efficient photocatalytic materials to produce hydrogen fuel under the exposure of light photons is still in the active stage. In a photoelectrochemical cell, it is required to create sufficient voltage between the anode and cathode to perform the water decomposition reaction. However, most of wide bandgap semiconductor materials are not able to respond to the visible part of the spectrum. Absorption of ultraviolet radiation alone is an undesirable property of photocatalyst operating in terrestrial conditions. One of the exciting ways to solve the above contradictions is the creation of photocatalytic systems consisting of a series of photocatalysts. That is why researchers try to use photochemical systems, where water can be decomposed using colloid particles without any external voltage. Many papers have reported [47] that hydrogen can be generated, even though the efficiency is very low. Evolution of oxygen is difficult because it requires process of four electrons and four H$^+$ transfers.

Under solar illumination, although photoexcited electrons and holes are produced, they simultaneously experience recombination and back reaction, which are competitive processes of photogeneration. Hence, most works focus only on half reactions where either H$_2$ or O$_2$ evolution is possible in the presence of sacrificial electron donor or acceptor.

C$_d$S/WO$_3$ photocatalysts produced a high hydrogen evolution rate of 369 µmol/gh with lactic acid as an electron donor [30]. Further modification of C$_d$S/WO$_3$ with Pd particles increased the hydrogen evolution rate to 2900 µmol/gh, 7.9-fold higher than for C$_d$S/WO$_3$.

Furthermore, the surface plasmon resonance (SPR) effect of non-stochiometric WO$_{3-x}$ was demonstrated [31] from C$_d$S/WO$_{3-x}$ heterostructures photocatalysts via photoinduced electron injection for hydrogen evolution. The non-elemental metal plasmonic material WO$_{3-x}$ has intense SPR in the visible/NIR region (Figure 4b). Free electrons in the conduction band of WO$_{3-x}$ can be generated from oxygen vacancies that are results of chemical reduction during synthesis. Further excitation of electrons can happen by SPR and then they can become hot electrons for the hydrogen generation as shown in Figure 4a. Photo-excited electrons on CdS inject into conduction band of WO$_{3-x}$, so that the SPR of the photocatalyst WO$_{3-x}$ is stable and some hot electrons participate in hydrogen evolution reactions (Figure 4c).

In addition, to choose the photoanode material for the half-reaction of water-splitting, attention should be paid to the selection of the electrolyte. For sulfide semiconductors and composites, a Na$_2$S/Na$_2$SO$_3$ mixture is used as an absorbing hole agent. In type II heterojunctions, for example, WO$_3$–NS/C$_d$S–NR, with high conductivity, WO$_3$ provides efficient charge collection and, therefore, reduces the rate of space charge recombination, which leads to the accumulation of holes in cadmium sulfide. An electrolyte based on Na$_2$S/Na$_2$SO$_3$ provides fast hole collection, which allows the half-reaction to occur without degradation of the photoanode [48]. The effect of some electrolyte solutions on the oxidative half-reactions of WO$_3$-based photoanodes was studied on [49]. James C.

Hill and Kyoung-Shin Choi studied photo-oxidative processes in chloride solutions, acetate solutions, phosphate solutions, perchlorate solutions, sulfate solutions and solutions with K^+ and Li^+ cations. The electrodeposited porous WO_3 layers were used as a photoanode. The results show that the presence of acetate and chloride ions suppressed the release of O_2. In a phosphate solution, the release of O_2 and the formation of peroxides was the main result of photooxidation. The oxidation of water in perchlorate electrolytes was accompanied by the release of O_2 and the formation of peroxides. In this case, the photocurrent density in such a system was lower in comparison with phosphate electrolytes. The authors also showed that cations have a significant effect on the efficiency of conversion of the photocurrent to O_2. For example, Li^+ ions adsorbed on the surface of WO_3 serve as blockers of water oxidation centers, while K^+ ions increase oxygen evolution in perchlorate, sulfate and phosphate solutions.

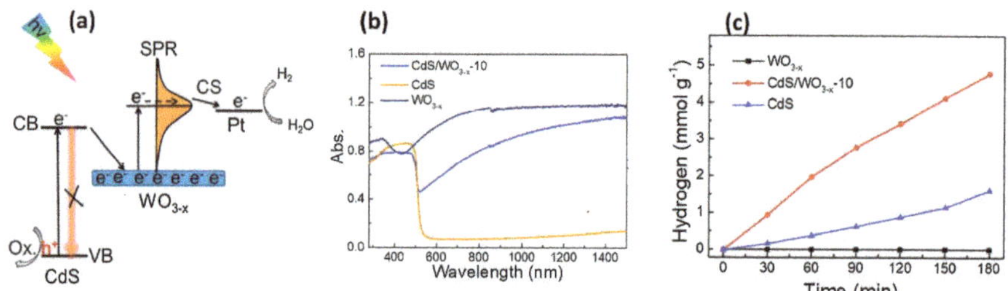

Figure 4. (**a**) Graphic illustration of charge transfer for CdS/WO_{3-x} composite; (**b**) DRS spectra of WO_{3-x}, CdS and CdS/WO_{3-x-10}; (**c**) hydrogen generation for CdS nanowires, WO_{3-x} and CdS/WO_{3-x-10} composites in 20 vol% lactic solution under illumination. Reproduced from [31], with permission from Elsevier, 2018.

The effect of tungsten trioxide layers on hydrogen reduction processes also demonstrates positive dynamics. When combining Cu_2O with WO_3, a semiconductor p–n junction is created and that generates the conditions for the rupture of photogenerated excitons. Thus, in the Cu_2O/WO_3 heterostructure, an enhancement of the half-reaction of reduction is demonstrated in comparison with the sole Cu_2O photocatalyst.

3.2. Z-Schemes

Z-scheme photocatalysts for overall water-splitting are a combined system involving two photon excitation processes (Figure 5) [50]. The two–photon excitation system was proposed by Bard et al. in 1979 which mimicked natural photosynthesis [51]. A Z-scheme is composed of one H_2 evolution photocatalyst and another O_2 evolution photocatalyst with electron mediator. Most of Z-scheme construction was demonstrated using Pt co-catalyst loaded $SrTiO_3$, TaON, $CaTa_2O_2$ N and $BaTa_2O_2N$ for hydrogen evolution and Pt/WO_3 for oxygen evolution. Photocatalytic activity depends on pH level, concentration of electron mediator and type of co-catalysts. For example, Hideki Kato showed that pH level affects photocatalytic activity of Z-scheme consisting of $Pt/SrTiO_3Rh–WO_3–FeCl_3$ system [52]. The pH value was adjusted using sulfuric and perchloric acids between 1.3 and 2.5. It was shown that the best photocatalytic activity was achieved at pH 2.4 and subsequent increasing of pH led to decrease of the activity. The sulfate ions-induced formation of $[Fe(H_2O)_5(SO_4)]^+$ species around pH 2.4. Under 48 h of solar illumination, the Z-scheme generated both 890 and 450 µmol H_2 and O_2, respectively.

Yugo Miseki reported a Z-scheme system with an oxygen evolution photocatalyst of $PtO_x/H–Cs–WO_3$ [53]. The Z-scheme water-splitting efficiency with $PtOx/H–Cs–WO_3$ was 3-fold higher than that of using PtO_x/WO_3. Adding Cs^+ ions to the PtO_x/WO_3 significantly improved the oxygen evolution rate. IO_3^- ion was used as an electron acceptor in this work. The apparent quantum yield at 420 nm was 20% which is the best index among photocatalysts using the IO_3^- redox.

Another Z-scheme, containing g-C_3N_4–WO_3 photocatalysts, demonstrated enhanced H_2 evolution [54]. The high photocatalytic activity is most likely due to direct electron transfer from WO_3 to g-C_3N_4 in the Z-scheme.

One Z-scheme system consisting of graphitic nitrite g-C_3N_4 and WO_3 nanocomposites modified with co-catalyst $Ni(OH)_x$ showed the highest hydrogen production rate of 576 μmol/(g.h). Photogenerated electrons and holes are efficiently separated by combination of g-C_3N_4 and $Ni(OH)_x$. The electron spin resonance (ESR) technique used DMPO (5,5-dimethyl-1-pyrroline N-oxide) as a trapping agent of $\bullet O_2^-$ and $\bullet OH$ radicals to register the change of the active oxidizing species in aqueous systems. As a result, the Z-scheme charge separation mechanism explained the high hydrogen production rate [55].

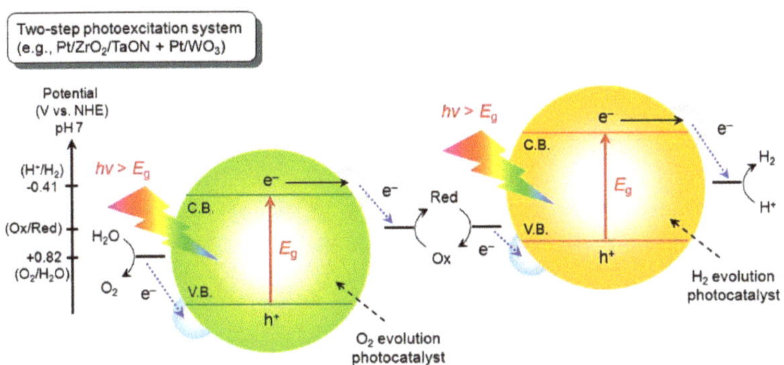

Figure 5. Schematic illustration of photocatalytic water-splitting by Z-system. C.B.—conduction band; V.B.—valence band; Eg—bandgap. Reproduced from [50], with permission from American Chemical Society, 2010.

Recently, construction of Z-scheme using ZnO–WO_{3-x} nanorods was successfully synthesized by the solvothermal method [56]. A high photocurrent value of 3.38 mA/cm^2 at 1.23 V vs. RHE, which is 3.02-fold higher than pure ZnO, was obtained by an effective Z-scheme charges-transfer process. Red shift of optical absorption and better electrochemical performances were achieved by decoration of ZnO nanorods with WO_{3-x} nanoparticles.

Sayama et al. reported [57] a stoichiometric production of H_2 and O_2 using a mixture of Pt-WO_3 and $PtSrTiO_3$ (Cr–Ta-doped) in NaI media. The Pt-loaded $SrTiO_3$ (Cr–Ta-doped) produced H_2 of 0.8 μmol h^{-1} from an aqueous NaI solution while the Pt-loaded WO_3 produced O_2 at an initial rate of 84 μmol h^{-1} in an aqueous $NaIO_3$ solution under visible light (l > 420 nm) separately. The H_2 evolution rate from the mixed photocatalyst system (1.8 μmol h^{-1}) was higher than that from Pt–$SrTiO_3$ (Cr–Ta-doped) alone (0.8 μmol h^{-1}), indicating that addition of the Pt–WO_3 effectively reduced the IO^{3-} ion to I$^-$.

Even though the band position of WO_3 is suitable for O_2 evolution, doping WO_3 with a metal can shift the energy level. Wang [58] studied electronic properties of WO_3 using density functional theory (DFT) calculations with a hybrid calculation. Replacing W by Mo and Cr in the lattice can modify the bandgap of WO_3 and improve absorption of visible light. The effect of replacing O atoms by S anions was simulated by substitution along the Z direction in the unit cell. The DFT results predicted that there is a decrease in energy gap (2.21 eV) as well as a conduction band shift up, which is beneficial for HER. The authors also tested the effect of doping WO_3 with Ti, Zr and Gf metals, resulting in a predicted upward shift of the conduction band like the case with S anions.

Mg-doped WO_3 photocatalysts [59] have been studied experimentally. The conduction band edge for p-type Mg-doped WO_3 was −2.7 eV vs. Saturated calomel electrode (SCE) at pH 12, which is more negative than the reduction potential of H_2. Hydrogen generation of 3 µmol/gh was achieved by doping WO_3 with Mg (5–10 wt%). Doping has also been done using other metals, including Mo [59].

4. Heterostructured WO_3 Nanocomposites for Photoelectrochemical Cell Systems

Photocatalytic activity of WO_3 depends on the crystal structure, morphology and surface areas. High surface areas of WO_3 usually increase the photo activity via providing more reaction sites. The certain morphology increases electron mobility, thus demonstrates better photocatalytic activity. For example, one dimensional WO_3 demonstrates relatively high photoactivity relative to nanoparticles. In the two dimensional WO_3 nanomaterials, it is very important to have optimum grain size which lead to high photoactivity. The crystal structure is critical for the photoactivity of WO_3. Furthermore, monoclinic structure of WO_3 offers different photocatalytic activity relative to other crystalline structure including tetragonal, etc.

Anodization [60,61], solvothermal [56], hydrothermal [62,63], spin coating [64], electrodeposition [65,66] and sol–gel [67] methods were used to fabricate different morphologies and structures.

4.1. Crystalline Structure

Many research efforts have been performed to investigate the effect of crystal structure on the tungsten photocatalytic activity. It was found that the monoclinic crystalline phase demonstrated stronger oxidation activity than other crystal phases such as hexagonal and orthorhombic. The monoclinic phase was found to be the most stable at room temperature [68–72]. Increase of the temperature gradually transformed $WO_3 \cdot 0.33H_2O$ from orthorhombic into anhydrous hexagonal and a final stable form monoclinic (Figure 6a). As the temperature transited from 400 to 500 °C, the color of the film turned into yellowish color obviously, which is corresponded to a red shift (Figure 6b). The photocurrent density increases until 500 °C, then it starts to decrease (Figure 6c). The monoclinic structure of WO_3 at 500 °C showed the highest photoelectrochemical performance, on the contrary the orthorhombic $WO_3 \cdot 0.33H_2O$ exhibited the lowest photocurrent density [68–70,73–75].

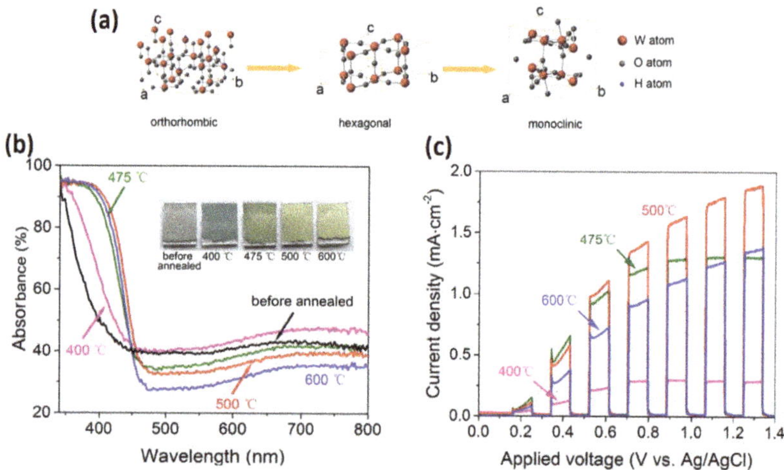

Figure 6. (**a**) Crystal-unit cells for orthorhombic $WO_3 \cdot 0.33H_2O$, hexagonal WO_3 and monoclinic WO_3; (**b**) absorption peaks of WO_3 films with/without heat treatment; (**c**) linear-sweep voltammetry of WO_3 photoanodes at different temperatures under chopped illumination. Reproduced from [68], with permission from American Chemical Society, 2016.

Nayak et al. [76] used combination of a facile precipitation and solvothermal methods to fabricate WO_3 nanowires. The precipitation method produced (Figure 7a–d) $WO_3 \cdot H_2O$ nanoplates with an orthorhombic phase, later the solvothermal approach was used to form WO_3 nanowires with a monoclinic phase (Figure 7e–h). The photocurrent density obtained from WO_3 monoclinic structure was 21-fold higher than that of $WO_3 \cdot H_2O$ orthorhombic phase. This enhancement was ascribed to the presence of different phases between $WO_3 \cdot H_2O$ nanoplates and WO_3 nanowires or the high crystallinity of WO_3 nanowires, which minimized the barrier of charge transfer at the interfacial charge and enhanced the PEC performance.

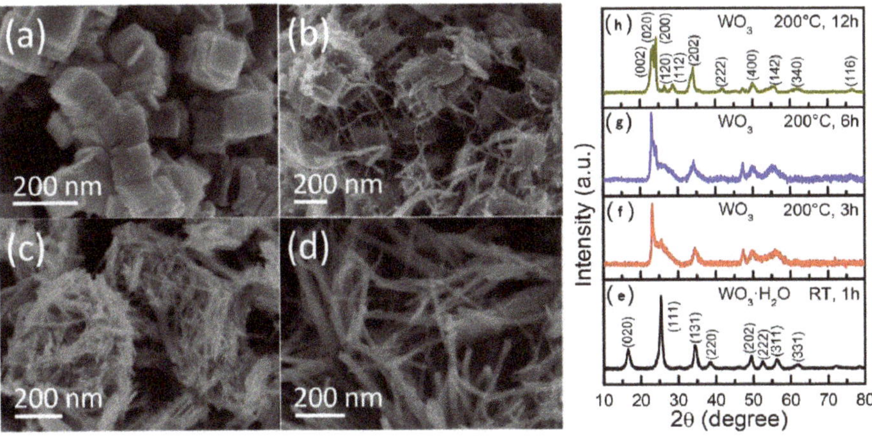

Figure 7. SEM images. (**a**) Stacked $WO_3 \cdot H_2O$ square nanoplates synthesized at room temperature, WO_3 nanowires evolving after (**b**) 3 h, (**c**) 6 h and (**d**) 12 h solvothermal treatment of $WO_3 \cdot H_2O$ nanoplates at 200 °C; (**e–h**) corresponding XRD patterns. Reproduced from [76], with permission from American Chemical Society, 2017.

The effects of crystal phase on the photocatalytic performance has been broadly explored. Park et al. [77] found that the annealing treatment reduced the surface disorder induced by water via improvement of the crystallinity or oxygen deficiencies of WO_3, led to enhancement of the PEC performance. Zeng's group explained the formation of peroxo species on the surface of $WO_3 \cdot H_2O$ as it has low degree of crystallinity. As the annealing temperature of WO_3 reached 500 °C, highly reactive (002) facets were formed to reduce defects, thus to minimize the recombination of electron–hole pairs [75]. The same conclusion was obtained by Su's group [72]. From the above studies, monoclinic WO_3 demonstrated higher PEC performance than that of as-prepared hydrated WO_3.

Recent investigations demonstrated that with surface engineering certain crystal planes possess preferences on the photoexcited electrons and holes, which lead to either preferential oxidation or preferential reduction reactions [78]. Furthermore, photo–electrochemical efficiency has been improved via exposing the high surface crystalline surface [79,80]. Among the three crystal planes or facets of WO_3 which are (200) with 1.43 J/m^2, (020) with 1.54 J/m^2 and (002) with (1.56 J/m^2) facet of WO_3, the crystal facet (002) showed preference for adsorbing the reaction species due to its highest surface energy [81]. Wang et al. confirmed this via DFT calculations [82]. The dangling O atoms of the weakest W–O bond on the (002) crystal plane of the monoclinic WO_3, offer plentiful active sites for H_2O and organic molecules through the hydrogen bond. Oxidization of water and degradation of organics on the (002) easily occur via consuming photo-excited holes and generating active oxygen species, which reduce the recombination of photogenerated carriers [83–85].

The morphology of the WO_3 films can be controlled by synthesis parameters such as synthesis time, temperature and the amount of the capping agent [75]. HRTEM study revealed that annealing

WO$_3$·H$_2$O plates transformed along with the (020) crystal face into WO$_3$ plates with preferentially (002) facet. At annealing 500 °C, the WO$_3$ showed 1.42 mA cm^{-2} at 1.23 V vs. RHE, which is relatively high current density. This is explained due to reduction of peroxo species on the surface of WO$_3$. The high energy crystal plane of WO$_3$ nanoplate enhanced PEC water-splitting. Zhang et al. [83] compared monoclinic WO$_3$ nanomultilayers which has preferable (002) facet with that of WO$_3$ nanorods and found that WO$_3$ nanomultilayers performed higher photocurrent densities than the WO$_3$ nanorods. These results were explained not by the specific surface area of WO$_3$ nanorods, but the presence of highly reactive (002) facets of WO$_3$, which contributed to the improved PEC water-splitting performance.

In addition, increasing studies have been made investigating the effect of the (002) crystal plane of 2D monoclinic WO$_3$ on PEC water-splitting.

To enhance the PEC water-splitting performance of WO$_3$, most of studies have been focused on engineering morphology, crystallinity, heterojunction, oxygen vacancy, doping and co-catalysts for enhancement of photocatalytic hydrogen evolution.

According to the crystalline structure of WO$_3$, it is confirmed that the monoclinic phase of WO$_3$ demonstrated higher OER than the hexagonal or orthorhombic phases since it is the most stable phase at room temperature and presence of highly reactive (002) facets.

4.2. Morphologic Effect

Various WO$_3$ nanomaterials with different morphologies including nanorods [86], nanoflake [87], nanotubes [88,89], nanoplates [90] and nanoparticles [91,92] were synthesized by various methods to provide active sites for catalysis. It was found that morphology change of WO$_3$ can significantly influence photocatalytic activity.

Ma and other authors [93] obtained nanoplates of WO$_3$ by topological method using Na$_2$WO$_4$ and HBF$_4$ and mentioned that intrinsic crystal lattice of tungsten acid plays important role to obtain morphology of final products. The crystal lattice of H$_2$WO$_4$ has (WO$_6$) octahedra layers with normal direction (010) and each layer is linked to each other via hydrogen bonds. That is why H$_2$WO$_4$ tends to form platelike nanocrystals with (010) direction. Another factor affecting morphology is addition of directing agents for nucleation and crystal growth. Interaction of H$_2$WO$_4$ crystal planes and HBF$_4$ can be reason of formation plate morphology of WO$_3$. Meng and others [94] synthesized hierarchical structure using citric acid C$_6$H$_8$O$_7$ and found that (–COOH) functional groups affect growth of nanoplate. They concluded [93,94] that uniform platelike morphology is favorable for gas sensing because it has more active sides for absorption of gas molecules. In addition, much work was done on WO$_3$ crystal growth using fluoroboric acid [95], polyethylene glycol (PEG) [96], polyvinyl alcohol (PVA) [87].

Strategy of increasing active sides for suitable absorption of light is a way to enhance photoelectrochemical performances of photocatalysts. For example, Jiao et al. demonstrated different morphology of tungsten trioxide hydrate (3WO$_3$·H$_2$O) films which grown by hydrothermal method using Na$_2$SO$_4$, (NH$_4$)$_2$SO$_4$ and CH$_3$COONH$_4$ as capping agents, respectively. Platelike, wedgelike and sheet like nanostructures can be obtained as shown in Figure 8 [97]. From Figure 8e it can be seen that sheet like nanostructures had the highest photocurrent density (1.15 mA/cm^2) under illumination and high photocatalytic activity for photodegradation of methanol. This was in good agreement with UV-vis absorbance spectroscopy results (Figure 8d). The authors believe that the reason for high current density of sheetlike morphology can be explained by the existence of small pores among sheets. This may be beneficial for accelerating the interface electron kinetics between the sheet and electrolyte due to its large active surface area.

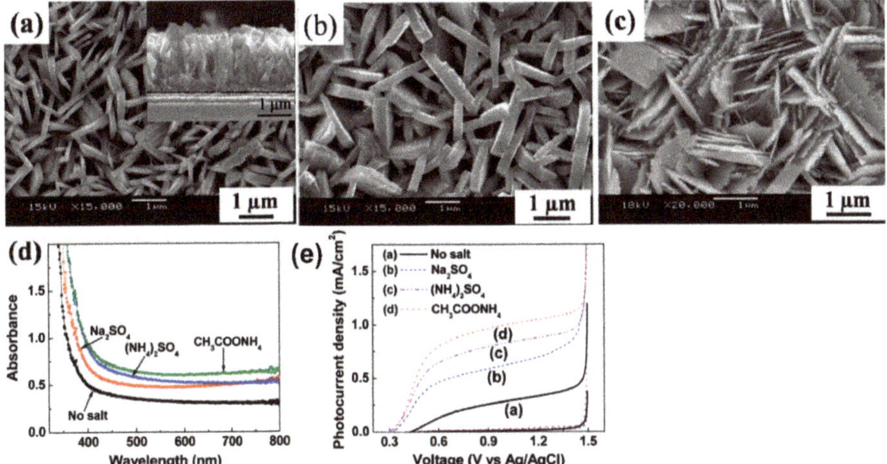

Figure 8. SEM images for WO$_3$ hydrate film with sheetlike nanostructures with (**a**) (NH$_4$)$_2$SO$_4$, (**b**) Na$_2$SO$_4$ and (**c**) CH$_3$COONH$_4$ as capping agent; (**d**) absorption spectra of WO$_3$ films; (**e**) current-potential scans of WO$_3$ films measured in darkness and under illumination in a 1-M H$_2$SO$_4$ electrolyte containing 0.1-M methanol. Reproduced from [97], with permission from American Chemical Society, 2011.

Davidne et al. [98] reviewed WO$_3$ nanostructures and studied the effect of morphology on photocatalytic activity for decomposition organic dyes. Nanostructures such as nanoplate, nanoneedle, nanorods and nanowire were obtained by hydrothermal method. It was found that photocatalytic efficiency has good correlation with band gap, crystalline phase, morphology and oxidation state. Nanoneedles with hexagonal structure showed the best photocatalytic efficiency in contrast to others.

Monoclinic nanorods showed higher photocurrent density than (2.09 mA/cm^2) nanoplates (1.61 mA/cm^2) in the hydrogen evolution reaction (HER) [96]. Some results concluded that 1D structures have high optical, electrical, photoconductivity properties and fast charge transportation [99–101]. Vertically oriented nanorods and nanoneedles have remarkable PEC results [65,66,102] due to high interfacial contact area which improves redox contact area and efficient light scattering.

However, some authors believed that 2D nanostructures like nanoplates have higher specific surface area than one-dimensional (1D) materials such as nanorods and nanowires. For example, Su and others demonstrated better photoelectrochemical characteristics and optical properties of WO$_3$ nanoflakes than WO$_3$ nanorods [87]. Hammad et al. [103] fabricated WO$_3$ nanorods (with a diameter 7 nm, length up to 700 nm) and WO$_3$ nanoplates with width 700 nm on fluorine-doped tin oxide (FTO) substrate via hydrothermal treatment. Results of electrochemical spectroscopy showed that nanoplates have better contact with substrate than nanorods which led to high photocurrent density of 400 µA/cm^2 over 350 µA/cm^2. Through changing concentration of HCl acid, Zhou et al. [104] synthesized perpendicularly oriented WO$_3$ nanorods and nanoplates at different amount of acid. WO$_3$ nanoplate arrays also showed a superior photocurrent density of 1 mA/cm^2 at 1.6 V vs. RHE than nanorods of 0.8 mA/cm^2. A high photocurrent density may be related with a long carrier diffusion length through 2D nanostructures comparing with 1D and efficient charge transportation and separation. Contradictory results between 2D and 1D nanostructures allow them to conclude that comparing different morphologies under different conditions do not give us true information. This is because one morphology can show different results depending on its morphologic parameter such as length, thickness and diameter.

In the aspect of photocatalytic efficiency evaluation in combination with morphology, WO$_3$–BiVO$_4$ nanostructures have been mostly studied. Lee and others [65] fabricated WO$_3$–BiVO$_4$, TiO$_2$–BiVO$_4$,

Fe$_2$O$_3$–BiVO$_4$ and SnO$_2$–BiVO$_4$ nanostructures and showed that PEC characteristics of bare WO$_3$ dramatically increased after modification with BiVO$_4$. The SEM and cross-sectional images of WO$_3$ nanorods coated with BiVO$_4$ are presented in Figure 9. They concluded that pairing WO$_3$ with BiVO$_4$ creates very promising photoanodes for water oxidation than others.

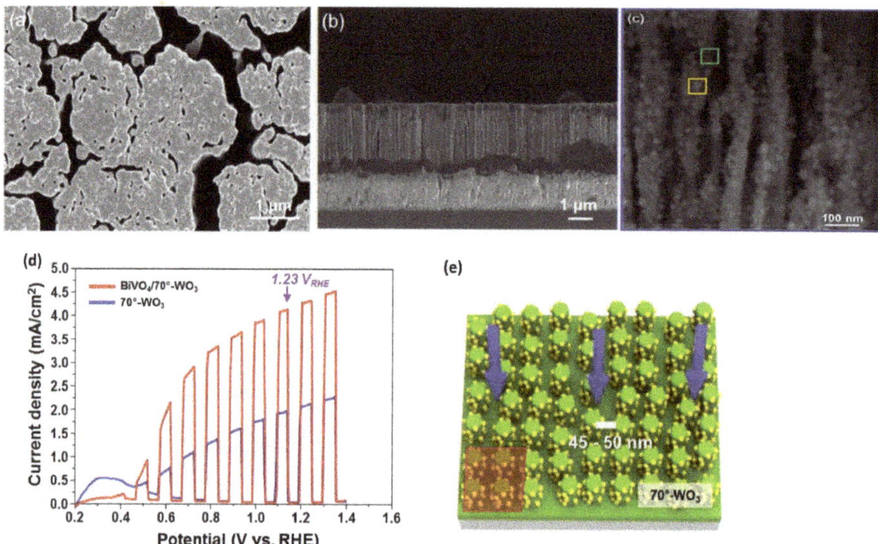

Figure 9. (**a,b**) Top and cross-section SEM images of the optimum BiVO4/70°-WO$_3$ nanorods; (**c**) expanded image of BiVO$_4$/WO$_3$ nanorods, (**d**) enhancement of photocurrent density of BiVO$_4$/WO$_3$ nanorods (70°) nanorods; (**e**) schematic illustration of BiVO$_4$/WO$_3$ nanorods. Reproduced from [65], with permission from Elsevier, 2016.

Improving PEC characteristics can be achieved by also using core–shell structures. Spatial separation of photogenerated charges between the core and shell is beneficial, however, excited charges stay inside and do not react with electrolyte. Nevertheless, fast transportation of charges to surface can diminish shell thickness [105]. Rao et al. [106] synthesized core–shell nanostructures of WO$_3$–BiVO$_4$ to improve light absorption and charge separation. A photocurrent and an incident photon-to-current conversion efficiency reached 3.1 mA/cm^2 and ~60% at 300–450 nm, respectively at 1.23 V vs. RHE under simulated sunlight.

Enhanced PEC performance was obtained by designing yolk-shell-shaped WO$_3$/BiVO$_4$ heterojunction which produced a photocurrent density of 2.3 mA/cm^2 with the highest value of ~5.0 mA/cm^2 after adding a Fe–Ni co-catalyst at a bias of 1.23 V vs. RHE under AM 1.5 illumination (100 mW/cm^2) [107]. These noticeable photocurrent density results demonstrated that core–shell structures may be potentially viable for photocatalytic applications.

Nanostructures with nanoporosity have shown a better PEC activity due to their large specific surface areas, relatively higher light absorption rate and excellent charge collection efficiency [108–110]. A high surface area of porous nanostructures makes them promising electrode materials for electrochemical surface reactions [111–113]. Furthermore, the nanoporous structure creates the depletion layer and reduced diffusion distance to the photoelectrodes/electrolyte interface, which diminish recombination of electrons and holes [114–116]. Song et al. [117] used versatile foaming-assisted electrospinning method to produce mesoporous WO$_3$ nanobelts which enhanced the PEC water-splitting performance compared with the as-prepared WO$_3$ nanofiber and WO$_3$ nanobelt samples. Shin et al. [114] used a laser ablation method to produce tree-like nanoporous WO$_3$ photoanode for a photoelectrochemical water–oxidation performance. Both SEM and TEM image in Figure 10, show 1D treelike morphology with a thickness ~3.2 µm and many clusters with nanoporous with average size of ~60 nm. The photocurrent density of treelike porous structures was 9-fold higher (1.8 mA/cm^2 at 1.23 V vs. RHE)) than dense WO_3. A quantum efficiency (QE) or incident photon-to-electron conversion efficiency (IPCE) was 70% at 350–400 nm.

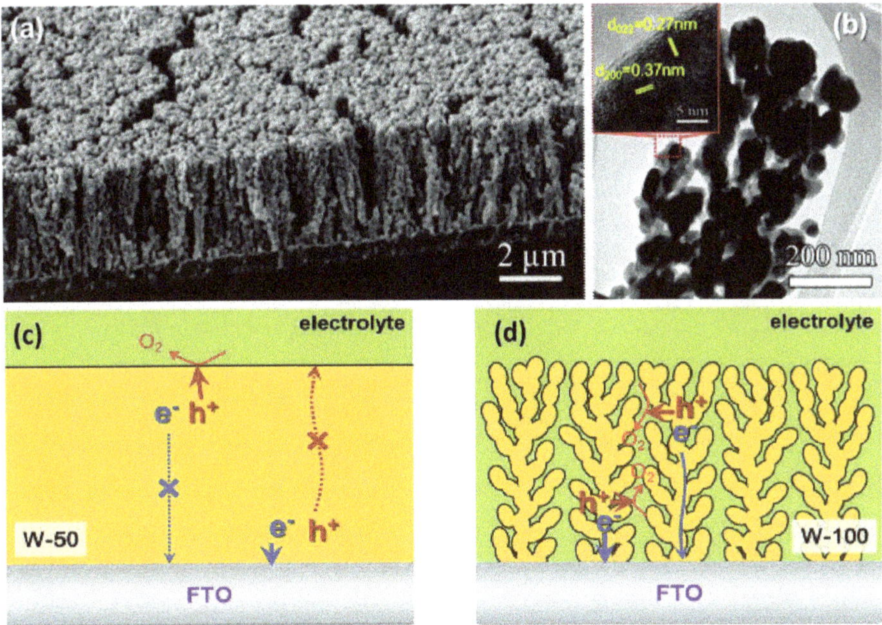

Figure 10. (**a**) SEM image of the WO_3 photoanode; (**b**) TEM image of the nanoporous WO_3 clusters, the inset: a high resolution TEM image; (**c**,**d**) schematic illustrations of charge transport/transfer processes. Reproduced from [114], with permission from Royal Society of Chemistry, 2015.

Fujimoto et al. [115] synthesized porous $BiVO_4$ using the auto combustion method. Adding oxidizing agent NH_4NO_3 and subsequent decomposition of organic additive after heating allowed to create porous film with small crystalline $BiVO_4$ nanoparticles during the synthesis. The optimized $WO_3/BiVO_4$ film produced a maximum IPCE value of 64% at 440 nm with photocurrent density of 3.43 mA cm^2 at 1.23 V vs. RHE (under one sun illumination).

Finally, it is obvious that different morphologies as a factor of synthesis method produce different PEC results and play an important role in configuration of water-splitting devices. However, it is hard to conclude that one morphology is more beneficial than others. In fact, other factors such as substrate on which the structure is grown, electrolyte, capping agents and, etc. may be cause of change in photocatalytic activity.

4.3. Binary Structures of Hierarchical Architectures Based on WO_3 Semiconductors

As mentioned above, single photocatalysts cannot satisfy all the requirements needed for water-splitting PEC systems. Therefore, scientists focus most of their attention on creating different kinds of heterostructure architectures from different various materials such as metal oxide/metal oxides [47,118–120], metal oxide/metals [92,121–123] and metal oxide/inorganic compounds [124] to create efficient systems for various spheres. Mixing several materials is a method commonly used to improve separation of charge carriers, photoelectrode stability, absorption of visible light, suitable carrier diffusion length and effective surface charge transfer.

4.3.1. Metal Oxide/Metal Oxide Binary Heterostructures

The sensitivity of TiO_2 can also be obtained by modification the surface using semiconductor photocatalyst with a smaller bandgap as WO_3 [125–127]. A WO_3/TiO_2 photocatalytic system was published in 1998 [128]. The photocatalytic activity of WO_3 coated with TiO_2 was tested for the oxidation of water using iron (III) acceptor. It was also found that iron (III) ion is preferred more than iron (II) ion as an electron acceptor for oxygen evolution.

Amorphous α-TiO_2 can be used to passivate the surface of WO_3 based nanostructured photoanodes. For example, Yang and et al. [129] demonstrated high photocurrent with 1.4 mA/cm^2 at 0.8 V in 0.1-M Na_2SO_4 electrolyte using WO_3 nanoflakes coated amorphous α-TiO_2 films. Passivation of WO_3 by α-TiO_2 was realized through the $O^{2-}W^{6+}$ bonding at contact surface between WO_3 and α-TiO_2. Hence, passivation of surface allows to decrease recombination and improve PEC oxidation.

Hierarchical WO_3/TiO_2 composites for hydrogen evolution was fabricated by Momeni [130] using the anodization method. TiO_2 nanotubes with a diameter ranged 80–110 nm were modified by WO_3. Controlling the concentration of the Na_2WO_4 solution allowed them to achieve the highest amount of H_2, with 2.14 mL/cm^2 under 120 min of solar illumination, which is approximately 3.02-fold higher than bare samples with TiO_2 nanotubes (0.71 mL/cm^2). It also showed increase of photocurrent value from 0.81 to 1.61 mA/cm^2 after modification proved the effectiveness of the coupled WO_3/TiO_2 system. The anodization method was also successfully used to prepare the hybrid WO_3/TiO_2 nanotube photoelectrodes [131] which showed better photo conversion efficiency, STH efficiency and H_2 generation.

Many other studies highlighted that coupled WO_3–TiO_2 systems have better characteristics. For example, the highest photocatalytic activity of nanocomposites particles for degradation of Rhodamine B [60,61], decomposition of 1,4-dichlorobenzene (DCB) aqueous solution [132] and azo dyes [133], for effective catalytic oxidation cyclopentene to glutaraldehyde [133] were obtained.

The effectiveness of photocatalysts was also attained by engineering morphology and specific surface area of the material since electron-hole transfer occurs on the surface. According to some studies, although electron–hole pairs can be generated in volume, they can annihilate before they reach the surface due to the low diffusion length.

Most of authors have shown that dual heterostructures of WO_3 and $BiVO_4$ are effective for driving water oxidation reactions [64,134,135]. $BiVO_4$ is an n-type semiconductor-like WO_3 which has a bandgap around 2.4 eV [26,136]. The theoretical solar-to-hydrogen efficiency using this material (9.8%) is more than that of WO_3 (4.8%)-based systems. Despite the fact that $BiVO_4$ is a direct semiconductor unlike WO_3, it has a poor charge transport and a short hole-diffusion length (~60 nm). Coupling WO_3 with $BiVO_4$ decreases recombination of photogenerated charge carriers and improves the efficiency of overall water-splitting systems. A dynamic of photogenerated carriers and effective charge separation of WO_3–$BiVO_4$ heterojunctions was explained by Grigioni using femtosecond transient absorption spectroscopy [137]. They determined the position of the WO_3 conduction bands (+0.41 V vs. RHE) and $BiVO_4$ (+0.02 V vs. RHE) by testing the photocatalytic reduction of thionine. The charge separation mechanism of $BiVO_4$–WO_3 system is shown in Figure 11a. When comparing the flat band position of $BiVO_4$ in the WO_3–$BiVO_4$ composite, a shift of 170 mV is observed. This shift was explained by electron equilibrium between the two materials due to a Fermi level shift. Photoelectrons transfer from $BiVO_4$ to WO_3 while holes localize in $BiVO_4$, so it is possible to separate photogenerated charge carriers spatially (Figure 11b).

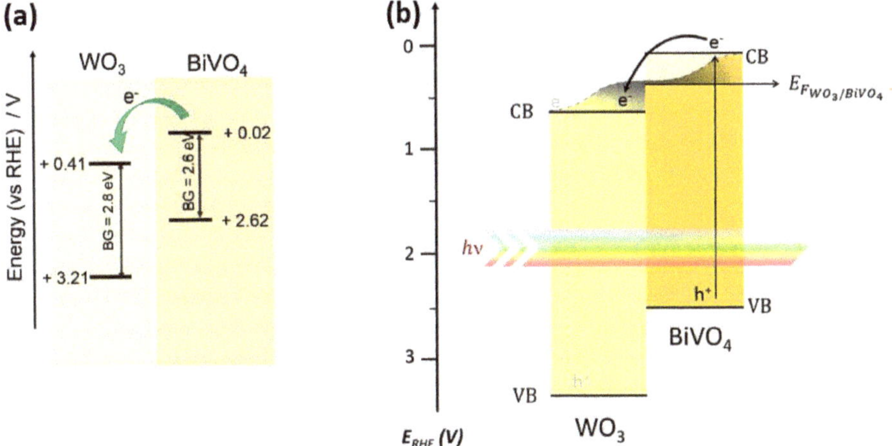

Figure 11. (a) Diagram of the band edge positions of pure WO$_3$ and BiVO$_4$ and (b) for a WO$_3$–BiVO$_4$ composite under solar irradiation. Reproduced from [137], with permission from American Chemical Society, 2015.

The morphology of WO$_3$ and the decoration method with BiVO$_4$ are also very important issues. Chae et al. synthesized mesoporous WO$_3$ films followed by a coating of BiVO$_4$ to research particle sizes and shapes, as well as the effects of the photoanode thickness. Large nanoplates showed a high injection efficiency while nanospheres enhanced the charge-separation efficiency [138]. Pihosh et al. synthesized WO$_3$–BiVO$_4$ vertically oriented nanorods by combining the glancing-angle deposition and normal physical sputtering techniques [66]. The photocurrent density achieved 3.1 mA/cm^2 at 1.23 V RHE under illumination of one sun. A nanopillar morphology of WO$_3$–BiVO$_4$ photoanodes fabricated by electrostatic spraying method also produced a photocurrent up to 3.2 mA/cm^2 [139]. An enhanced photocurrent density of 4.55 mA/cm^2 was achieved by using a WO$_3$–BiVO$_4$ photoanode [65]. Deposition of BiVO$_4$ nanodots on WO$_3$ nanorods had an increased photon to hydrogen efficiency of 80% at 1.23 V vs. RHE, which is higher than the theoretical efficiency for bare BiVO$_4$. Rao et al. fabricated WO$_3$–BiVO$_4$ core shell nanowires and showed that the photoanodes demonstrated a $\eta_{arb} \times \eta_{sep}$ up to 53%. A combination of BiVO$_4$ with more conductive WO$_3$ leads to effective charge carrier separation and the photocurrent achieved 3.1 mA/cm^2 at 1.23 V vs. RHE.

Iron is an abundant and important metal in the earth's crust, so its use is considered economically viable. Oxidation of iron can lead to formation of the known hematite phase α-Fe$_2$O$_3$ which has semiconductor properties. It has good stability in most electrolytes pH > 3 and has a narrow bandgap (~2.2 eV) which can absorb 40% of the solar spectrum. Although hematite electrodes are well studied for PEC system, photoconversion efficiency is still lower than the theoretical value due to low hole mobility (~2–4 nm). Moreover, poor electrical conductivity, high recombination rate of electron-hole pairs [140] and the slow kinetics of oxygen evolution [141] limits its use. Some studies focus on binary heterostructures with WO$_3$–hematite α-Fe$_2$O$_3$ photoanodes [142]. A photocurrent of 1.66 mV/cm^2 was observed at 1.23 V RHE, while the photon to current efficiency was 73.7% at 390 nm. Schematic illustration of WO$_3$-Fe$_2$O$_3$ composite nanosheets and bandgaps are shown in Figure 12a. The optical absorption measured by a UV-vis diffuse reflectance spectroscopy was found to be improved for the composite WO$_3$ and Fe$_2$O$_3$ material (Figure 12b). Luo published enhanced electrochemical characteristics of a WO$_3$@Fe$_2$O$_3$ photoelectrode compared to bare WO$_3$ and Fe$_2$O$_3$ [143]. Effective photoelectrochemical splitting of seawater with Fe$_2$O$_3$/WO$_3$ nanorods was achieved by Li et al. [144]. Although optical absorption is promising, the photocurrent of Fe$_2$O$_3$/WO$_3$-based photoanodes is still low.

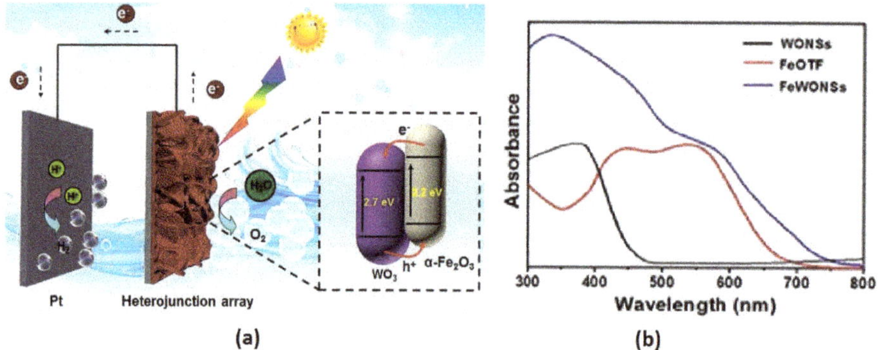

Figure 12. (**a**) WO_3–α–Fe_2O_3 nanocomposites and its band diagram for PEC water-splitting (**b**) UV-vis absorption spectra of $WO_3@Fe_2O_3$. Reproduced from ref [142], with permission from John Wiley and Sons, 2016.

4.3.2. Metaloxide/Inorganic Compounds Heterostructures

The heterostructures formed from WO_3 and sulfur components have narrow bandgaps. For example, antimony sulfide, Sb_2S_3, (1.7–1.9 eV), bismuth sulfite, Bi_2S_3, (~1.3 eV) and tungsten disulfide, WS_2, (~1.3 eV) [145] are very effective. For example, Zhang [146] synthesized WO_3/Sb_2S_3 heterostructures via a simple hydrothermal method to improve PEC performances. Tungsten trioxide nanorods and nanoplates were synthesized by controlling the concentration of acid and tungsten precursor along with subsequent growth of Sb_2S_3 nanoparticles. It was demonstrated, that WO_3/Sb_2S_3 heterostructures have better electrochemical and optical characteristics than pristine WO_3.

A high photocurrent of 5.95 mV/cm² was achieved using a three-dimensional WO_3/Bi_2S_3 heterojunction [147]. Bi_2S_3 is also a n-type semiconductor with bandgap 1.3 eV and has more negative conduction band edge than WO_3. A WO_3/Bi_2S_3 heterojunction was fabricated by combining of hydrothermal method, SILAR (successive ionic layer absorption and reaction) process and CBD (chemical bath deposition). Relatively high light absorption, small electron transfer impedance and high charge carrier were proved by UV-vis, EIS (Electrochemical Impedance Spectroscopy) and Mott-Shottky methods.

Despite the fact that the WO_3 bandgap energy is not suitable for hydrogen evolution, it is still useful for solving problems such as high electron-hole recombination rates and poor electrical conductivity of some photocathodes [145]. A study of $WO_3@WS_2$ core–shell nanostructures fabricated by plasma assisted sublimation was published by Kumar et al. [145]. The highest achieved cathodic photocurrent was 16.2 mA/cm² for WS_2 at 0.3 V vs. RHE. Sulfurization of the WO_3 surface forms a WS_2 layer with a rich defect structure, resulting in a high catalytic activity.

4.3.3. Metal Oxide/Plasmon Particle Systems

The plasmonic effect induced by noble metal particles plays an important role in decorating photoelectrodes. Photoactivity of photocatalysts can be increased by enhancing light scattering and SPR [148,149]. Moreover, noble metals play a role as a co-catalyst for OER due to good electrical contact between the metal and semiconductor [150,151]. Altering the surface properties of WO_3 photoanodes with plasmonic nanoparticles Au and Ag has shown enhanced visible light absorption and high photocurrent density [152]. Hu showed a high faradic efficiency of 94% for $WO_3@Au$ composites [149]. Enhanced photocurrent density and morphology of heterostructure is presented in Figure 13a,b. Modified WO_3 by plasmonic Ag and Pt nanoparticles showed enhanced photocurrent of 1.13 mA/cm² at 1.23 V vs. RHE under AM 1.5G illumination in a 0.2 M Na_2SO_4 solution, which is nearly 3.32 times that of bare WO_3 [153]. The photocurrent density for binary systems are represented in Table 1.

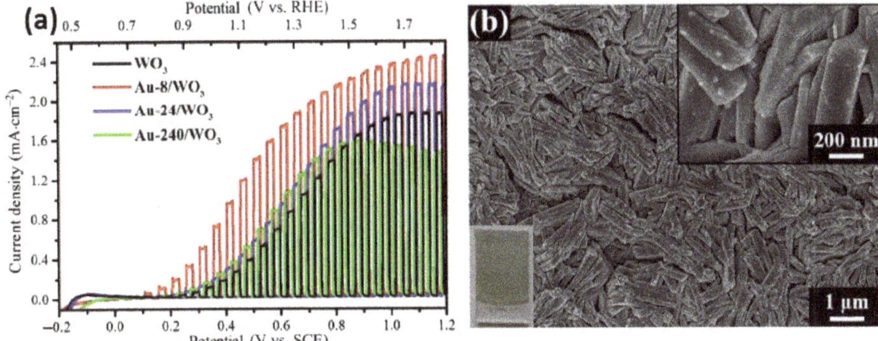

Figure 13. (**a**) Current -potential curves of $WO_3@Au$ composite with $HAuCl_4$ concentrations of 8, 24 and 240 umol in 0.1-M Na_2SO_4 electrolyte and (**b**) SEM images of $WO_3@Au$ composites. Reproduced from ref [149] with permission from Springer Nature, 2016.

Table 1. Photocurrent densities of binary heterostructures.

Photoanodes	Methods	Morphology	Electrolyte	Potential	P (mW/cm²)	J (mA/cm²)/STH (%)/Gas Evolution (mL/cm²)	Ref
WO_3/TiO_2	Solvothermal	Nanoflake	0.1-M Na_2SO_4	0.8 vs. SCE (1.45 vs. RHE)	100	1.4	[129]
WO_3/TiO_2	Anodization	Nanotubes	1-M NaOH	0.7 vs. Ag/AgCl (1.7 RHE)	100	1.6	[130]
WO_3/TiO_2	Anodization	Nanotubes	1-M KOH	0.6 V vs. SCE (1.62 RHE)	100	2 /3.1%/16.2	[131]
WO_{3-x}/ZnO	Solvothermal method	Nanorods	1-M Na_2SO_4	1.23 vs. RHE	100	3.38	[56]
$WO_3/BiVO_4$	Glancing-angle deposition/ electrodeposition	Vertically oriented nanorods	0.5-M Na_2SO_4	1.23 vs. RHE	100	3.1	[66]
$WO_3/BiVO_4$	Electrostaticspraying method	Nanotextured pillar	0.5-M Na_2SO_4	0.7 V vs. Ag/AgCl (1.44 vs. RHE)	100	2.1	[139]
$WO_3/BiVO_4$	Layer-by-layer	Film	0.5-M Na_2SO_4	1.23 vs. Ag/AgCl (1.817 vs. RHE)	100	2.78	[134]
$WO_3/BiVO_4$	Spin coating	Film	0.5-M Na_2SO_4	1.23 V vs. Ag/AgCl 1.817 vs. RHE	100	1.2	[64]
$WO_3/BiVO_4$	Pulsed electrodeposition	Nanorods	0.1-M Na_2SO_3	1.23 vs. RHE	100	4.55	[65]
$WO_3/BiVO_4$	Anodic oxidation	Nanoporous film	0.1-M KH_2PO_4	0.6 V vs. Ag/AgCl (1.21 vs. RHE)	100	02.01	[135]
$WO_3/BiVO_4$	Electrospinning	Nanofibers	0.5-M Na_2SO_4	1.23 vs. RHE	100	2.8	[154]
WO_3/Fe_2O_3	Solvothermal	Nanosheets	0.5-M Na_2SO_4	1.23 vs. RHE	100	1.66	[142]

Table 1. *Cont.*

Photoanodes	Methods	Morphology	Electrolyte	Potential	P (mW/cm²)	J (mA/cm²)/STH (%)/Gas Evolution (mL/cm²)	Ref
WO_3/Fe_2O_3	Hydrothermal	Nanorods	0.1-M Na_2SO_4	1.23 vs. RHE	100	1	[144]
WO_3/Fe_2O_3	Sol–gel	Film	0.1-M NaOH	1.23 vs. RHE	100	0.7	[67]
WO_3/Sb_2S_3	Hydrothermal	Nanoplate/ nanorods	1-M H_2SO_4	0.8 vs. RHE	100	1.79	[146]
WO_3/Bi_2S_3	CBD	Nanorods/ nanoplates	0.1-M Na_2SO_3	0.9 vs. RHE	100	5.95	[147]
WO_3/Au	Hydrothermal	Nanoplate	0.1-M Na_2SO_4	1.23 vs. RHE	100	1.5	[149]

4.4. Ternary Systems for Efficient Water Decomposition

A typical heterojunction between two dissimilar semiconductors comes to equilibrium without any external electric field. The result shown above is a potential difference that appears at the interface as an internal electric field. This field accelerates charge carrier drift and decreases the number of electron–hole recombination, improving the semiconductor's photocatalytic activity [155]. The exploitation of multilayer structures in photocatalysis is considered more beneficial. An illustration of the advantages of a cascade transition of charge carriers is well explained in numerous works [156] and [157]. The authors of [156] work investigated the photocatalytic properties of a composite of CdS, TiO_2 and tungsten trioxide. Since, TiO_2 has the conduction band edge which is between conduction band edges of CdS and WO_3, in such ternary composites electronic transitions are cascading. After generation of excitons in CdS, electrons easily migrate to TiO_2 and WO_3 of potential difference as shown in Figure 14.

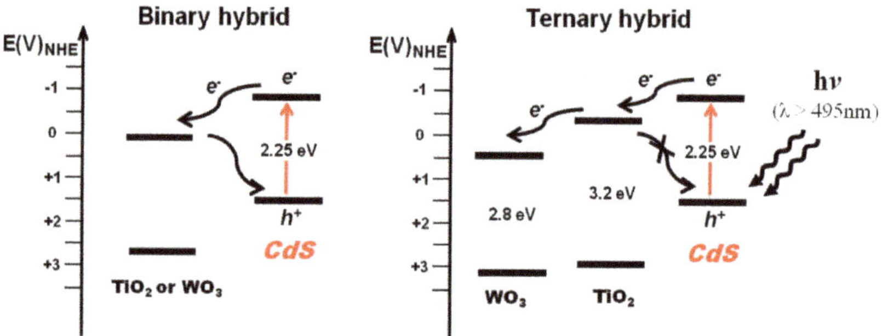

Figure 14. Schematic representation of a comparison of electronic transition processes in binary and ternary hybrid. Reproduced from [156], with permission from American Chemical Society, 2011.

In addition to ternary composites made only of semiconductor materials, there are also ternary structures based on semiconductors and metal complexes. An example of such ternary photocatalysts can be photonic heterostructure of $CdS–Au–WO_3$. Cui et al. reported, that Au nanoparticles deposited between WO_3 and CdS leads to form heterostructure which had photocatalytic properties superior to similar two-phase systems. It was thought that such amelioration caused by a synergistic integration of photon absorption with effective electron transfer in the heterostructure [62]. The use of doping elements—or modification by particles such as CdS quantum dots [158] in usual two-component heterojunctions—is also referred to as a ternary structure. Doping elements such as Yb [159] and Mo have been shown to suppress charge carrier recombination during photocatalysis [160], improving the efficiency of reactions. In ref. [161], high photocurrent density was achieved by doping active materials to make bilayers $WO_3/MoBiVO_4$ ($BiV_{0.95}Mo_{0.05}O_4$). The photocurrent of the Mo-doped content increased by a factor of 3 and 1.5 relative to pristine photoanodes based on WO_3 and $BiVO_4/WO_3$, respectively. Another promising way to create performable photocatalysts is a combination of a catalyst with heterostructures. Some examples of such formations are materials which obtained by the deposition of catalytic coatings NiOOH [162] and FeOOH [163] on the surface of two-phase WO_3 structures. The deposited materials suppress both the degradation of the photoactive material and the Faradic losses [164], accelerating the photoelectrochemical reaction processes. Shouli Bai et al. also combined catalyst layers with a heterojunction by depositing NiFe bimetallic complexes onto a WO_3/Fe_2O_3 surface [165]. Their strategy increased the photocurrent density of the ternary photoanode system to 3.0 mA cm^{-2}, which, according to Shouli Bai, is 5 and 7-fold higher relative to that of pristine WO_3 and α-Fe_2O_3 structures, respectively. The role of the catalyst here is to improve the absorption of holes accumulated on the electrode surface. In another study, the deposition of CoPd bimetallic nanoparticles onto the surface of WO_3/α-Fe_2O_3 photoanodes causes a cathode shift of the initial potential, increasing the photocurrent density from 0.15 to 0.5 mA/cm^2 during water oxidation at 1.23 V relative to RHE when illuminated with AM 1.5 G [166]. Substitution of iron oxide with cadmium sulfide in heterostructures based on WO_3 also makes it possible to sensitively increase the photoresponse of electrodes during water decomposition. In [167] work preparation of such ternary compositions conducted via three simple hydrothermal, impregnation and photo-assisted deposition steps. Thus, authors obtained rodlike structures with a performance of photocurrent at the level of 5.85 mA/cm^2 at 1.23 V (vs. RHE). Sun with co-workers explain this phenomenon by creation a larger built-in potential at interface WO_3/CdS formed via impregnating appropriate CdS onto surface of WO_3. This drives transport of electrons from CdS to WO_3 with improvement of exciton separation. In this case, not the entire charge is torn well enough. Part of the charge recombines due to the weak involvement of holes in the valence band of cadmium sulfide in the process of water oxidation. Decoration surface of WO_3/CdS heterojunctions with Co-Pi co-catalyst advances the transfer kinetics of charge advanced which is positive to suppression of charge recombination. In this case, the mechanism of improving charge transfer to the sites of redox half-reactions is also achieved by adding phosphate anions to the electrolyte. In fact, the use of various electrolytes, such as glycerol-water mixture [168], in its effect on the parameters of the transferred charges between photocatalytic coatings and a split liquid. Varieties of compositions of working electrolytes and a list of structural heterojunctions, as well as geometric schemes that receive the influence and influence of all this on the photoresponse of PEC systems are shown in Table 2. It is known that tungsten trioxide is widely used as the primary semiconductor material in three transient systems in photocatalysis. Therefore, numerous of recent works dedicated to photoinduced dye degradation processes [169] and the direct splitting of water [170,171] were carried out with exploiting WO_3. Liu formed an effective three-component photoanode based on tungsten trioxide nanosheets synthesized by hydrothermal method and decorated with $Zn_xBi_2S_{3+x}$ quantum dots via layer-by-layer adsorption [172]. In obtained core/shell structure of $Zn_xBi_2S_{3+x}/WO_3$, surface $Zn_xBi_2S_{3+x}$ served as a protective layer for tungsten trioxide. Comparable photocatalytic studies conducted under visible light irradiation in 0.1-M Na_2S and 0.1-M Na_2SO_3 aqueous solution at pH ~9 showed that $Zn_xBi_2S_{3+x}/WO_3$ composite has higher photocatalytic activity then Bi_2S_3/WO_3.structure. The photocurrent density was 7 mA/cm^2 at a bias voltage of -0.1 V. This attributed to the involvement of ZnS nanostructures with high photocatalytic properties [173]. Designing semiconducting heterocomposites via a surface–interface engineering approach showed high effectiveness for enhancing exciton separation/transportability and photoelectrochemical features. The photoactive layer of 2D $BiVO_4$-layer/WO_3 array modified with cobalt phosphate studied for oxygen evolution and showed 1.8 mA/cm^2 at 1.23 V vs. RHE in a phosphate buffer under an AM1.5G sun. The result is 5 and 12-fold higher than that of bare WO_3 and $BiVO_4$ photoanodes, respectively [174]. Morphology of WO_3 nanostructures affects to charge separation ability in the active layer and to charge collection efficiency in the $WO_3/BiVO_4$ heterojunction. The low-dimensional nanosphere WO_3 layer showed higher photocatalytic efficiency than the platelike or rodlike ones [71].

There are three different component systems with different morphology in which the hierarchy of the band structure observed for effective water-splitting. The main types of morphology include multi-heterojunction-based photocatalysts composed of WO_3 nanorods, Pt nanoparticles and TiO_2 nanoparticles [175], hierarchical heterostructures with core and double shells [176], rectangular $AgIn(WO_4)_2$ nanotubes which showed excellent photocatalytic properties for decomposing water to evolve H_2 [177] and linked porous structures such as WO_3/porous–$BiVO_4$/FeOOH [178]. Special attention should be paid to the three-component system made on spiral WO_3 nanostructures decorated with doped Mo and $BiVO_4$ nanoparticles [179]. Based on the assertions that 1D structures charge carriers go straight to the terminals or half-reaction centers [180], Xinjian Shi et al. used a spring morphology with an increased surface area while retaining the properties of 1D structures. As a result of the successive deposition of these structures onto conductive glass, a spiral heterocomposite WO_3/(W, Mo)–$BiVO_4$ with a length of 5.5 μm was obtained and the photocurrent density exceeded 3.9 mA/cm^2. The process of making triple transitions is possible using various techniques, such as electrochemical reduction-oxidation technology [181], a hybrid synthetic method, including electrodeposition and hydrothermal treatment [182], a solvothermal-calcination process [183], an electrostatic-driven self-assembly correlated with ion-exchange process [162] and a polymer complex method [184]. Jun Lv et al. obtained photoactive $LiCr(WO_4)_2$. After heat treatment at 700 °C for 5 h, crystal structures of tungstate were obtained, consisting of alternating layers of WO_6 and CrO_6 octahedra lying on the YZ plane. The WO_6 octahedra are connected by separating edges, leading to the appearance of unrelated zigzag rows along the Y-axis. On the other hand, CrO_6 octahedra not related to each other. Photocatalytic tests of $LiCr(WO_4)_2$ showed that the release of H_2 proceeds upon irradiation with visible light up to 540 nm [184]. Reaching the rest of the visible spectrum remains the goal. Ji Hyun Baek et al. developed a photoanode based on $BiVO_4$/WO_3/SnO_2 using a simple method of applying layers on conductive glass to obtain a thin double-heterojunction coating on the order of 320 nm. A characteristic feature of $BiVO_4$/WO_3/SnO_2 is the large bandwidth of visible light, which allowed the researchers to combine the photoanode with a perovskite solar cell into a tandem PEC system. This allowed the reaction to split water without applying a bias voltage [26]. In general, the development of PEC cells using independent absorbents of incident light is a promising direction, so the next section will deal tandem hydrogen production systems.

4.5. WO_3-Based Tandem PEC Cells

Two strategies can produce tandem cells for photoinduced water-splitting. One strategy used is to increase the capture of photons is a tandem system of a photoanode and photocathode with n- and p-conductivities of active layers, respectively. The splitting of water under light irradiation in this type of PEC cell occurs as a result water oxidation by photogenerated holes on the photoanode surface. Meanwhile, reduction to molecular hydrogen is initiated by electrons on the photocathode surface. At the same time, negative charge carriers generated in the photoanode are directed under the action of the field, to positively charged carriers in the photocathode material, where they recombine. Robert Coridan et al. investigated the photocatalytic properties of Si/WO_3 heterojunctions and Si/ITO/WO_3 arrays in a core-shell manner [185]. The operation of the tandem structure depends not only on the bandgap of the semiconductors used but also on the alignment of the strip edge and the state of the electrical connection between photo absorbents. When a mismatch of conductivity levels and valence levels of semiconductors included in tandem circuits occurs, low efficiencies of PEC hydrogen evolution cells [186]. A similar effect was observed in the work of Heli Wang et al. in which they combined n-type tungsten trioxide and hematite nanorod metal oxides with p-GaInP$_2$. It was found that even when photo electrodes are illuminated with a source with a power of 1 W/cm^2, a photocurrent appears but the density values of which are rather weak. This is due to low electron mobility of in the hematite layers, short hole-diffusion length, and insufficient potential difference between the levels of the conduction band and valence semiconductors, which help reduce charge recombination [187]. Geometrical optimization of the morphology of the active layers of photoelectrodes refers to an increase the density of short circuit current [188]. Investigating the PEC properties of the tandem structure of WO$_3$/Si, Zhuo Xing et al. concluded that it was necessary to add an intermediate layer between p and n semiconductors to reduce the number of recombinations of photoinduced charges. In [189], metal W served as an intermediate layer, resulting in a WO$_3$/W/Si ternary structure demonstrated an increase in the photogenerated current density by a factor of 10 compared to the WO$_3$/Si structure.

Another possible way for general water-splitting without assistance is to combine photoelectrodes with photovoltaic cells to form a tandem PEC/PV cell. In one study [190], organic molecules were used as sensitizers in a tandem PEC, which is a powerful strategy for designing hydrogen evolution systems since they allow large-scale modification of photoelectrodes by adjusting the dye redox potentials or redox mediators.

In the tandem devices shown in Figure 15, the BiVO$_4$/WO$_3$ photoelectrode absorbs short-wavelength photons and more extended wavelength absorbs by a dye/TiO$_2$ electrode [185–190]. This method offers better concession between device performance, complexity and stability [3]. In addition to scientific methods, when choosing materials, morphology and hierarchy of architecture, engineering aspects related to the spatial and reciprocal arrangement of the physical elements of tandem structures are also important. For efficient use of incident photons, Pihosh Y. and colleagues produced a PEC-PV tandem system based on WO$_3$ NRs/BiVO$_4$+CoPi photoanode and an AlGaAsP solar cell, which were placed at 45° relative to each other using a V-shaped stand [102]. This design allows the passage of reflected light from the photoanode to the surface of the solar cell.

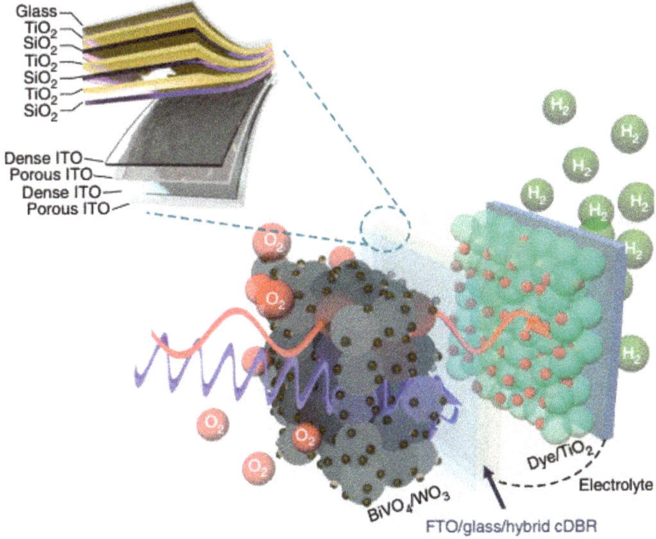

Figure 15. Schematic illustration of a PEC tandem system based on BiVO$_4$/WO$_3$ and DSSC. Reproduced from [191], with permission from Springer Nature, 2016.

Thus, the characteristic differences between ternary systems and binary systems are the improvement of photocatalytic properties and corrosion resistance. As described above, tandem structures provide an operating mode for photoelectrochemical processes in a wide range of the radiation spectrum. It increases the number of components of the hydrogen evolution cells, which leads to a complication of the assembly of heterostructures and to a high cost of the obtained layers. Therefore, when choosing which or used components of tertiary composites, pay attention to postprocessing, including thermal.

Table 2. Photocurrent densities of ternary heterostructures.

Photocatalytic Material	Methods	Morphology	Electrolyte	Potential	Irradiation	Photocurrent Density	Ref
WO_3–Pt–CdS	Combination of wet-chemical, photodeposition and hydrothermal techniques	Hollow microspheres composed of small crystallites	$0.5\ M$ Na_2SO_4	0.5 V vs. Ag/AgCl (1.82 V vs. RHE)	Vis light	$0.16\ \mu A/cm^2$	[192]
$SnO_2/WO_3/$ $BiVO_4$	Combination of electron beam deposition and metal organic decomposition technique	Planar film	0.5-M Na_2SO_3	1.23 vs. RHE	100 mW/cm^2	$2.01\ mA/cm^2$	[193]
$WO_3/C_3N_4//CoO_x$	Combination of a hydrothermal method with wet impregnation	film		1.23 V vs. Ag/AgCl (1.8 V vs. RHE)	100 mW/cm^2	$5.76\ mA/cm^2$	[170]
$CuWO_4$–WO_3	electrodeposition	film	0.1-M KH_2PO_4	0.618 V vs. Ag/AgCl (1.23 vs. RHE)	100 mW/cm^2	$0.3\ mA/cm^2$	[194]
$WO_3/(Er,\ W):BiVO_4$	spray coating	monoclinic clinobisvanite structure	0.1-M K_2HPO_4	1.23 V vs. RHE	100 mW/cm^2	$4.1 \pm 0.19\ mA\ cm^{-2}$	[195]
$WO_3/(Er,\ W):BiVO_4$	spray coating	monoclinic clinobisvanite structure	0.1-M K_2HPO_4	2.3 V vs. RHE	100 mW/cm^2	$7.2 \pm 0.39\ mA\ cm^{-2}$	[195]
$TiO_2/WO_3/$ $BiVO_4$	hydrothermal	brochosomes-like	0.5-M Na_2SO_4	0.35 V vs. RHE	100 mW/cm^2	$3.13\ mA\ cm^{-2}$	[196]
$WO_3/$ $Fe_2O_3/Co(OH)$	electrospray deposition	worm-like nanobars	0.1-M NaOH	1.23 vs. RHE		$0.62\ mA\ cm^{-2}$	[197]
Ag-functionalized $CuWO_4/WO_3$	electrophoretic deposition	thin film	potassium phosphate buffer solution	0.62 V vs. Ag/AgCl (1.23 V vs. RHE)		$0.205\ mA\ cm^{-2}$	[198]
$CuWO_4/BiVO_4$ with Co-Pi	drop-casting and thermal annealing method	nanoflakes	1.0 M of Na_2SO_4 with 0.1 M of sodium phosphate buffer (pH = 7)	1.23 V vs. RHE	100 mW/cm^2	$2.25\ mA\ cm^{-2}$	[199]
$BiVO_4/WO_3/SnO_2$ connected with perovskite solar cell tandem device	Spin-coating	triple-layer planar film	pH 7 phosphate buffer electrolyte	1.23 V vs. RHE	100 mW/cm^2	$3.1\ mA/cm^2$	[26]
$ZnWO_4/WO_3$	Piezo-dispensing	Spot Arrays	0.1-M Na_2SO_4 at pH 7	0.7 V vs. Ag/AgCl (1.31 V vs. RHE)		$0.75\ mA/cm^2$	[200]
b-$Cu_2V_2O_7/WO_3$	Seeded-growth approach		0.1-M sodium borate buffer (pH 8.2) containing 0.1-M Na_2SO_3	1.23 V vs. RHE	100 mW/cm^2	$0.45\ mA\ cm^{-2}$	[201]
$CaMn_2O_4/WO_3$	Spin-coating	Thin film	0.5-M Na_2SO_4 solution (pH 6)	1.09 V vs. RHE		$1.5 \times 10^{-3}\ mA\ cm^{-2}$	[202]

Table 2. *Cont.*

Photocatalytic Material	Methods	Morphology	Electrolyte	Potential	Irradiation	Photocurrent Density	Ref
Pt/WO$_3$/Ag	Hydrothermal method, chemical bath, photoassisted electrodeposition	Nanorods			100 mW/cm^2	1.13 mA/cm^2	[153]
WO$_3$/CdS/NiOOH	hydrothermal method, successive ionic layer adsorption and reaction, photo-assisted electrodeposition	Nanorods	d 0.2-M Na$_2$SO$_4$-0.1-M Na$_2$SO$_3$	1.23 V vs. RHE		1.5–2 mA/cm^2	[203]
ZnWO$_4$/WO$_3$	hydrothermal	Nanorods	0.5 M Na$_2$SO$_4$	1.23 V vs. RHE	100 mW/cm^2	1.87 mA cm^{-2}	[204]
WO$_3$/BiVO$_4$/ZnO	drop-casting method, atomic layer deposition	Nanosheets	0.5-M Na$_2$SO$_4$	1.23 V vs. RHE	100 mW/cm^2	2.5–3.00 mA cm^{-2}	[205]
Au-surface/BiVO$_4$/WO$_3$/Au-bottom	hydrothermal, sol–gel spin-coated,	Nanospheres	0.5 M Na$_2$SO$_4$	1.23 V vs. RHE		1.31 mA/cm^2	[63]
WO$_3$/C@CoO	hydrothermal process and immersion method	Octopus tentacles-like	1.0-M KOH	55 mV (vs. RHE)		10 mA cm^{-2}	[206]
WO$_3$@ZnWO$_4$@ZnO	layer deposition technique and hydrothermal process	nanosheets	mixed aqueous solution of 0.35-M Na2S and 0.25-M NaSO3 (pH = 13.4)	1.23 V vs. RHE	100 mW/cm^2	~1.57 mA/cm^2	[207]
WO$_3$/rGO/Sb$_2$S$_3$	chemical bath deposition	nanoplates	0.5-M Na$_2$SO$_4$ (pH~7)	1.23 V vs. RHE		1.20 mA/cm^2	[208]
Cu$_2$O/CuO/WO$_3$	Electrodeposition, spin-coating	Thin film		0 V vs. RHE		−1.9 mA/cm^2	[209]
WO3/BiVO4/Co-Pi	Electrodeposition	composite inverse opals	0.5-M Na$_2$SO$_4$	1.4 V versus Ag/AgCl (0.67 V vs. RHE)	100 mW cm^{-2}	4.5 mA cm^{-2}	[210]
WO$_3$/BiVO$_4$/TiO$_2$	Spin-coating, wet chemistry	platelike	0.1-M Na$_2$SO$_4$	1.23 V vs. RHE	100 mW/cm^2	~3.9 mA/cm^2	[211]
TiO$_2$/WO$_3$/Pt	Electrospinning technique	fibers	0.2-M Na$_2$SO$_4$			15–20×10^{-3} mA/cm^2	[212]
TiO$_2$-TiCl$_4$-WO$_3$	Hydrothermal method + Electrodeposition	nanorods	KOH	1.23 V vs NHE	100 mW/cm^2	3.86 mW/cm^2	[213]

5. Conclusions and Outlook

The goal of research in the field of photoinduced decomposition of water is to develop high performance photocatalytic systems with high STH efficiency. The transmission of the photocatalytic systems from the field of laboratory research to the large scale production is a key point. The principle of using semiconductor coatings based on tungsten trioxide for PEC cells is justified by the economic aspects associated with the low cost of the material, as well as with its physicochemical properties. Using nanotechnology and nanomaterials is a suitable method for addressing several of the issues listed above. Metal oxide nanoparticles can be obtained by a wide range of physical and chemical methods. They can be classified as top-down and bottom-up methods. Top-down approaches rely on physical processes, such as abrasion or ball milling. Nano powders produced in this method usually exhibit wide distribution sizes and their size, shape and morphology are difficult to control. In addition, possible structural and surface impurities can have a significant effect on surface chemistry and the catalytic properties of nanomaterials. Low-dimensional structures are most advantageous from the point of view of effective absorption of light with the generation of charge carriers, migration of charge carriers to the surfaces of the material, which fit over the exciton lifetime, as well as possessing a fairly significant semiconductor/electrolyte contact area. The mixing of semiconductors (i.e., the formation of composites) is also an accepted strategy for the development of photocatalysts that respond to radiation in the visible range. This strategy is based on a hierarchical architecture for connecting a wide-gap semiconductor with a narrow-gap semiconductor with a more negative level of the conduction band. Thus, the conduction of electrons can be introduced from a narrow-gap semiconductor into a wide band semiconductor, leading to better absorption for the mixed photocatalyst. An additional advantage of using composite semiconductor photocatalysts are that it reduces carrier recombination by facilitating electron transfer crossing interface of particles. In photocatalyst composites, semiconductor particles stay in electronic contact individually. For a successful combination of semiconductors, the certain requirements are needed to be met: the conductivity level of the narrow-gap semiconductor should be more negative than the level of the wide-gap semiconductor; the position of the conductivity level of the wide-gap semiconductor should be more negative than the recovery potential; electron injection should be quick. All of the steps can improve the characteristics of the material, as well as eliminate the influence of its shortcomings on the process of splitting water under the action of light. In any case, a review of the literature in this area indicates a special level of development in the field of photoelectrochemistry for hydrogen evolution using active materials from pure tungsten trioxide or in various compositions with it. However, so far, the complete and qualitative decomposition of water and the generation of hydrogen under the influence of sunlight has a low rate, which indicates insufficient feasibility of industrial use of existing technologies. Based on the current trend towards the creative and experimental activity of researchers in this direction, the authors of this article express deep confidence in the imminent achievement of quantum efficiency of PEC systems sufficient for universal use in human life in the near future.

Author Contributions: N.N. designed and supervised the whole manuscript. He wrote the whole manuscript together with Y.S., A.M. and N.B.; Y.S. contributed to write Sections 2 and 4.1; A.M. wrote Sections 3, 4.2 and 4.3. N.B. wrote Sections 4.4 and 4.5; all authors wrote introduction, conclusion and outlook together. All authors have read and agreed to the published version of the manuscript.

Funding: This research was partially funded by Nazarbayev University FGRG grant number SEDS2020 016 and the Young Scientist Program of the Committee of Science of the Ministry of Education and Science of the Republic of Kazakhstan, grant number AP08052381.

Acknowledgments: The authors would like to express special thanks to Aygul Nuraje from BTHS for proof-reading and editing the manuscript. N.N. greatly acknowledge financial supported from Nazarbayev University. N.B. thanks the Young Scientist Program of the Ministry of Science and Education of the Republic of Kazakhstan for financial support.

Conflicts of Interest: The authors declare no potential conflict of interest.

References

1. Gratzel, M. Photoelectrochemical cells. *Nature* **2001**, *414*, 338–344. [CrossRef] [PubMed]
2. Sathre, R.; Scown, C.D.; Morrow, W.R.; Stevens, J.C.; Sharp, I.D.; Ager, J.W.; Walczak, K.; Houle, F.A.; Greenblatt, J.B. Life-cycle net energy assessment of large-scale hydrogen production via photoelectrochemical water splitting. *Energy Environ. Sci.* **2014**, *7*, 3264–3278. [CrossRef]
3. Prevot, M.S.; Sivula, K. Photoelectrochemical tandem cells for solar water splitting. *J. Phys. Chem. C* **2013**, *117*, 17879–17893. [CrossRef]
4. Wang, G.; Wang, H.; Ling, Y.; Tang, Y.; Yang, X.; Fitzmorris, R.C.; Wang, C.; Zhang, J.Z.; Li, Y. Hydrogen-treated TiO$_2$ nanowire arrays for photoelectrochemical water splitting. *Nano Lett.* **2011**, *11*, 3026–3033. [CrossRef]
5. Zhang, Z.; Zhang, L.; Hedhili, M.N.; Zhang, H.; Wang, P. Plasmonic gold nanocrystals coupled with photonic crystal seamlessly on TiO$_2$ nanotube photoelectrodes for efficient visible light photoelectrochemical water splitting. *Nano Lett.* **2012**, *13*, 14–20. [CrossRef]
6. Zhang, Z.; Hossain, M.F.; Takahashi, T. Photoelectrochemical water splitting on highly smooth and ordered TiO$_2$ nanotube arrays for hydrogen generation. *Int. J. Hydrog. Energy* **2010**, *35*, 8528–8535. [CrossRef]

7. Sivula, K.; Le Formal, F.; Grätzel, M. Solar water splitting: Progress using hematite (α-Fe_2O_3) photoelectrodes. *ChemSusChem* **2011**, *4*, 432–449. [CrossRef]

8. Cesar, I.; Kay, A.; Gonzalez Martinez, J.A.; Grätzel, M. Translucent thin film Fe_2O_3 photoanodes for efficient water splitting by sunlight: Nanostructure-directing effect of Si-doping. *J. Am. Chem. Soc.* **2006**, *128*, 4582–4583. [CrossRef]

9. Liu, Y.; Xu, Z.; Yin, M.; Fan, H.; Cheng, W.; Lu, L.; Song, Y.; Ma, J.; Zhu, X. Enhanced photoelectrocatalytic performance of α-Fe_2O_3 thin films by surface plasmon resonance of Au nanoparticles coupled with surface passivation by atom layer deposition of Al_2O_3. *Nanoscale Res. Lett.* **2015**, *10*, 374. [CrossRef]

10. Wang, D.; Li, R.; Zhu, J.; Shi, J.; Han, J.; Zong, X.; Li, C. Photocatalytic water oxidation on $BiVO_4$ with the electrocatalyst as an oxidation cocatalyst: Essential relations between electrocatalyst and photocatalyst. *J. Phys. Chem. C* **2012**, *116*, 5082–5089. [CrossRef]

11. Jo, W.J.; Jang, J.W.; Kong, K.-j.; Kang, H.J.; Kim, J.Y.; Jun, H.; Parmar, K.; Lee, J.S. Phosphate doping into monoclinic BiVO4 for enhanced photoelectrochemical water oxidation activity. *Angew. Chem.* **2012**, *124*, 3201–3205. [CrossRef]

12. Wang, D.; Jiang, H.; Zong, X.; Xu, Q.; Ma, Y.; Li, G.; Li, C. Crystal facet dependence of water oxidation on $BiVO_4$ sheets under visible light irradiation. *Chem. A Eur. J.* **2011**, *17*, 1275–1282. [CrossRef]

13. Steinfeld, A. Solar hydrogen production via a two-step water-splitting thermochemical cycle based on Zn/ZnO redox reactions. *Int. J. Hydrog. Energy* **2002**, *27*, 611–619. [CrossRef]

14. Wolcott, A.; Smith, W.A.; Kuykendall, T.R.; Zhao, Y.; Zhang, J.Z. Photoelectrochemical study of nanostructured ZnO thin films for hydrogen generation from water splitting. *Adv. Funct. Mater.* **2009**, *19*, 1849–1856. [CrossRef]

15. Maeda, K.; Domen, K. Solid solution of GaN and ZnO as a stable photocatalyst for overall water splitting under visible light. *Chem. Mater.* **2010**, *22*, 612–623. [CrossRef]

16. Tacca, A.; Meda, L.; Marra, G.; Savoini, A.; Caramori, S.; Cristino, V.; Bignozzi, C.A.; Pedro, V.G.; Boix, P.P.; Gimenez, S. Photoanodes based on nanostructured WO_3 for water splitting. *ChemPhysChem* **2012**, *13*, 3025–3034. [CrossRef]

17. Hameed, A.; Gondal, M.; Yamani, Z. Effect of transition metal doping on photocatalytic activity of WO_3 for water splitting under laser illumination: Role of 3d-orbitals. *Catal. Commun.* **2004**, *5*, 715–719. [CrossRef]

18. Enesca, A.; Duta, A.; Schoonman, J. Study of photoactivity of tungsten trioxide (WO_3) for water splitting. *Thin Solid Films* **2007**, *515*, 6371–6374. [CrossRef]

19. Pala, R.A.; Leenheer, A.J.; Lichterman, M.; Atwater, H.A.; Lewis, N.S. Measurement of minority-carrier diffusion lengths using wedge-shaped semiconductor photoelectrodes. *Energy Environ. Sci.* **2014**, *7*, 3424–3430. [CrossRef]

20. Coridan, R.H.; Arpin, K.A.; Brunschwig, B.S.; Braun, P.V.; Lewis, N.S. Photoelectrochemical behavior of hierarchically structured Si/WO_3 core–shell tandem photoanodes. *Nano Lett.* **2014**, *14*, 2310–2317. [CrossRef]

21. Wang, Y.; Shi, H.; Cui, K.; Zhang, L.; Ge, S.; Jinghua, Y. Reversible electron storage in tandem photoelectrochemical cell for light driven unassisted overall water splitting. *Appl. Catal. B Environ.* **2020**, *275*, 119094. [CrossRef]

22. Li, Y.; Zhang, W.; Qiu, B. Enhanced surface charge separation induced by Ag nanoparticles on WO_3 photoanode for photoelectrochemical water splitting. *Chem. Lett.* **2020**, *49*, 741–744. [CrossRef]

23. Jun, J.; Ju, S.; Moon, S.; Son, S.; Huh, D.; Liu, Y.; Kim, K.; Lee, H. The optimization of surface morphology of Au nanoparticles on WO_3 nanoflakes for plasmonic photoanode. *Nanotechnology* **2020**, *31*, 204003. [CrossRef]

24. Huang, J.; Zhang, Y.; Ding, Y. Rationally designed/constructed CoO_x/WO_3 anode for efficient photoelectrochemical water oxidation. *Acs Catal.* **2017**, *7*, 1841–1845. [CrossRef]

25. Wang, S.L.; Mak, Y.L.; Wang, S.; Chai, J.; Pan, F.; Foo, M.L.; Chen, W.; Wu, K.; Xu, G.Q. Visible–Near-Infrared-Light-Driven Oxygen Evolution Reaction with Noble-Metal-Free WO_2–WO_3 Hybrid Nanorods. *Langmuir* **2016**, *32*, 13046–13053. [CrossRef] [PubMed]

26. Baek, J.H.; Kim, B.J.; Han, G.S.; Hwang, S.W.; Kim, D.R.; Cho, I.S.; Jung, H.S. $BiVO_4$/WO_3/SnO_2 Double-Heterojunction photoanode with enhanced charge separation and visible-transparency for bias-free solar water-splitting with a perovskite solar cell. *ACS Appl. Mater. Interfaces* **2017**, *9*, 1479–1487. [CrossRef] [PubMed]

27. Liu, Y.; Wygant, B.R.; Mabayoje, O.; Lin, J.; Kawashima, K.; Kim, J.-H.; Li, W.; Li, J.; Mullins, C.B. Interface engineering and its effect on WO3-based photoanode and tandem cell. *ACS Appl. Mater. Interfaces* **2018**, *10*, 12639–12650. [CrossRef]

28. Lee, W.J.; Shinde, P.S.; Go, G.H.; Ramasamy, E. Ag grid induced photocurrent enhancement in WO_3 photoanodes and their scale-up performance toward photoelectrochemical H_2 generation. *Int. J. Hydrogen Energy* **2011**, *36*, 5262–5270. [CrossRef]

29. Zhang, J.; Zhu, G.; Liu, W.; Xi, Y.; Golosov, D.; Zavadski, S.; Melnikov, S. 3D core-shell WO_3@ α-Fe_2O_3 photoanode modified by ultrathin FeOOH layer for enhanced photoelectrochemical performances. *J. Alloys Compd.* **2020**, 154992. [CrossRef]

30. Zhang, L.J.; Li, S.; Liu, B.K.; Wang, D.J.; Xie, T.F. Highly efficient CdS/WO_3 photocatalysts: Z-scheme photocatalytic mechanism for their enhanced photocatalytic H_2 evolution under visible light. *ACS Catal.* **2014**, *4*, 3724–3729. [CrossRef]

31. Lou, Z.; Zhu, M.; Yang, X.; Zhang, Y.; Whangbo, M.-H.; Li, B.; Huang, B. Continual injection of photoinduced electrons stabilizing surface plasmon resonance of non-elemental-metal plasmonic photocatalyst CdS/WO_3-x for efficient hydrogen generation. *Appl. Catal. B Environ.* **2018**, *226*, 10–15. [CrossRef]

32. Islam, S.Z.; Reed, A.; Wanninayake, N.; Kim, D.Y.; Rankin, S.E. Remarkable enhancement of photocatalytic water oxidation in N_2/Ar plasma treated, mesoporous TiO_2 films. *J. Phys. Chem. C* **2016**, *120*, 14069–14081. [CrossRef]

33. Islam, S.Z.; Rankin, S.E. Hydrazine-based synergistic Ti (III)/N doping of surfactant-templated TiO_2 thin films for enhanced visible light photocatalysis. *Mater. Chem. Phys.* **2016**, *182*, 382–393. [CrossRef]

34. Hou, D.; Hu, X.; Ho, W.; Hu, P.; Huang, Y. Facile fabrication of porous Cr-doped $SrTiO_3$ nanotubes by electrospinning and their enhanced visible-light-driven photocatalytic properties. *J. Mater. Chem. A* **2015**, *3*, 3935–3943. [CrossRef]

35. Kudo, A.; Miseki, Y. Heterogeneous photocatalyst materials for water splitting. *Chem. Soc. Rev.* **2009**, *38*, 253–278. [CrossRef]

36. Takanabe, K. Solar water splitting using semiconductor photocatalyst powders. In *Solar Energy for Fuels*; Springer: New York, NY, USA, 2015; pp. 73–103.

37. Takanabe, K. Photocatalytic water splitting: Quantitative approaches toward photocatalyst by design. *ACS Catal.* **2017**, *7*, 8006–8022. [CrossRef]

38. Fujishima, A.; Honda, K. Electrochemical photolysis of water at a semiconductor electrode. *Nature* **1972**, *238*, 37–38. [CrossRef]

39. Youngblood, W.J.; Lee, S.-H.A.; Maeda, K.; Mallouk, T.E. Visible light water splitting using dye-sensitized oxide semiconductors. *Acc. Chem. Res.* **2009**, *42*, 1966–1973. [CrossRef]

40. Maeda, K.; Eguchi, M.; Lee, S.-H.A.; Youngblood, W.J.; Hata, H.; Mallouk, T.E. Photocatalytic hydrogen evolution from hexaniobate nanoscrolls and calcium niobate nanosheets sensitized by ruthenium (II) bipyridyl complexes. *J. Phys Chem. C* **2009**, *113*, 7962–7969. [CrossRef]

41. Islam, S.Z.; Reed, A.; Kim, D.Y.; Rankin, S.E. N_2/Ar plasma induced doping of ordered mesoporous TiO_2 thin films for visible light active photocatalysis. *Microporous Mesoporous Mater.* **2016**, *220*, 120–128. [CrossRef]

42. Jafari, T.; Moharreri, E.; Amin, A.S.; Miao, R.; Song, W.; Suib, S.L. Photocatalytic water splitting—the untamed dream: A review of recent advances. *Molecules* **2016**, *21*, 900. [CrossRef] [PubMed]

43. Wang, Q.; Hisatomi, T.; Jia, Q.; Tokudome, H.; Zhong, M.; Wang, C.; Pan, Z.; Takata, T.; Nakabayashi, M.; Shibata, N. Scalable water splitting on particulate photocatalyst sheets with a solar-to-hydrogen energy conversion efficiency exceeding 1%. *Nat. Mater.* **2016**, *15*, 611–615. [CrossRef] [PubMed]

44. Iwase, A.; Ng, Y.H.; Ishiguro, Y.; Kudo, A.; Amal, R. Reduced graphene oxide as a solid-state electron mediator in Z-scheme photocatalytic water splitting under visible light. *J. Am. Chem. Soc.* **2011**, *133*, 11054–11057. [CrossRef] [PubMed]

45. Sayama, K.; Mukasa, K.; Abe, R.; Abe, Y.; Arakawa, H. A new photocatalytic water splitting system under visible light irradiation mimicking a Z-scheme mechanism in photosynthesis. *J. Photochem. Photobiol. A Chem.* **2002**, *148*, 71–77. [CrossRef]

46. Matsumoto, Y.; Unal, U.; Tanaka, N.; Kudo, A.; Kato, H. Electrochemical approach to evaluate the mechanism of photocatalytic water splitting on oxide photocatalysts. *J. Solid State Chem.* **2004**, *177*, 4205–4212. [CrossRef]

47. Afroz, K.; Moniruddin, M.; Bakranov, N.; Kudaibergenov, S.; Nuraje, N. A heterojunction strategy to improve the visible light sensitive water splitting performance of photocatalytic materials. *J. Mater. Chem. A* **2018**, *6*, 21696–21718. [CrossRef]

48. Wang, Z.; Yang, G.; Tan, C.K.; Nguyen, T.D.; Tok, A.I.Y. Amorphous TiO_2 coated hierarchical WO_3 Nanosheet/CdS Nanorod arrays for improved photoelectrochemical performance. *Appl. Surface Sc.* **2019**, *490*, 411–419. [CrossRef]

49. Hill, J.C.; Choi, K.-S. Effect of Electrolytes on the Selectivity and Stability of n-type WO_3 Photoelectrodes for Use in Solar Water Oxidation. *J. Phys. Chem. C* **2012**, *116*, 7612–7620. [CrossRef]

50. Maeda, K.; Domen, K. Photocatalytic water splitting: Recent progress and future challenges. *J. Phys. Chem. Lett.* **2010**, *1*, 2655–2661. [CrossRef]

51. Bard, A.J. Photoelectrochemistry and heterogeneous photo-catalysis at semiconductors. *J. Photochem.* **1979**, *10*, 59–75. [CrossRef]

52. Kato, H.; Sasaki, Y.; Iwase, A.; Kudo, A. Role of iron ion electron mediator on photocatalytic overall water splitting under visible light irradiation using Z-scheme systems. *Bull. Chem. Soc. Jpn.* **2007**, *80*, 2457–2464. [CrossRef]

53. Miseki, Y.; Fujiyoshi, S.; Gunji, T.; Sayama, K. Photocatalytic water splitting under visible light utilizing $I_3{-}/I{-}$ and $IO_3{-}/I{-}$ redox mediators by Z-scheme system using surface treated PtO x/WO_3 as O_2 evolution photocatalyst. *Catal. Sci. Technol.* **2013**, *3*, 1750–1756. [CrossRef]

54. Yu, W.; Chen, J.; Shang, T.; Chen, L.; Gu, L.; Peng, T. Direct Z-scheme g-C_3N_4/WO_3 photocatalyst with atomically defined junction for H_2 production. *Appl. Catal. B Environ.* **2017**, *219*, 693–704. [CrossRef]

55. He, K.; Xie, J.; Luo, X.; Wen, J.; Ma, S.; Li, X.; Fang, Y.; Zhang, X. Enhanced visible light photocatalytic H_2 production over Z-scheme g-C_3N_4 nansheets/WO_3 nanorods nanocomposites loaded with Ni (OH)x cocatalysts. *Chin. J. Catal.* **2017**, *38*, 240–252. [CrossRef]

56. Chen, Y.; Wang, L.; Gao, R.; Zhang, Y.-C.; Pan, L.; Huang, C.; Liu, K.; Chang, X.-Y.; Zhang, X.; Zou, J.-J. Polarization-Enhanced direct Z-scheme ZnO-WO_3-x nanorod arrays for efficient piezoelectric-photoelectrochemical Water splitting. *Appl. Catal. B Environ.* **2019**, *259*, 118079. [CrossRef]

57. Sayama, K.; Mukasa, K.; Abe, R.; Abe, Y.; Arakawa, H. Stoichiometric water splitting into H_2 and O_2 using a mixture of two different photocatalysts and an $IO_3{-}/I{-}$ shuttle redox mediator under visible light irradiation. *Chem. Commun.* **2001**, 2416–2417. [CrossRef]

58. Wang, F.; Di Valentin, C.; Pacchioni, G. Doping of WO_3 for photocatalytic water splitting: Hints from density functional theory. *J. Phys. Chem. C* **2012**, *116*, 8901–8909. [CrossRef]

59. Hwang, D.W.; Kim, J.; Park, T.J.; Lee, J.S. Mg-doped WO_3 as a novel photocatalyst for visible light-induced water splitting. *Catal. Lett.* **2002**, *80*, 53–57. [CrossRef]

60. Higashimoto, S.; Ushiroda, Y.; Azuma, M. Electrochemically assisted photocatalysis of hybrid WO_3/TiO_2 films: Effect of the WO_3 structures on charge separation behavior. *Top. Catal.* **2008**, *47*, 148–154. [CrossRef]

61. Ke, D.; Liu, H.; Peng, T.; Liu, X.; Dai, K. Preparation and photocatalytic activity of WO3/TiO_2 nanocomposite particles. *Mater. Lett.* **2008**, *62*, 447–450. [CrossRef]

62. Cui, X.F.; Wang, Y.J.; Jiang, G.Y.; Zhao, Z.; Xu, C.M.; Wei, Y.C.; Duan, A.J.; Liu, J.; Gao, J.S. A photonic crystal-based CdS-Au-WO_3 heterostructure for efficient visible-light photocatalytic hydrogen and oxygen evolution dagger. *Rsc Adv.* **2014**, *4*, 15689–15694. [CrossRef]

63. Chen, B.; Zhang, Z.; Baek, M.; Kim, S.; Kim, W.; Yong, K. An antenna/spacer/reflector based Au/$BiVO_4$/WO_3/Au nanopatterned photoanode for plasmon-enhanced photoelectrochemical water splitting. *Appl. Catal. B Environ.* **2018**, *237*, 763–771. [CrossRef]

64. Chatchai, P.; Murakami, Y.; Kishioka, S.-Y.; Nosaka, A.Y.; Nosaka, Y. Efficient photocatalytic activity of water oxidation over WO_3/$BiVO_4$ composite under visible light irradiation. *Electrochim. Acta* **2009**, *54*, 1147–1152. [CrossRef]

65. Lee, M.G.; Kim, D.H.; Sohn, W.; Moon, C.W.; Park, H.; Lee, S.; Jang, H.W. Conformally coated BiVO4 nanodots on porosity-controlled WO_3 nanorods as highly efficient type II heterojunction photoanodes for water oxidation. *Nano Energy* **2016**, *28*, 250–260. [CrossRef]

66. Pihosh, Y.; Turkevych, I.; Mawatari, K.; Asai, T.; Hisatomi, T.; Uemura, J.; Tosa, M.; Shimamura, K.; Kubota, J.; Domen, K. Nanostructured WO_3/$BiVO_4$ photoanodes for efficient photoelectrochemical water splitting. *Small* **2014**, *10*, 3692–3699. [CrossRef]

67. Müller, A.; Kondofersky, I.; Folger, A.; Fattakhova-Rohlfing, D.; Bein, T.; Scheu, C. Dual absorber Fe_2O_3/WO_3 host-guest architectures for improved charge generation and transfer in photoelectrochemical applications. *Mater. Res. Express* **2017**, *4*, 016409. [CrossRef]

68. Feng, X.; Chen, Y.; Qin, Z.; Wang, M.; Guo, L. Facile fabrication of sandwich structured WO_3 nanoplate arrays for efficient photoelectrochemical water splitting. *ACS Appl. Mater. Interfaces* **2016**, *8*, 18089–18096. [CrossRef]

69. Kalanur, S.S.; Hwang, Y.J.; Chae, S.Y.; Joo, O.S. Facile growth of aligned WO_3 nanorods on FTO substrate for enhanced photoanodic water oxidation activity. *J. Mater. Chem. A* **2013**, *1*, 3479–3488. [CrossRef]

70. Fan, X.; Gao, B.; Wang, T.; Huang, X.; Gong, H.; Xue, H.; Guo, H.; Song, L.; Xia, W.; He, J. Layered double hydroxide modified WO_3 nanorod arrays for enhanced photoelectrochemical water splitting. *Appl. Catal. A General* **2016**, *528*, 52–58. [CrossRef]

71. Chae, S.Y.; Lee, C.S.; Jung, H.; Joo, O.-S.; Min, B.K.; Kim, J.H.; Hwang, Y.J. Insight into charge separation in $WO_3/BiVO_4$ heterojunction for solar water splitting. *ACS Appl. Mater. Interfaces* **2017**, *9*, 19780–19790. [CrossRef]

72. Su, J.; Zhang, T.; Wang, L. Engineered WO_3 nanorods for conformal growth of $WO_3/BiVO_4$ core–shell heterojunction towards efficient photoelectrochemical water oxidation. *J. Mater. Sci. Mater. Electron.* **2017**, *28*, 4481–4491. [CrossRef]

73. Yang, J.; Li, W.; Li, J.; Sun, D.; Chen, Q. Hydrothermal synthesis and photoelectrochemical properties of vertically aligned tungsten trioxide (hydrate) plate-like arrays fabricated directly on FTO substrates. *J. Mater. Chem.* **2012**, *22*, 17744–17752. [CrossRef]

74. Amano, F.; Li, D.; Ohtani, B. Fabrication and photoelectrochemical property of tungsten (VI) oxide films with a flake-wall structure. *Chem. Commun.* **2010**, *46*, 2769–2771. [CrossRef] [PubMed]

75. Zeng, Q.; Li, J.; Bai, J.; Li, X.; Xia, L.; Zhou, B. Preparation of vertically aligned WO_3 nanoplate array films based on peroxotungstate reduction reaction and their excellent photoelectrocatalytic performance. *Appl. Catal. B Environ.* **2017**, *202*, 388–396. [CrossRef]

76. Nayak, A.K.; Sohn, Y.; Pradhan, D. Facile green synthesis of WO_3 H_2O nanoplates and WO_3 nanowires with enhanced photoelectrochemical performance. *Cryst. Growth Design* **2017**, *17*, 4949–4957. [CrossRef]

77. Park, M.; Seo, J.H.; Song, H.; Nam, K.M. Enhanced visible light activity of single-crystalline WO_3 microplates for photoelectrochemical water oxidation. *J. Phys. Chem. C* **2016**, *120*, 9192–9199. [CrossRef]

78. Yang, H.G.; Sun, C.H.; Qiao, S.Z.; Zou, J.; Liu, G.; Smith, S.C.; Cheng, H.M.; Lu, G.Q. Anatase TiO_2 single crystals with a large percentage of reactive facets. *Nature* **2008**, *453*, 638–641. [CrossRef] [PubMed]

79. Wang, X.; Liu, G.; Wang, L.; Pan, J.; Lu, G.Q.M.; Cheng, H.-M. TiO_2 films with oriented anatase {001} facets and their photoelectrochemical behavior as CdS nanoparticle sensitized photoanodes. *J. Mater. Chem.* **2011**, *21*, 869–873. [CrossRef]

80. Zou, J.-P.; Wu, D.-D.; Luo, J.; Xing, Q.-J.; Luo, X.-B.; Dong, W.-H.; Luo, S.-L.; Du, H.-M.; Suib, S.L. A strategy for one-pot conversion of organic pollutants into useful hydrocarbons through coupling photodegradation of MB with photoreduction of CO_2. *Acs Catal.* **2016**, *6*, 6861–6867. [CrossRef]

81. Zhang, D.; Wang, S.; Zhu, J.; Li, H.; Lu, Y. WO_3 nanocrystals with tunable percentage of (0 0 1)-facet exposure. *Appl. Catal. B Environ.* **2012**, *123*, 398–404. [CrossRef]

82. Wang, S.; Chen, H.; Gao, G.; Butburee, T.; Lyu, M.; Thaweesak, S.; Yun, J.-H.; Du, A.; Liu, G.; Wang, L. Synergistic crystal facet engineering and structural control of WO_3 films exhibiting unprecedented photoelectrochemical performance. *Nano Energy* **2016**, *24*, 94–102. [CrossRef]

83. Zhang, J.; Zhang, P.; Wang, T.; Gong, J. Monoclinic WO3 nanomultilayers with preferentially exposed (002) facets for photoelectrochemical water splitting. *Nano Energy* **2015**, *11*, 189–195. [CrossRef]

84. Dong, P.; Hou, G.; Xi, X.; Shao, R.; Dong, F. WO 3-based photocatalysts: Morphology control, activity enhancement and multifunctional applications. *Environ. Sci. Nano* **2017**, *4*, 539–557. [CrossRef]

85. Zheng, J.Y.; Song, G.; Hong, J.; Van, T.K.; Pawar, A.U.; Kim, D.Y.; Kim, C.W.; Haider, Z.; Kang, Y.S. Facile fabrication of WO_3 nanoplates thin films with dominant crystal facet of (002) for water splitting. *Cryst. Growth Des.* **2014**, *14*, 6057–6066. [CrossRef]

86. Navarro, J.R.; Mayence, A.; Andrade, J.; Lerouge, F.; Chaput, F.; Oleynikov, P.; Bergstrom, L.; Parola, S.; Pawlicka, A. WO_3 nanorods created by self-assembly of highly crystalline nanowires under hydrothermal conditions. *Langmuir* **2014**, *30*, 10487–10492. [CrossRef]

87. Su, J.; Feng, X.; Sloppy, J.D.; Guo, L.; Grimes, C.A. Vertically aligned WO₃ nanowire arrays grown directly on transparent conducting oxide coated glass: Synthesis and photoelectrochemical properties. *Nano Lett.* **2011**, *11*, 203–208. [CrossRef]

88. Rao, P.M.; Cho, I.S.; Zheng, X. Flame synthesis of WO₃ nanotubes and nanowires for efficient photoelectrochemical water-splitting. *Proc. Combust. Inst.* **2013**, *34*, 2187–2195. [CrossRef]

89. Koo, W.-T.; Choi, S.-J.; Kim, N.-H.; Jang, J.-S.; Kim, I.-D. Catalyst-decorated hollow WO₃ nanotubes using layer-by-layer self-assembly on polymeric nanofiber templates and their application in exhaled breath sensor. *Sens. Actuators B Chem.* **2016**, *223*, 301–310. [CrossRef]

90. Chen, D.; Hou, X.; Wen, H.; Wang, Y.; Wang, H.; Li, X.; Zhang, R.; Lu, H.; Xu, H.; Guan, S. The enhanced alcohol-sensing response of ultrathin WO₃ nanoplates. *Nanotechnology* **2009**, *21*, 035501. [CrossRef]

91. Enferadi-Kerenkan, A.; Ello, A.S.; Do, T.-O. Synthesis, organo-functionalization, and catalytic properties of tungsten oxide nanoparticles as heterogeneous catalyst for oxidative cleavage of oleic acid as a model fatty acid into diacids. *Ind. Eng. Chem. Res.* **2017**, *56*, 10639–10647. [CrossRef]

92. Abdullin, K.A.; Kalkozova, Z.K.; Markhabayeva, A.A.; Dupre, R.; Moniruddin, M.; Nuraje, N. Core–Shell (W@ WO₃) Nanostructure to Improve Electrochemical Performance. *ACS Appl. Energy Mater.* **2018**, *2*, 797–803. [CrossRef]

93. Ma, J.; Zhang, J.; Wang, S.; Wang, T.; Lian, J.; Duan, X.; Zheng, W. Topochemical preparation of WO₃ nanoplates through precursor H₂WO₄ and their gas-sensing performances. *J. Phys. Chem. C* **2011**, *115*, 18157–18163. [CrossRef]

94. Meng, D.; Wang, G.; San, X.; Song, Y.; Shen, Y.; Zhang, Y.; Wang, K.; Meng, F. Synthesis of WO₃ flower-like hierarchical architectures and their sensing properties. *J. Alloys Compd.* **2015**, *649*, 731–738. [CrossRef]

95. Adhikari, S.; Sarkar, D. Hydrothermal synthesis and electrochromism of WO₃ nanocuboids. *RSC Adv.* **2014**, *4*, 20145–20153. [CrossRef]

96. Ham, D.J.; Phuruangrat, A.; Thongtem, S.; Lee, J.S. Hydrothermal synthesis of monoclinic WO₃ nanoplates and nanorods used as an electrocatalyst for hydrogen evolution reactions from water. *Chem. Eng. J.* **2010**, *165*, 365–369. [CrossRef]

97. Jiao, Z.; Wang, J.; Ke, L.; Sun, X.W.; Demir, H.V. Morphology-tailored synthesis of tungsten trioxide (hydrate) thin films and their photocatalytic properties. *ACS Appl. Mater. Interfaces* **2011**, *3*, 229–236. [CrossRef]

98. Nagy, D.; Nagy, D.; Szilágyi, I.M.; Fan, X. Effect of the morphology and phases of WO 3 nanocrystals on their photocatalytic efficiency. *RSC Adv.* **2016**, *6*, 33743–33754. [CrossRef]

99. Sieb, N.R.; Wu, N.-c.; Majidi, E.; Kukreja, R.; Branda, N.R.; Gates, B.D. Hollow metal nanorods with tunable dimensions, porosity, and photonic properties. *Acs Nano* **2009**, *3*, 1365–1372. [CrossRef]

100. Lu, X.; Wang, G.; Zhai, T.; Yu, M.; Gan, J.; Tong, Y.; Li, Y. Hydrogenated TiO₂ nanotube arrays for supercapacitors. *Nano Lett.* **2012**, *12*, 1690–1696. [CrossRef]

101. Kalanoor, B.S.; Seo, H.; Kalanur, S.S. Recent developments in photoelectrochemical water-splitting using WO₃/BiVO₄ heterojunction photoanode: A review. *Mater. Sci. Energy Technol.* **2018**, *1*, 49–62.

102. Pihosh, Y.; Turkevych, I.; Mawatari, K.; Uemura, J.; Kazoe, Y.; Kosar, S.; Makita, K.; Sugaya, T.; Matsui, T.; Fujita, D.; et al. Photocatalytic generation of hydrogen by core-shell WO₃/BiVO₄ nanorods with ultimate water splitting efficiency. *Sci. Rep.* **2015**, *5*, 11141. [CrossRef]

103. Hammad, A.; El-Bery, H.M.; El-Shazly, A.; Elkady, M. Effect of WO₃ morphological structure on its photoelectrochemical properties. *Int. J. Electrochem. Sci* **2018**, *13*, 362–372. [CrossRef]

104. Zhou, J.; Lin, S.; Chen, Y.; Gaskov, A. Facile morphology control of WO₃ nanostructure arrays with enhanced photoelectrochemical performance. *Appl. Surface Sci.* **2017**, *403*, 274–281. [CrossRef]

105. Perera, D.; Lorek, R.; Khnayzer, R.S.; Moroz, P.; O'Connor, T.; Khon, D.; Diederich, G.; Kinder, E.; Lambright, S.; Castellano, F.N. Photocatalytic activity of core/shell semiconductor nanocrystals featuring spatial separation of charges. *J. Phys. Chem. C* **2012**, *116*, 22786–22793. [CrossRef]

106. Rao, P.M.; Cai, L.; Liu, C.; Cho, I.S.; Lee, C.H.; Weisse, J.M.; Yang, P.; Zheng, X. Simultaneously efficient light absorption and charge separation in WO₃/BiVO₄ core/shell nanowire photoanode for photoelectrochemical water oxidation. *Nano Lett.* **2014**, *14*, 1099–1105. [CrossRef]

107. Jin, B.; Jung, E.; Ma, M.; Kim, S.; Zhang, K.; Kim, J.I.; Son, Y.; Park, J.H. Solution-processed yolk–shell-shaped WO₃/BiVO₄ heterojunction photoelectrodes for efficient solar water splitting. *J. Mater. Chem. A* **2018**, *6*, 2585–2592. [CrossRef]

108. Li, Z.; Zheng, G.; Wang, J.; Li, H.; Wu, J.; Du, Y. Refining waste hardmetals into tungsten oxide nanosheets via facile method. *J. Nanoparticle Res.* **2016**, *18*, 98. [CrossRef]

109. Tan, Y.; Wang, H.; Liu, P.; Shen, Y.; Cheng, C.; Hirata, A.; Fujita, T.; Tang, Z.; Chen, M. Versatile nanoporous bimetallic phosphides towards electrochemical water splitting. *Energy Environ. Sci.* **2016**, *9*, 2257–2261. [CrossRef]

110. Kim, H.; Hwang, D.; Kim, Y.; Lee, J. Highly donor-doped (110) layered perovskite materials as novel photocatalysts for overall water splitting. *Chem. Commun.* **1999**, 1077–1078. [CrossRef]

111. Lee, C.Y.; Wang, L.; Kado, Y.; Killian, M.S.; Schmuki, P. Anodic nanotubular/porous hematite photoanode for solar water splitting: Substantial effect of iron substrate purity. *ChemSusChem* **2014**, *7*, 934–940. [CrossRef]

112. Troitskaia, I.; Gavrilova, T.; Atuchin, V. Structure and micromorphology of titanium dioxide nanoporous microspheres formed in water solution. *Phys. Procedia* **2012**, *23*, 65–68. [CrossRef]

113. Fan, X.; Liu, Y.; Chen, S.; Shi, J.; Wang, J.; Fan, A.; Zan, W.; Li, S.; Goddard, W.A.; Zhang, X.-M. Defect-enriched iron fluoride-oxide nanoporous thin films bifunctional catalyst for water splitting. *Nat. Commun.* **2018**, *9*, 1–11. [CrossRef] [PubMed]

114. Shin, S.; Han, H.S.; Kim, J.S.; Park, I.J.; Lee, M.H.; Hong, K.S.; Cho, I.S. A tree-like nanoporous WO$_3$ photoanode with enhanced charge transport efficiency for photoelectrochemical water oxidation. *J. Mater. Chem. A* **2015**, *3*, 12920–12926. [CrossRef]

115. Fujimoto, I.; Wang, N.; Saito, R.; Miseki, Y.; Gunji, T.; Sayama, K. WO3/BiVO4 composite photoelectrode prepared by improved auto-combustion method for highly efficient water splitting. *Int. J. Hydrogen Energy* **2014**, *39*, 2454–2461. [CrossRef]

116. Choi, K.-S. Shape effect and shape control of polycrystalline semiconductor electrodes for use in photoelectrochemical cells. *J. Phys. Chem. Lett.* **2010**, *1*, 2244–2250. [CrossRef]

117. Song, K.; Gao, F.; Yang, W.; Wang, E.; Wang, Z.; Hou, H. WO$_3$ mesoporous nanobelts towards efficient photoelectrocatalysts for water splitting. *ChemElectroChem* **2018**, *5*, 322–327. [CrossRef]

118. Markhabayeva, A.A.; Moniruddin, M.; Dupre, R.; Abdullin, K.A.; Nuraje, N. Designing of WO$_3$@ Co$_3$O$_4$ Heterostructure to Enhance Photoelectrochemical Performances. *J. Phys. Chem. A* **2020**, *124*, 486–491. [CrossRef]

119. Kronawitter, C.X.; Vayssieres, L.; Shen, S.; Guo, L.; Wheeler, D.A.; Zhang, J.Z.; Antoun, B.R.; Mao, S.S. A perspective on solar-driven water splitting with all-oxide hetero-nanostructures. *Energy Environ. Sci.* **2011**, *4*, 3889–3899. [CrossRef]

120. Moniruddin, M.; Afroz, K.; Shabdan, Y.; Bizri, B.; Nuraje, N. Hierarchically 3D assembled strontium titanate nanomaterials for water splitting application. *Appl. Surface Sci.* **2017**, *419*, 886–892. [CrossRef]

121. Shabdan, Y.; Ronasi, A.; Coulibaly, P.; Moniruddin, M.; Nuraje, N. Engineered core-shell nanofibers for electron transport study in dye-sensitized solar cells. *AIP Adv.* **2017**, *7*, 065008. [CrossRef]

122. Abdullin, K.A.; Azatkaliev, A.; Gabdullin, M.; Kalkozova, Z.K.; Mukashev, B.; Serikkanov, A. Preparation of Nanosized Tungsten and Tungsten Oxide Powders. *Phys. Solid State* **2018**, *60*, 2634–2639. [CrossRef]

123. Lisitsyna, L.; Denisov, G.; Dauletbekova, A.; Karipbayev, Z.T.; Markhabaeva, A.; Vaganov, V.; Lisitsyn, V.M.; Akilbekov, A. Luminescence of LiF crystals doped with uranium. *Proc. J. Phys. Conf. Ser.* **2017**, *830*, 012156. [CrossRef]

124. Moniruddin, M.; Oppong, E.; Stewart, D.; McCleese, C.; Roy, A.; Warzywoda, J.; Nuraje, N. Designing CdS-Based Ternary Heterostructures Consisting of Co-Metal and CoOx Cocatalysts for Photocatalytic H$_2$ Evolution under Visible Light. *Inorg. Chem.* **2019**, *58*, 12325–12333. [CrossRef] [PubMed]

125. De Tacconi, N.R.; Chenthamarakshan, C.; Rajeshwar, K.; Pauporté, T.; Lincot, D. Pulsed electrodeposition of WO$_3$–TiO$_2$ composite films. *Electrochem. Commun.* **2003**, *5*, 220–224. [CrossRef]

126. Smith, W.; Wolcott, A.; Fitzmorris, R.C.; Zhang, J.Z.; Zhao, Y. Quasi-core-shell TiO$_2$/WO$_3$ and WO$_3$/TiO$_2$ nanorod arrays fabricated by glancing angle deposition for solar water splitting. *J. Mater. Chem.* **2011**, *21*, 10792–10800. [CrossRef]

127. Bae, S.W.; Ji, S.M.; Hong, S.J.; Jang, J.W.; Lee, J.S. Photocatalytic overall water splitting with dual-bed system under visible light irradiation. *Int. J. Hydrogen Energy* **2009**, *34*, 3243–3249. [CrossRef]

128. Ohno, T.; Tanigawa, F.; Fujihara, K.; Izumi, S.; Matsumura, M. Photocatalytic oxidation of water on TiO$_2$-coated WO$_3$ particles by visible light using Iron (III) ions as electron acceptor. *J. Photochem. Photobiol. A Chem.* **1998**, *118*, 41–44. [CrossRef]

129. Yang, M.; He, H.; Zhang, H.; Zhong, X.; Dong, F.; Ke, G.; Chen, Y.; Du, J.; Zhou, Y. Enhanced photoelectrochemical water oxidation on WO_3 nanoflake films by coupling with amorphous TiO_2. *Electrochim. Acta* **2018**, *283*, 871–881. [CrossRef]

130. Momeni, M.M.; Ghayeb, Y.; Davarzadeh, M. Single-step electrochemical anodization for synthesis of hierarchical WO_3–TiO_2 nanotube arrays on titanium foil as a good photoanode for water splitting with visible light. *J. Electroanal. Chem.* **2015**, *739*, 149–155. [CrossRef]

131. Lai, C.W.; Sreekantan, S. Preparation of hybrid WO_3–TiO_2 nanotube photoelectrodes using anodization and wet impregnation: Improved water-splitting hydrogen generation performance. *Int. J. Hydrogen Energy* **2013**, *38*, 2156–2166. [CrossRef]

132. Kwon, Y.T.; Song, K.Y.; Lee, W.I.; Choi, G.J.; Do, Y.R. Photocatalytic behavior of WO_3-loaded TiO_2 in an oxidation reaction. *J. Catal.* **2000**, *191*, 192–199. [CrossRef]

133. Sajjad, A.K.L.; Shamaila, S.; Tian, B.; Chen, F.; Zhang, J. Comparative studies of operational parameters of degradation of azo dyes in visible light by highly efficient WOx/TiO_2 photocatalyst. *J. Hazard. Mater.* **2010**, *177*, 781–791. [CrossRef]

134. Hong, S.J.; Lee, S.; Jang, J.S.; Lee, J.S. Heterojunction $BiVO_4/WO_3$ electrodes for enhanced photoactivity of water oxidation. *Energy Environ. Sci.* **2011**, *4*, 1781–1787. [CrossRef]

135. Xia, L.; Bai, J.; Li, J.; Zeng, Q.; Li, X.; Zhou, B. A highly efficient $BiVO_4/WO_3/W$ heterojunction photoanode for visible-light responsive dual photoelectrode photocatalytic fuel cell. *Appl. Catal. B Environ.* **2016**, *183*, 224–230. [CrossRef]

136. Santato, C.; Odziemkowski, M.; Ulmann, M.; Augustynski, J. Crystallographically oriented mesoporous WO_3 films: Synthesis, characterization, and applications. *J. Am. Chem. Soc.* **2001**, *123*, 10639–10649. [CrossRef] [PubMed]

137. Grigioni, I.; Stamplecoskie, K.G.; Selli, E.; Kamat, P.V. Dynamics of photogenerated charge carriers in $WO_3/BiVO_4$ heterojunction photoanodes. *J. Phys. Chem. C* **2015**, *119*, 20792–20800. [CrossRef]

138. Pattengale, B.; Ludwig, J.; Huang, J. Atomic insight into the W-doping effect on carrier dynamics and photoelectrochemical properties of $BiVO_4$ photoanodes. *J. Phys. Chem. C* **2016**, *120*, 1421–1427. [CrossRef]

139. Mali, M.G.; Yoon, H.; Kim, M.-w.; Swihart, M.T.; Al-Deyab, S.S.; Yoon, S.S. Electrosprayed heterojunction $WO_3/BiVO_4$ films with nanotextured pillar structure for enhanced photoelectrochemical water splitting. *Appl. Phys. Lett.* **2015**, *106*, 151603. [CrossRef]

140. Cesar, I.; Sivula, K.; Kay, A.; Zboril, R.; Graätzel, M. Influence of feature size, film thickness, and silicon doping on the performance of nanostructured hematite photoanodes for solar water splitting. *J. Phys. Chem. C* **2008**, *113*, 772–782. [CrossRef]

141. Sadtler, B.; Demchenko, D.O.; Zheng, H.; Hughes, S.M.; Merkle, M.G.; Dahmen, U.; Wang, L.-W.; Alivisatos, A.P. Selective facet reactivity during cation exchange in cadmium sulfide nanorods. *J. Am. Chem. Soc.* **2009**, *131*, 5285–5293. [CrossRef]

142. Li, Y.; Zhang, L.; Liu, R.; Cao, Z.; Sun, X.; Liu, X.; Luo, J. WO_3@ α-Fe_2O_3 Heterojunction arrays with improved photoelectrochemical behavior for neutral ph water splitting. *ChemCatChem* **2016**, *8*, 2765–2770. [CrossRef]

143. Luo, W.; Yu, T.; Wang, Y.; Li, Z.; Ye, J.; Zou, Z. Enhanced photocurrent–voltage characteristics of WO_3/Fe_2O_3 nano-electrodes. *J. Phys. D Appl. Phys.* **2007**, *40*, 1091. [CrossRef]

144. Li, Y.; Feng, J.; Li, H.; Wei, X.; Wang, R.; Zhou, A. Photoelectrochemical splitting of natural seawater with α-Fe_2O_3/WO_3 nanorod arrays. *Int. J. Hydrogen Energy* **2016**, *41*, 4096–4105. [CrossRef]

145. Kumar, P.; Singh, M.; Reddy, G.B. Core-Shell WO_3-WS_2 Nanostructured Thin Films via Plasma Assisted Sublimation and Sulfurization. *ACS Appl. Nano Mater.* **2019**, *2*, 1691–1703. [CrossRef]

146. Zhang, J.; Liu, Z.; Liu, Z. Novel WO_3/Sb_2S_3 heterojunction photocatalyst based on WO_3 of different morphologies for enhanced efficiency in photoelectrochemical water splitting. *ACS Appl. Mater. Interfaces* **2016**, *8*, 9684–9691. [CrossRef]

147. Wang, Y.; Tian, W.; Chen, L.; Cao, F.; Guo, J.; Li, L. Three-dimensional WO_3 nanoplate/Bi_2S_3 nanorod heterojunction as a highly efficient photoanode for improved photoelectrochemical water splitting. *ACS Appl. Mater. Interfaces* **2017**, *9*, 40235–40243. [CrossRef]

148. Wang, P.; Huang, B.; Dai, Y.; Whangbo, M.-H. Plasmonic photocatalysts: Harvesting visible light with noble metal nanoparticles. *Phys. Chem. Chem. Phys.* **2012**, *14*, 9813–9825. [CrossRef]

149. Hu, D.; Diao, P.; Xu, D.; Wu, Q. Gold/WO_3 nanocomposite photoanodes for plasmonic solar water splitting. *Nano Res.* **2016**, *9*, 1735–1751. [CrossRef]

150. Warren, S.C.; Thimsen, E. Plasmonic solar water splitting. *Energy Environ. Sci.* **2012**, *5*, 5133–5146. [CrossRef]

151. Thimsen, E.; Le Formal, F.; Gratzel, M.; Warren, S.C. Influence of plasmonic Au nanoparticles on the photoactivity of Fe_2O_3 electrodes for water splitting. *Nano Lett.* **2010**, *11*, 35–43. [CrossRef]

152. Singh, T.; Müller, R.; Singh, J.; Mathur, S. Tailoring surface states in WO_3 photoanodes for efficient photoelectrochemical water splitting. *Appl. Surface Sci.* **2015**, *347*, 448–453. [CrossRef]

153. Li, Y.Z.; Liu, Z.; Guo, Z.; Ruan, M.; Li, X.; Liu, Y. Efficient WO_3 photoanode modified by Pt layer and plasmonic Ag for enhanced charge separation and transfer to promote photoelectrochemical performances. *ACS Sustain. Chem. Eng.* **2019**, *7*, 12582–12590.

154. Xu, S.; Fu, D.; Song, K.; Wang, L.; Yang, Z.; Yang, W.; Hou, H. One-dimensional $WO_3/BiVO_4$ heterojunction photoanodes for efficient photoelectrochemical water splitting. *Chem. Eng. J.* **2018**, *349*, 368–375. [CrossRef]

155. Wu, P.D.; Liu, Z.F.; Ruan, M.N.; Guo, Z.G.; Zhao, L. Cobalt-phosphate modified Fe-Zn0.2Cd0.8S/CuSbS$_2$ heterojunction photoanode with multiple synergistic effect for enhancing photoelectrochemical water splitting. *Appl. Surface Sci.* **2019**, *476*, 716–723. [CrossRef]

156. Kim, H.I.; Kim, J.; Kim, W.; Choi, W. Enhanced Photocatalytic and Photoelectrochemical Activity in the Ternary Hybrid of $CdS/TiO_2/WO_3$ through the Cascadal Electron Transfer. *J. Phys. Chem. C* **2011**, *115*, 9797–9805. [CrossRef]

157. Zeng, Q.Y.; Bai, J.; Li, J.H.; Li, L.S.; Xia, L.G.; Zhou, B.X.; Sun, Y.G. Highly-stable and efficient photocatalytic fuel cell based on an epitaxial $TiO_2/WO_3/W$ nanothorn photoanode and enhanced radical reactions for simultaneous electricity production and wastewater treatment. *Appl. Energy* **2018**, *220*, 127–137. [CrossRef]

158. Liu, C.H.; Qiu, Y.Y.; Zhang, J.; Liang, Q.; Mitsuzaki, N.; Chen, Z.D. Construction of CdS quantum dots modified g-C_3N_4/ZnO heterostructured photoanode for efficient photoelectrochemical water splitting. *J. Photochem. Photobiol. A Chem.* **2019**, *371*, 109–117. [CrossRef]

159. Liew, S.L.; Zhang, Z.; Goh, T.W.G.; Subramanian, G.S.; Seng, H.L.D.; Hor, T.S.A.; Luo, H.K.; Chi, D.Z. Yb-doped WO_3 photocatalysts for water oxidation with visible light. *Int. J. Hydrogen Energy* **2014**, *39*, 4291–4298. [CrossRef]

160. Subramanyam, P.; Vinodkumar, T.; Nepak, D.; Deepa, M.; Subrahmanyam, C. Mo-doped $BiVO_4$@reduced graphene oxide composite as an efficient photoanode for photoelectrochemical water splitting. *Catal. Today* **2019**, *325*, 73–80. [CrossRef]

161. Zhang, K.; Shi, X.J.; Kim, J.K.; Park, J.H. Photoelectrochemical cells with tungsten trioxide/Mo-doped $BiVO_4$ bilayers. *Phys. Chem. Chem. Phys.* **2012**, *14*, 11119–11124. [CrossRef]

162. Tian, L.; Yang, X.F.; Cui, X.K.; Liu, Q.Q.; Tang, H. Fabrication of dual direct Z-scheme g-$C_3N_4/MoS_2/Ag_3PO_4$ photocatalyst and its oxygen evolution performance. *Appl. Surface Sci.* **2019**, *463*, 9–17. [CrossRef]

163. Kim, J.Y.; Youn, D.H.; Kang, K.; Lee, J.S. Highly Conformal Deposition of an Ultrathin FeOOH Layer on a Hematite Nanostructure for Efficient Solar Water Splitting. *Angew. Chem. Int. Ed.* **2016**, *55*, 10854–10858. [CrossRef] [PubMed]

164. Lhermitte, C.R.; Verwer, J.G.; Bartlett, B.M. Improving the stability and selectivity for the oxygen-evolution reaction on semiconducting WO_3 photoelectrodes with a solid-state FeOOH catalyst. *J. Mater. Chem. A* **2016**, *4*, 2960–2968. [CrossRef]

165. Bai, S.L.; Yang, X.J.; Liu, C.Y.; Xiang, X.; Luo, R.X.; He, J.; Chen, A.F. An Integrating Photoanode of WO_3/Fe_2O_3 Heterojunction Decorated with NiFe-LDH to Improve PEC Water Splitting Efficiency. *Acs Sustain. Chem. Eng.* **2018**, *6*, 12906–12913. [CrossRef]

166. Davi, M.; Ogutu, G.; Schrader, F.; Rokicinska, A.; Kustrowski, P.; Slabon, A. Enhancing Photoelectrochemical Water Oxidation Efficiency of WO_3/alpha-Fe_2O_3 Heterojunction Photoanodes by Surface Functionalization with CoPd Nanocrystals. *Eur. J. Inorg. Chem.* **2017**, *2017*, 4267–4274. [CrossRef]

167. Sun, J.; Sun, L.; Yang, X.; Bai, S.; Luo, R.; Li, D.; Chen, A. Photoanode of coupling semiconductor heterojunction and catalyst for solar PEC water splitting. *Electrochim. Acta* **2020**, *331*, 135282. [CrossRef]

168. Tahir, M.; Siraj, M.; Tahir, B.; Umer, M.; Alias, H.; Othman, N. Au-NPs embedded Z–scheme WO_3/TiO_2 nanocomposite for plasmon-assisted photocatalytic glycerol-water reforming towards enhanced H2 evolution. *Appl. Surface Sci.* **2020**, *503*, 144344. [CrossRef]

169. Zhu, Z.F.; Yan, Y.; Li, J.Q. Preparation of flower-like $BiOBr-WO_3-Bi_2WO_6$ ternary hybrid with enhanced visible-light photocatalytic activity. *J. Alloys Compd.* **2015**, *651*, 184–192. [CrossRef]

170. Hou, Y.; Zuo, F.; Dagg, A.P.; Liu, J.K.; Feng, P.Y. Branched WO_3 nanosheet array with layered C_3N_4 heterojunctions and CoOx Nanoparticles as a flexible photoanode for efficient photoelectrochemical water oxidation. *Adv. Mater.* **2014**, *26*, 5043–5049. [CrossRef]

171. Yan, H.J.; Tian, C.G.; Wang, L.; Wu, A.P.; Meng, M.C.; Zhao, L.; Fu, H.G. Phosphorus-modified tungsten nitride/reduced graphene oxide as a high-performance, non-noble-metal electrocatalyst for the hydrogen evolution reaction. *Angew. Chem. Int. Ed.* **2015**, *54*, 6325–6329. [CrossRef]

172. Liu, C.J.; Yang, Y.H.; Li, W.Z.; Li, J.; Li, Y.M.; Shi, Q.L.; Chen, Q.Y. Highly Efficient Photoelectrochemical Hydrogen Generation Using ZnxBi2S3+x Sensitized Platelike WO_3 Photoelectrodes. *Acs Appl. Mater. Interfaces* **2015**, *7*, 10763–10770. [CrossRef] [PubMed]

173. Hu, J.S.; Ren, L.L.; Guo, Y.G.; Liang, H.P.; Cao, A.M.; Wan, L.J.; Bai, C.L. Mass production and high photocatalytic activity of ZnS nanoporous nanoparticles. *Angew. Chem. Int. Ed.* **2005**, *44*, 1269–1273. [CrossRef] [PubMed]

174. Zhang, X.L.; Wang, X.; Wang, D.F.; Ye, J.H. Conformal $BiVO_4$-Layer/WO_3-Nanoplate-Array Heterojunction Photoanode Modified with Cobalt Phosphate Cocatalyst for Significantly Enhanced Photoelectrochemical Performances. *Acs Appl. Mater. Interfaces* **2019**, *11*, 5623–5631. [CrossRef] [PubMed]

175. Wang, C.H.; Zhang, X.T.; Yuan, B.; Wang, Y.X.; Sun, P.P.; Wang, D.; Wei, Y.A.; Liu, Y.C. Multi-heterojunction photocatalysts based on WO3 nanorods: Structural design and optimization for enhanced photocatalytic activity under visible light. *Chem. Eng. J.* **2014**, *237*, 29–37. [CrossRef]

176. Ran, L.; Yin, L.W. Ternary Hierarchical Cu_7S_4/TiO_2/CoCr-LDH Heterostructured Nanorod Arrays with Multiphase Reaction Interfaces for More Efficient Photoelectrochemical Water Splitting. *Adv. Mater. Interfaces* **2019**, *6*, 14. [CrossRef]

177. Song, S.Y.; Zhang, Y.; Xing, Y.; Wang, C.; Feng, J.; Shi, W.D.; Zheng, G.L.; Zhang, H.J. Rectangular AgIn(WO4)(2) nanotubes: A promising photoelectric material. *Adv. Funct. Mater.* **2008**, *18*, 2328–2334. [CrossRef]

178. Ma, Z.Z.; Hou, H.L.; Song, K.; Fang, Z.; Wang, L.; Gao, F.M.; Yang, Z.B.; Tang, B.; Yang, W.Y. Ternary WO_3/Porous-$BiVO_4$/FeOOH Hierarchical Architectures: Towards Highly Efficient Photoelectrochemical Performance. *Chemelectrochem* **2018**, *5*, 3660–3667. [CrossRef]

179. Shi, X.J.; Choi, Y.; Zhang, K.; Kwon, J.; Kim, D.Y.; Lee, J.K.; Oh, S.H.; Kim, J.K.; Park, J.H. Efficient photoelectrochemical hydrogen production from bismuth vanadate-decorated tungsten trioxide helix nanostructures. *Nat. Commun.* **2014**, *5*, 8. [CrossRef]

180. Beermann, N.; Vayssieres, L.; Lindquist, S.E.; Hagfeldt, A. Photoelectrochemical studies of oriented nanorod thin films of hematite. *J. Electrochem. Soc.* **2000**, *147*, 2456–2461. [CrossRef]

181. Zhang, L.; Huang, Y.; Dai, C.H.; Liang, Q.M.; Yang, P.; Yang, H.H.; Yan, J.H. Constructing ZnO/$ZnCr_2O_4$@TiO_2-NTA Nanocomposite for Photovoltaic Conversion and Photocatalytic Hydrogen Evolution. *J. Electron. Mater.* **2019**, *48*, 1724–1729. [CrossRef]

182. Cai, J.J.; Li, S.; Qin, G.W. Interface engineering of Co_3O_4 loaded $CaFe_2O_4$/Fe_2O_3 heterojunction for photoelectrochemical water oxidation. *Appl. Surface Sci.* **2019**, *466*, 92–98. [CrossRef]

183. Yu, C.L.; Chen, F.Y.; Zeng, D.B.; Xie, Y.; Zhou, W.Q.; Liu, Z.; Wei, L.F.; Yang, K.; Li, D.H. A facile phase transformation strategy for fabrication of novel Z-scheme ternary heterojunctions with efficient photocatalytic properties. *Nanoscale* **2019**, *11*, 7720–7733. [CrossRef] [PubMed]

184. Lv, J.; Zhao, Z.Y.; Li, Z.S.; Ye, J.H.; Zou, Z.G. Preparation and photocatalytic property of LiCr(WO4)(2). *J. Alloys Compd.* **2009**, *485*, 346–350. [CrossRef]

185. Coridan, R.H.; Shaner, M.; Wiggenhorn, C.; Brunschwig, B.S.; Lewis, N.S. Electrical and Photoelectrochemical Properties of WO_3/Si Tandem Photoelectrodes. *J. Phys. Chem. C* **2013**, *117*, 6949–6957. [CrossRef]

186. Wang, H.L.; Turner, J.A. Characterization of Hematite Thin Films for Photoelectrochemical Water Splitting in a Dual Photoelectrode Device. *J. Electrochem. Soc.* **2010**, *157*, F173–F178. [CrossRef]

187. Wang, H.L.; Deutsch, T.; Turner, J.A. Direct water splitting under visible light with nanostructured hematite and WO_3 photoanodes and a GaInP2 photocathode. *J. Electrochem. Soc.* **2008**, *155*, F91–F96. [CrossRef]

188. Fountaine, K.T.; Atwater, H.A. Mesoscale modeling of photoelectrochemical devices: Light absorption and carrier collection in monolithic, tandem, Si vertical bar WO_3 microwires. *Opt. Express* **2014**, *22*, A1453–A1461. [CrossRef]

189. Xing, Z.; Shen, S.H.; Wang, M.; Ren, F.; Liu, Y.; Zheng, X.D.; Liu, Y.C.; Xiao, X.H.; Wu, W.; Jiang, C.Z. Efficient enhancement of solar-water-splitting by modified "Z-scheme" structural WO_3-W-Si photoelectrodes. *Appl. Phys. Lett.* **2014**, *105*, 143902. [CrossRef]

190. Sherman, B.D.; Sheridan, M.V.; Wee, K.R.; Marquard, S.L.; Wang, D.G.; Alibabaei, L.; Ashford, D.L.; Meyer, T.J. A Dye-Sensitized Photoelectrochemical Tandem Cell for Light Driven Hydrogen Production from Water. *J. Am. Chem. Soc.* **2016**, *138*, 16745–16753. [CrossRef]

191. Shi, X.; Jeong, H.; Oh, S.J.; Ma, M.; Zhang, K.; Kwon, J.; Choi, I.T.; Choi, I.Y.; Kim, H.K.; Kim, J.K.; et al. Unassisted photoelectrochemical water splitting exceeding 7% solar-to-hydrogen conversion efficiency using photon recycling. *Nat. Commun.* **2016**, *7*, 11943. [CrossRef]

192. Akple, M.S.; Chimmikuttanda, S.P. A ternary Z-scheme WO$_3$-Pt-CdS composite for improved visible-light photocatalytic H-2 production activity. *J. Nanoparticle Res.* **2018**, *20*, 16. [CrossRef]

193. Bhat, S.S.M.; Lee, S.A.; Suh, J.M.; Hong, S.-P.; Jang, H.W. Triple Planar Heterojunction of SnO$_2$/WO$_3$/BiVO$_4$ with Enhanced Photoelectrochemical Performance under Front Illumination. *Appl. Sci. Basel* **2018**, *8*, 1765. [CrossRef]

194. Yourey, J.E.; Kurtz, J.B.; Bartlett, B.M. Water Oxidation on a CuWO$_4$-WO$_3$ Composite Electrode in the Presence of Fe(CN)(6) (3-): Toward Solar Z-Scheme Water Splitting at Zero Bias. *J. Phys. Chem. C* **2012**, *116*, 3200–3205. [CrossRef]

195. Prasad, U.; Prakash, J.; Gupta, S.K.; Zuniga, J.; Mao, Y.B.; Azeredo, B.; Kannan, A.N.M. Enhanced Photoelectrochemical Water Splitting with Er- and W-Codoped Bismuth Vanadate with WO$_3$ Heterojunction-Based Two-Dimensional Photoelectrode. *Acs Appl. Mater. Interfaces* **2019**, *11*, 19029–19039. [CrossRef] [PubMed]

196. Pan, Q.; Zhang, H.; Yang, Y.; Cheng, C. 3D Brochosomes-Like TiO$_2$/WO$_3$/BiVO$_4$ Arrays as Photoanode for Photoelectrochemical Hydrogen Production. *Small* **2019**, *15*, 924. [CrossRef]

197. Ma, W.; Wu, X.; Huang, K.; Wang, M.; Fu, R.; Chen, H.; Feng, S. A Co(OH)(x) nanolayer integrated planar WO$_3$/Fe$_2$O$_3$ photoanode for efficient photoelectrochemical water splitting. *Sustain. Energy Fuels* **2019**, *3*, 2135–2141. [CrossRef]

198. Salimi, R.; Alvani, A.A.S.; Mei, B.T.; Naseri, N.; Du, S.F.; Mul, G. Ag-Functionalized CuWO$_4$/WO$_3$ nanocomposites for solar water splitting. *New J. Chem.* **2019**, *43*, 2196–2203. [CrossRef]

199. Peng, B.; Xia, M.Y.; Li, C.; Yue, C.S.; Diao, P. Network Structured CuWO$_4$/BiVO$_4$/Co-Pi Nanocomposite for Solar Water Splitting. *Catalysts* **2018**, *8*, 663. [CrossRef]

200. Leonard, K.C.; Nam, K.M.; Lee, H.C.; Kang, S.H.; Park, H.S.; Bard, A.J. ZnWO$_4$/WO$_3$ Composite for Improving Photoelectrochemical Water Oxidation. *J. Phys. Chem. C* **2013**, *117*, 15901–15910. [CrossRef]

201. Scarongella, M.; Gadiyar, C.; Strach, M.; Rimoldi, L.; Loiudice, A.; Buonsanti, R. Assembly of -Cu$_2$V$_2$O$_7$/WO$_3$ heterostructured nanocomposites and the impact of their composition on structure and photoelectrochemical properties. *J. Mater. Chem. C* **2018**, *6*, 12062–12069. [CrossRef]

202. Li, K.Z.; Zhang, C.; Liu, A.J.; Chu, D.M.; Zhang, C.Y.; Yang, P.; Du, Y.K.; Huang, J. Mesoporous tungsten oxide modified by nanolayered manganese-calcium oxide as robust photoanode for solar water splitting. *J. Colloid Interface Sci.* **2018**, *516*, 145–152. [CrossRef] [PubMed]

203. Li, Y.T.; Liu, Z.F.; Zhang, J.; Guo, Z.G.; Xin, Y.; Zhao, L. 1D/0D WO$_3$/CdS heterojunction photoanodes modified with dual co-catalysts for efficient photoelectrochemical water splitting. *J. Alloy. Compd.* **2019**, *790*, 493–501. [CrossRef]

204. Cui, Y.; Pan, L.; Chen, Y.; Afzal, N.; Ullah, S.; Liu, D.Y.; Wang, L.; Zhang, X.W.; Zou, J.J. Defected ZnWO4-decorated WO$_3$ nanorod arrays for efficient photoelectrochemical water splitting. *Rsc. Adv.* **2019**, *9*, 5492–5500. [CrossRef]

205. Ma, Z.Z.; Song, K.; Wang, L.; Gao, F.M.; Tang, B.; Hou, H.L.; Yang, W.Y. WO$_3$/BiVO$_4$ Type-II Heterojunction Arrays Decorated with Oxygen-Deficient ZnO Passivation Layer: A Highly Efficient and Stable Photoanode. *Acs Appl. Mater. Interfaces* **2019**, *11*, 889–897. [CrossRef]

206. Lv, Y.; Liu, Y.K.; Chen, C.M.; Wang, T.H.; Zhang, M. Octopus tentacles-like WO$_3$/C@CoO as high property and long life-time electrocatalyst for hydrogen evolution reaction. *Electrochim. Acta* **2018**, *281*, 1–8. [CrossRef]

207. Yuan, K.P.; Cao, Q.; Li, X.Y.; Chen, H.Y.; Deng, Y.H.; Wang, Y.Y.; Luo, W.; Lu, H.L.; Zhang, D.W. Synthesis of WO$_3$@ZnWO$_4$@ZnO-ZnO hierarchical nanocactus arrays for efficient photoelectrochemical water splitting. *Nano Energy* **2017**, *41*, 543–551. [CrossRef]

208. Lin, H.S.; Lin, L.Y. Improving Visible-light Responses and Electric Conductivities by Incorporating Sb$_2$S$_3$ and Reduced Graphene Oxide in a WO$_3$ Nanoplate Array for Photoelectrochemical Water Oxidation. *Electrochim. Acta* **2017**, *252*, 235–244. [CrossRef]

209. Jamali, S.; Moshaii, A. Improving photo-stability and charge transport properties of Cu_2O/CuO for photo-electrochemical water splitting using alternate layers of WO_3 or $CuWO_4$ produced by the same route. *Appl. Surface Sci.* **2017**, *419*, 269–276. [CrossRef]

210. Zhang, H.F.; Zhou, W.W.; Yang, Y.P.; Cheng, C.W. 3D WO3/BiVO4/Cobalt Phosphate Composites Inverse Opal Photoanode for Efficient Photoelectrochemical Water Splitting. *Small* **2017**, *13*, 8. [CrossRef]

211. Kalanur, S.S.; Yoo, I.H.; Park, J.; Seo, H. Insights into the electronic bands of $WO_3/BiVO_4/TiO_2$, revealing high solar water splitting efficiency. *J. Mater. Chem. A* **2017**, *5*, 1455–1461. [CrossRef]

212. Gao, H.Q.; Zhang, P.; Hu, J.H.; Pan, J.M.; Fan, J.J.; Shao, G.S. One-dimenSional Z-scheme $TiO_2/WO3/Pt$ heterostructures for enhanced hydrogen generation. *Appl. Surface Sci.* **2017**, *391*, 211–217. [CrossRef]

213. Peng, X.N.; He, C.; Liu, Q.Y.; Wang, X.N.; Wang, H.B.; Zhang, Y.J.; Ma, Q.C.; Zhang, K.; Han, Y.B.; Wang, H. Strategic Surface Modification of TiO_2 nanorods by WO_3 and $TiCl_4$ for the Enhancement in Oxygen Evolution Reaction. *Electrochim. Acta* **2016**, *222*, 1112–1119. [CrossRef]

 *nanomaterials*

Review

Recent Developments of TiO$_2$-Based Photocatalysis in the Hydrogen Evolution and Photodegradation: A Review

Baglan Bakbolat [1,2], **Chingis Daulbayev** [1,2,3,*], **Fail Sultanov** [1,2,*], **Renat Beissenov** [3,4,*], **Arman Umirzakov** [4], **Almaz Mereke** [4], **Askhat Bekbaev** [5,*] and **Igor Chuprakov** [6,7]

1   Faculty of Chemistry and Chemical Technology, al-Farabi Kazakh National University, Almaty 050000, Kazakhstan; baglan.bakbolat@mail.ru
2   Laboratory of Energy-Intensive and Nanomaterials, Institute of Combustion Problems, Almaty 050000, Kazakhstan
3   Department of Engineering Physics, Satbayev University, Almaty 050013, Kazakhstan
4   Laboratory of EPR Spectroscopy Named after Y.V. Gorelkinsky, LLP "Institute of Physics and Technology", Almaty 050032, Kazakhstan; arman_umirzakov@mail.ru (A.U.); mereke.almaz@mail.ru (A.M.)
5   Faculty of Physics and Technology, al-Farabi Kazakh National University, Almaty 050000, Kazakhstan
6   Frank Laboratory of Neutron Physics, Joint Institute for Nuclear Research (JINR), Dubna 141980, Russia; chupa@nf.jinr.ru
7   Faculty of Physics and Technical Sciences, L.N. Gumilyov Eurasian National University, Astana 010000, Kazakhstan
*   Correspondence: chingis.daulbayev@kaznu.kz (C.D.); sultanov.fail@kaznu.kz (F.S.); r.beissenov@satbayev.university (R.B.); bekbaev.ashat2@kaznu.kz (A.B.)

Received: 7 August 2020; Accepted: 29 August 2020; Published: 9 September 2020

Abstract: The growth of industrialization, which is forced to use non-renewable energy sources, leads to an increase in environmental pollution. Therefore, it is necessary not only to reduce the use of fossil fuels to meet energy needs but also to replace it with cleaner fuels. Production of hydrogen by splitting water is considered one of the most promising ways to use solar energy. TiO$_2$ is an amphoteric oxide that occurs naturally in several modifications. This review summarizes recent advances of doped TiO$_2$-based photocatalysts used in hydrogen production and the degradation of organic pollutants in water. An intense scientific and practical interest in these processes is aroused by the fact that they aim to solve global problems of energy conservation and ecology.

Keywords: titanium dioxide; photocatalysis; hydrogen evolution; photodegradation

---

1. Introduction

At present, hydrogen is regarded as the fuel of the future. Compared to carbon fuel, hydrogen is considered a renewable and environmentally friendly source of energy. There are various methods for producing hydrogen on an industrial scale. However, all known methods are characterized by high energy consumption, which makes the process of hydrogen production on a large scale disadvantageous. The production of hydrogen by photocatalytic water splitting is technologically simple, and the outcoming gases are environmentally friendly.

TiO$_2$ is a wide-bandgap semiconductor. In nature, TiO$_2$ is usually found in three different crystalline structures: rutile, anatase, and brookite. TiO$_2$ in anatase form is the most widespread photocatalyst for hydrogen evolution [1]. However, it cannot be used in the spectrum of visible light, since its bandgap (Eg) for different crystalline phases (anatase—3.2 eV, rutile—3.0 eV, and brookite—3.3 eV) is in the UV region. Also, the efficiency of photocatalysis in addition to the bandgap depends on many other factors [2]. In the 1970s, Fujishima and Honda studied the photoelectrochemical splitting of water, where a TiO$_2$-based electrode demonstrated the ability to split water under the influence of ultraviolet radiation [3]. Photocatalysis is a complex reaction consisting of the processes starting from light absorption to generate charge carriers up to surface catalytic reactions due to which gas is formed (Figure 1). This process only requires a photocatalyst (TiO$_2$), which is not consumed during the entire process, water, sunlight or ultraviolet radiation. During the absorption of light ($\geq E_g$) by the photocatalyst, an excited electron (e_{CB}^-) transfers from the valence band into the conduction band. This transition of the electron leads to the generation of a positively charged carrier-hole (h_{VB}^+) in the valence band (Equation (1)). Also, these charge carriers can recombine among themselves (Equation (2)) [4,5].

$$TiO_2 \xrightarrow{h\nu \geq E_g} h_{VB}^+ + e_{CB}^- \tag{1}$$

$$h_{VB}^+ + e_{CB}^- \xrightarrow{recombination} Energy \tag{2}$$

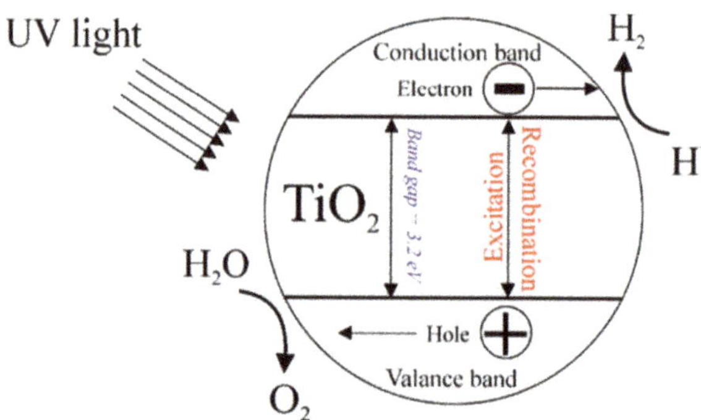

Figure 1. The mechanism of photocatalytic water splitting on TiO$_2$ based photocatalyst.

Another important application of TiO$_2$-based photocatalysts is based on their ability to discolor and completely decompose organic dyes contained in water. In addition, there are a number of works in which TiO$_2$-based photocatalysts were used to neutralize harmful to the atmosphere gases [6–9]. Chemical stability, easy accessibility, nontoxicity and the ability to oxidize under the influence of radiation, allow TiO$_2$-based photocatalysts to solve the main global problems associated with pollution of the environment and need for renewable energy [10,11]. Currently, TiO$_2$ as a photocatalyst is commercially produced in powder form. The most common brand of TiO$_2$ based photocatalyst are P-25 Degussa/Evonik, TiO$_2$ nanofibers from Kertak, TiO$_2$ Millennium PC-500, TiO$_2$ Hombikat UV-100, Sigma Aldrich TiO$_2$, TiO$_2$ PK-10, and P90 Aeroxide.

This paper aims to provide a brief overview of articles published starting from 2017 on new research and developments of TiO$_2$ based photocatalysts with significant advances. In this review, attention is also paid to the study on the mechanism of photocatalytic processes, factors affecting the activity of photocatalysts, and new techniques used to increase the activity of TiO$_2$ based photocatalysts during the splitting of water with the evolution of hydrogen and decomposition of organic compounds utilized for water purification.

2. Factors Affecting the Efficiency of Photocatalysis and Techniques Used to Improve the Efficiency of TiO$_2$-Based Photocatalysts

2.1. Lifetime of Photogenerated Charge Carriers

The activity of photocatalysts strongly depends on the lifetime of photogenerated electron-hole pairs. An important role is played by the rate at which charge carriers can reach the surface of the photocatalysts. The results of spectroscopic studies show that the time intervals between redox reactions or recombination involving charge carriers are extremely short, resulting in a significant reduction of the photocatalytic activity of TiO_2. In the case of recombination of charge carriers in a sufficiently fast interval (<0.1 ns), the photocatalytic activity of the semiconductor is not observed. For example, the lifetime of electron-hole pairs of ~250 ns (TiO_2) is considered relatively long [12]. Thus, it can be concluded that a high recombination rate and barriers, that prevent the transfer of charge carriers to the semiconductor surface, reduce photoactivity, despite the high concentration of initially photogenerated pairs. In this regard, to avoid their recombination, it becomes necessary to use cocatalysts in order to increase the lifetimes of electrons and holes.

2.2. The Particle Size of Photocatalyst

Compared to microparticles, TiO_2 nanoparticles have, generally, a higher photocatalytic activity [13,14]. This is due to the small diameter of the nanoparticles, in which the charge requires minimal effort to transfer to the surface. If the particle size decreases, the distance that photogenerated electrons and holes need to travel to the surface where the reactions take place is reduced, thereby reducing the probability of recombination. For TiO_2 photocatalyst microparticles, the penetration depth of UV rays is limited and amounts to about ~100 nm. This means that the inner part of the TiO_2 photocatalyst microparticle remains in a passive state. [15]. Figure 2 shows the scheme of light absorption by nano- and microparticles of TiO_2. This is one of the reasons for the increased interest in nanosized particles of TiO_2.

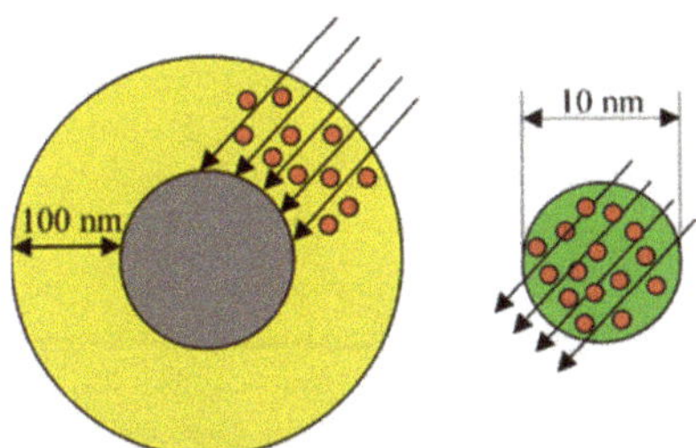

Figure 2. Light absorption by micro- and nanoparticles of TiO_2.

As shown in Figure 2, a decrease in the particle size of the photocatalyst to nanoscale facilitates the absorption of light by the entire volume of particles. However, there is a limitation regarding the minimum sizes to which it is desirable to reduce the particles of the photocatalyst, due to the onset of quantum effects. They become significant at particle sizes less than 2 nm for both anatase and rutile, and finally, this leads to a change in the bandgap. An increase in the size of the photocatalyst crystals leads to a decrease in the recombination of the electron-hole pair at the defects of the crystal lattice and to an increase in photocatalytic activity. For example, in [16], nanoparticles with a size of 25 nm were found to be more productive than nanoparticles of 15 nm. On the other hand, the bandgap is directly proportional to the size of the photocatalyst crystals. That is why it is necessary to find the optimal crystal size of TiO_2 and control it in the process of its obtaining.

To increase the photoactivity of TiO_2 in the visible region of the spectrum, the spectral region of its absorption should be expanded. There are several approaches for sensitizing TiO_2 to visible light: doping with cationic and anionic elements or metal nanoparticles. Elements of the 3d- and 2p-groups are often used as additives to reduce the value of the bandgap of the photocatalyst [17,18].

2.3. Doping with Cations

The essence of cationic doping is the introduction of metal cations into the crystal structure of TiO_2 at the position of Ti^{4+} ions. Rare-earth, noble, and transition metal cations can be used [19]. Doping with cations significantly expands the absorption spectrum of TiO_2, increases the redox potential of the formed radicals, and increases quantum efficiency by reducing the degree of recombination of electrons and holes. The nature and concentration of the dopant change the charge distribution on the TiO_2 surface and affects the process of photo corrosion and photocatalytic activity [20]. However, an increase in the absorption of visible light does not always lead to an increase in the activity of the photocatalyst. As a result of doping with cations, a certain number of defects appear in the TiO_2 structure, which can act as charge recombination centers, this leading to a decrease in photocatalytic activity even under the influence of UV light.

In [21], Kryzhitsky et al. show the change in the activity of photocatalytic properties of rutile and anatase forms of nanocrystalline TiO_2 depending on the nature of metal-based dopants. According to the results of the study, it is found that doping does not significantly change the bandgap of rutile, while in the case of doping anatase with iron and chromium, its bandgap narrows significantly. As a result of doping with metals, the photocatalytic activity of anatase (A) increases in the following order: A < A/Co < A/Cu < A/Fe. In the case of rutile (R), its photocatalytic activity decreases in the following order: R > R/Co > R/Cu > R/Fe > R/Cr. According to the authors, the decrease in the photoactivity of TiO_2 may be associated with the inhibitory effect of impurity cations.

2.4. Doping with Anionic Elements

Over the past few years, it has been shown that TiO_2 samples doped with nonmetallic elements (nitrogen, carbon, sulfur, boron, phosphorus, and fluorine) in the anionic positions of TiO_2, demonstrate high photoactivity in the UV and visible regions of the solar spectrum [22,23]. Among all the anions, carbon, nitrogen, and fluorine caused the most significant interest [24–26]. The substitution of oxygen atoms to carbon leads to the formation of new levels (C2p) above the ceiling of the valence band of TiO_2 (O2p), which reduces the bandgap and shifts the absorption spectrum. The inclusion of carbon in TiO_2 can also lead to the formation of carbon compounds on the surface of the photocatalyst, which acts as absorption centers of visible radiation [27].

Doping with nitrogen atoms is the most popular way to improve the photocatalytic performance of TiO_2. Introduction of nitrogen into the TiO_2 structure contributes to a significant shift of the absorption spectrum into the visible region of the solar spectrum, a change in the refractive index, an increase in hardness, electrical conductivity, elastic modulus, and photocatalytic activity in regard to visible light [28,29]. Upon substitution of anions, a new level is formed above the valence band of TiO_2 [30]. As shown in Figure 3, the presence of nitrogen leads to a change of the bandgap E_{g1} (TiO_2) > E_{g2} (N-doped TiO_2), thus contributing to the absorption of photons of light with lower energy.

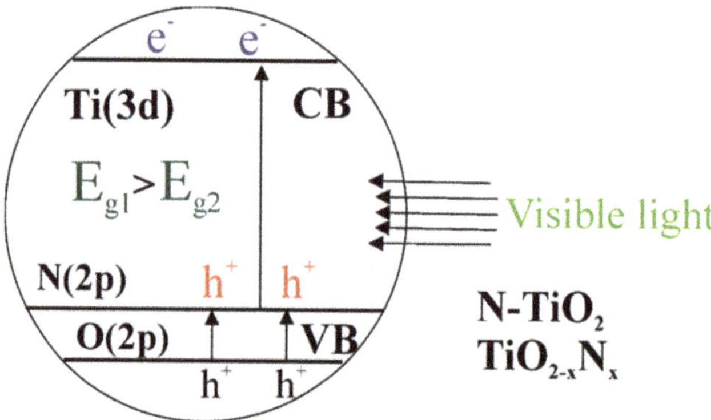

Figure 3. Changes in the band gap of TiO_2 upon doping with nitrogen atoms.

Doping with nitrogen in the oxygen position is difficult since the ionic radius of nitrogen (1.71 Å) is much larger than that for oxygen (1.4 Å). Another auspicious element in the anionic positions of

TiO_2 is fluorine atoms [31]. Unlike nitrogen, fluorine atoms easily replace oxygen due to the close ion radius (1.33 Å for F^- and 1.4 Å for O^{2-}). The increase in photocatalytic activity is mainly associated with an improvement in the degree of crystallinity of TiO_2 due to doping with fluorine [32]. It has been determined that crystallinity and the specific surface area also affect the photoactivity of TiO_2 [33,34]. The crystalline modification of TiO_2 in comparison with amorphous TiO_2 has significantly fewer defects, which reduce the possibility of recombination processes and contributes to the efficient movement of photogenerated charge carriers in the semiconductor. Since redox reactions occur on the surface of TiO_2, one of the main requirements for photocatalysts is the presence of a developed specific surface area. However, the presence of a developed specific surface area implies a large number of defects in the structure and a low degree of crystallinity and, as a result, reduces photocatalytic activity. Therefore, to increase photocatalytic activity, it is important to find a balance between the above factors.

2.5. Doping/Loading with Metal Nanoparticles

The application of metal nanoparticles is another alternative approach to the modification of photocatalysts. A review of the recent literature shows that metals (Co, Pt, Ag, Au, Pd, Ni, Cu, Eu, Fe, etc.) significantly increase the photocatalytic activity of TiO_2 [35–37]. The low location of the Fermi level of these metals compared to TiO_2 can lead to the movement of electrons from the TiO_2 structure to metal particles deposited on its surface. This helps to avoid the recombination of charge carriers, since the holes remain in the valence band of TiO_2. This is also beneficial for avoiding the recombination of charge carriers since the holes remain in the valence band of TiO_2. A number of conducted investigations indicate that the properties of these photocatalysts depend on the dispersion of metal particles [38,39]. Enhanced photocatalytic properties of metals appear when their size decreases <2.0 nm [40]. Despite the foregoing, too high concentration of metal particles can block the surface of TiO_2 and prevent the absorption of photons, leading to a decrease in the efficiency of the photocatalyst.

3. The Utilization of Photocatalysts Based on TiO_2

3.1. Hydrogen Evolution

Hydrogen can be produced by using nanoscale TiO_2 based photocatalysts with various morphologies in the form of nanowires, nanospheres, nanorods, nanotubes, and nanosheets [41]. Table 1 lists some TiO_2 nanocomposites with different structures, as well as their photocatalytic characteristics.

Table 1. TiO_2-based photocatalysts with different structures utilized for hydrogen evolution under splitting water mixtures.

Photocatalyst	Structure	Light Source	Sacrificial Agents	Evolution H_2	Ref.
Au/TiO_2 Pt/TiO_2	microspheres	15 W fluorescent tubes (λ_{max} = 365 nm)	25% vol. methanol	1118 µmol h^{-1} 2125 µmol h^{-1}	[42]
TiO_2	nanofibers	300 W Xenon lamp	10% vol. methanol	3200 µmol h^{-1} g^{-1}	[43]
Cu/TiO_2	nanorods	300 W Xe lamp (λ > 300 nm)	20% vol. methanol	1023.8 µmol h^{-1}	[44]
TiO_2/WO_3/Au	nanofibers	300 W Xe arc lamp	35% vol. methanol	269.63 µmol h^{-1}	[45]
M/TiO_2/rGO M = Au or Pt	nanoparticles	300 W Xenon lamp (λ > 300 nm)	20% vol. methanol	670 µmol h^{-1}	[46]
MoSe$_2$/TiO_2	nanoparticles	Xe arc lamp (PLS-SXE300)	10% vol. methanol	4.9 µmol h^{-1}	[47]
BCN-TiO_2	nanosheets& nanoparticles	300 W xenon lamp with a UV-cutoff filter ($\lambda \geq$ 420 nm)	20% vol. triethanolamine	68.54 µmol h^{-1} g^{-1}	[48]
TiO_2/C_3N_4	double-shell microtubes	300 W xenon lamp	20% vol. methanol	10.1 mmol h^{-1} g^{-1}	[49]
ZnS@g-C_3N_4/TiO_2	nanospheres	300 W Xenon lamp (λ > 400 nm)	10% vol. triethanolamine	422 µmol h^{-1} g^{-1}	[50]

P. Melián et al. [42] demonstrated that loading of nanosized TiO_2 microspheres with Au and Pt metals increases the hydrogen evolution twice. In the case of using Au, the maximum hydrogen evolution was determined at 1.5 wt.% content of Au in the sample with the yield of hydrogen 1118 µmol h^{-1}, while for Pt its optimal content in the sample was 0.27 wt.% with the yield of hydrogen

2125 µmol h^{-1}. The excess of dopants may result in decrease of photocatalyst activity due to possible complete coverage of TiO_2 surface, thus, hindering the light to be absorbed. A literature review also showed that the difference in the optimal ratio for each metal could be associated with the formation of recombination centers on the semiconductor surface by metal particles [51,52].

A positive effect on the rate of H_2 production was found when using sacrificial agents acting as electron donors (hole scavengers) during photoreforming, in which the hydroxyl radical is consumed by the sacrificial agents. In general, there are two types of sacrificial reagents: organic and inorganic based electron donors. Among organic electron donors, the most effective are water–alcohol mixtures, in particular, methanol > ethanol > ethylene glycol > glycerol [53–59]. However, an increase in the concentration of the sacrificial agent does not always lead to an increase in the yield of hydrogen. Y.-K. Park et al. [60] showed that the rate of hydrogen evolution also increases depending on the concentration of methanol (Figure 4a). At low concentrations, the rate of hydrogen formation in solutions is proportional to the concentration of methanol, while at higher concentrations, it approaches to a constant value [61]. Nevertheless, after adding a certain amount of methanol and ethanol, a further increase in its concentration leads to a decrease in the rate of hydrogen evolution (Figure 4b). The yield of hydrogen during photoreforming has a maximum output during 80–90 min and after it decreases. This is due to the formation of a significant amount of methane and ethane, during which photogenerated electrons (e^-_{CB}) and holes (h^+_{VB}) are consumed [54].

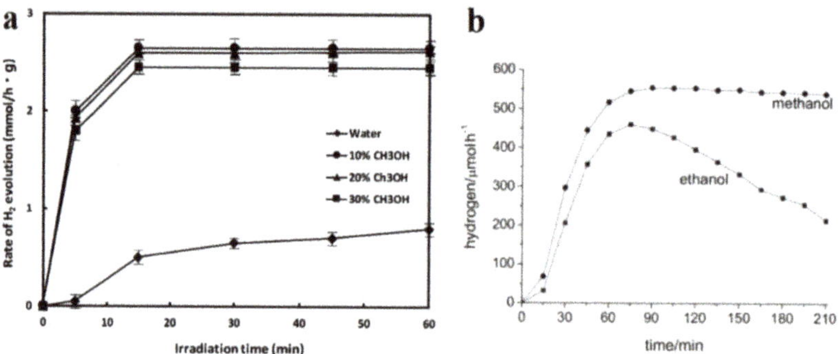

Figure 4. (a) Rate of hydrogen evolution from photocatalysis of aqueous methanol solution on Ag/TiO$_2$ photocatalysts. This figure is reprinted from [60], with permission from Elsevier, 2020; (b) H$_2$ production patterns for 24.47 M (100% *v/v*) of methanol and 17.06 M (100% *v/v*) of ethanol. This figure is reprinted from [54], with permission from Elsevier, 2015.

The amount of dopant also has a significant effect on the efficiency of light absorption by the photocatalyst and on its photocatalytic activity. For example, Udayabhanu et al. [62] prepared Cu-TiO$_2$/CuO nanocomposites containing different amount of Cu. The color of the obtained samples, depending on the concentration of Cu-TiO$_2$/CuO (CUT) from 1 to 4 mol% (the samples were as named CUT 1, CUT 2, CUT 3, and CUT 4) changes from light green to dark green (Figure 5). To evaluate the photocatalytic activity of hydrogen production, scientists compared the activities of obtained nanocomposites with conventional TiO$_2$. Sunlight was used as the source of radiation, and glycerol was chosen as a sacrificial agent. According to the results, the CUT 3 sample based on Cu-TiO$_2$/CuO possessed the highest yield of hydrogen (10.453 µmol h^{-1} g^{-1} of H$_2$ under sunlight and 4.714 µmol h^{-1} g^{-1} of H$_2$ under visible light). Nevertheless, this type of photocatalyst is effective only for hydrogen production, since it has shown low efficiency in the decomposition of organic dye and metal detoxification in water.

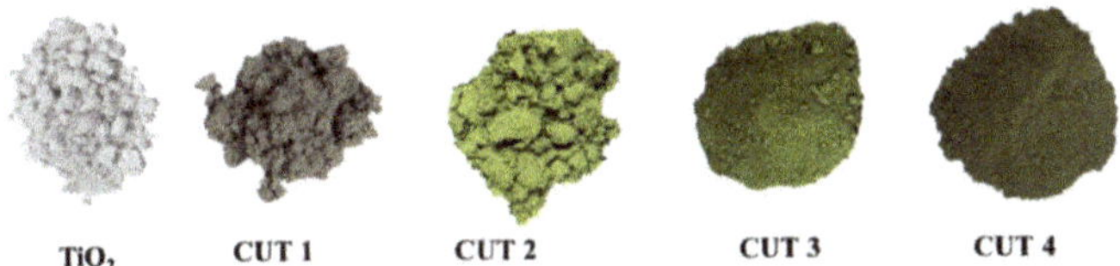

Figure 5. Colour of synthesized pristine and Cu-TiO$_2$/CuO composite nanopowders. This figure is reprinted from [62], with permission from Elsevier, 2020.

The photocatalyst efficiency is affected not only by the nature of the alloying element but also by its concentration. For example, X. Xing et al. in [63] demonstrated the dependence of the yield of hydrogen on the light intensity and the concentration of the dopant. From Figure 6a, it is clear that the increase of concentrations of each photocatalyst (pure TiO$_2$ and Au/TiO$_2$) results in increase of the rate of hydrogen production. However, the hydrogen generation rate for Au/TiO$_2$ photocatalyst at the same light intensity is 18–21 times higher than that for pure TiO$_2$. The explanation to this is that Au can limit charge carriers' recombinations and, what is more, the visible light absorption of photocatalyst is enhanced by the localized surface plasmon resonance effect of Au nanosized particles. The influence of the light intensity (from 1 to 9 kW/m^2) and the duration of exposure on the rate of hydrogen generation for Au/TiO$_2$ nanoparticles with a concentration of 1 g/L is also shown in Figure 6b.

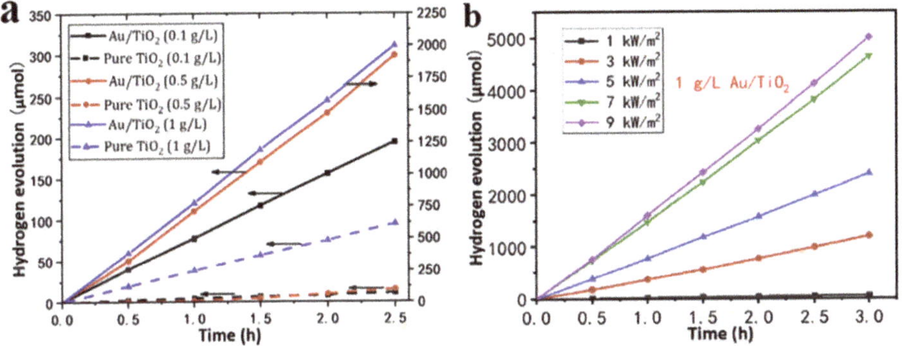

Figure 6. (**a**) Hydrogen evolution of pure TiO$_2$ nanoparticles and Au/TiO$_2$ nanoparticles at the same light intensity of 5 kW/m^2; (**b**) Hydrogen evolution in 1–9 kW/m^2 light intensities of 1 g/L Au/TiO$_2$ solutions. Both figures are reprinted from [63], with permission from Elsevier, 2020.

The photocatalytic properties of the material can be improved by creating composites based on different photocatalysts. For example, E.-C. Su et al. [64] obtained a composite photocatalyst based on Pt/N-TiO$_2$/SrTiO$_3$-TiO$_2$ in the form of nanotubes using a two-stage hydrothermal process (SrTiO$_3$ is also a photocatalyst with a bandgap of 3.2 eV with a perovskite-type structure [65]). The obtained results showed that this composite photocatalyst is able to operate under sunlight with the rate of hydrogen evolution up to 3873 µmol/h/g.

In practice, photocatalytic reactions are mainly carried out at room temperature. It is found that increasing temperature has a positive effect on the activity of some photocatalysts, which makes it relevant to develop new photocatalysts based on the anatase form of TiO$_2$ with a thermostable phase.

3.2. Photodegradation

Organic dyes are used on a large scale in modern industries. Wastewater polluted with such substances subsequently leads to environmental problems [66,67]. This is due to the fact that most organic dyes consist of biodegradable aromatic structures and azo-groups. Adsorption and catalytic oxidation remain the most effective among the methods for treating wastewater from dyes [68–70]. For photocatalytic degradation of organic compounds, in most cases, TiO_2, in the doped form, is used.

Graphene has a large specific surface area and electrical conductivity. The presence of such properties in graphene is of interest to scientists in the preparation of a graphene/nano-TiO_2-based photocatalyst [71–73]. After irradiation of a TiO_2-based photocatalyst, electrons move into the graphene structure. Based on the author's results [74], it helps to avoid recombination between charge carriers. TiO_2/graphene-based photocatalyst can be obtained by creating either a chemical bond between them or via the approach of their mechanically mixing [75,76]. Due to existence of chemical bond, the electron is transferred unhindered from the photocatalyst to graphene, reducing the probability of recombination. This is the main explanation for the increased activity of the photocatalyst with graphene.

Composites based on TiO_2/graphene attract scientists' attention not only as highly efficient photocatalysts due to light absorption in a wide range and charge separation, but also taking in account its high adsorption capacity to pollutants. However, TiO_2/graphene-based composite produced by the hydrothermal method cannot serve as an effective photocatalyst. The reason is the agglomeration of graphene layers, which adversely affects the adsorption and photocatalytic properties. The solution to this problem is described in [77], where graphene-based aerogel was used as an auxiliary material for TiO_2 [78,79]. Aerogel was obtained by thermal reduction of graphene oxide. Table 2 lists some reports on the use of composite TiO_2/graphene based photocatalysts to remove various organic compounds (pollutants and dyes) from water.

Table 2. Removal percentage of some organic pollutants by TiO_2/graphene based photocatalysts.

Photocatalyst	Organic Pollutant	Light Source	Irradiation Time	Efficiency	Ref.
TiO_2@rGO	2,4,6 trichlorophenol	Mercury lamp (11 W)	180 min	90%	[80]
TiO_2/Fe_3O_4/GO	Methylene blue	Halogen lamp (500 W)	90 min	76%	[81]
GO/TiO_2 nanotubes	Perfluorooctanoic acid	UV lamp (8 W)	240 min	97%	[82]
N-TiO_2/Ag_3PO_4@GO	Acid Blue 25	Halogen bulb (250 W)	20 min	98%	[83]
Ag and rGO modified TiO_2	Tetrabromobisphenol A	Xenon light (500 W)	80 min	99.6%	[84]
N-doped graphene/TiO_2	Bisphenol A	Mercury lamp (300 W)	60 min	100%	[85]
3D polyaniline/TiO_2/rGO hydrogel	BPA	Mercury lamp (500 W)	40 min	100%	[86]

Before testing photocatalytic activity, it was necessary to saturate the materials in the dark conditions. If saturation is not performed at dark conditions, a decrease in the concentration of the target pollutant or the color intensity change of the used in experiment dye will be associated not only with photocatalysis, but also with adsorption and photocatalytic degradation, so the effectiveness of the photocatalyst will be incorrect. X. Sun et al. [77] compared two samples of a photocatalyst of the same mass: hydrothermally obtained TiO_2-reduced graphene oxide (rGO), and aerogel based on TiO_2-rGO (Figure 7a). As can be seen, despite the equal mass of both samples, the TiO_2-rGO based aerogel has a larger specific volume than that of the second sample. For further characterization of their adsorption, both samples were used to adsorb methylene blue in the dark. During the experiment, the absorption intensity was analyzed. The results showed that it took 2 min for TiO_2-rGO based aerogel to achieve adsorption saturation, while for TiO_2-rGO powder, it took more than 10 min (Figure 7b). Figure 7c,d demonstrates the color change of methylene blue, which proves the high absorption coefficient of visible light by TiO_2-rGO aerogels. According to the authors, the adsorption rate also affects the efficiency of photocatalysts.

Photocatalytic activity directly depends on the number of active centers on the surface of the photocatalyst. Photolithography is a block of technological processes of photochemical technology aimed at creating the relief in the film, as well as a film of metal deposited on a substrate [87,88]. Using this method, a group of scientists [89] managed to obtain TiO_2 films with lattice, square, and hexagonal structures (Figure 8) and investigate the influence of a such surface textures on the photocatalysis. Obtained results showed that the activity of photocatalyst in the form of a film is not improved by increasing the values of specific surface area. Surface texture also has an effect on mass transfer during photocatalysis.

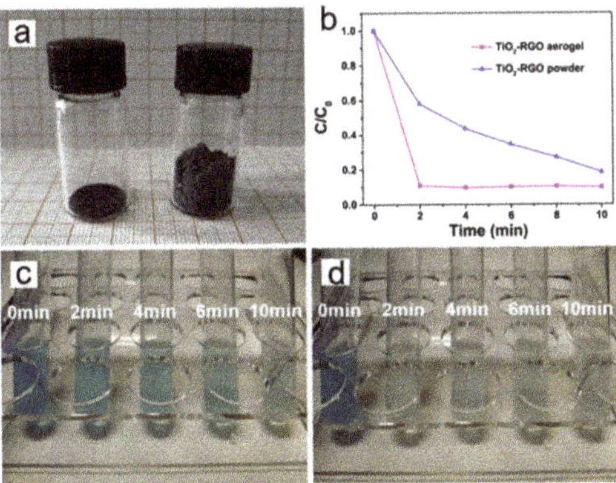

Figure 7. (**a**) Photograph of 80 mg TiO$_2$-reduced graphene oxide (rGO) powder (left) and aerogel (right); (**b**) methylene blue (MB) dark adsorption results of the TiO$_2$-rGO aerogel/powder; color changing of MB solution during dark adsorption with TiO$_2$-rGO powder (**c**), and aerogel (**d**). All figures are reprinted from [77], with permission from Elsevier, 2020.

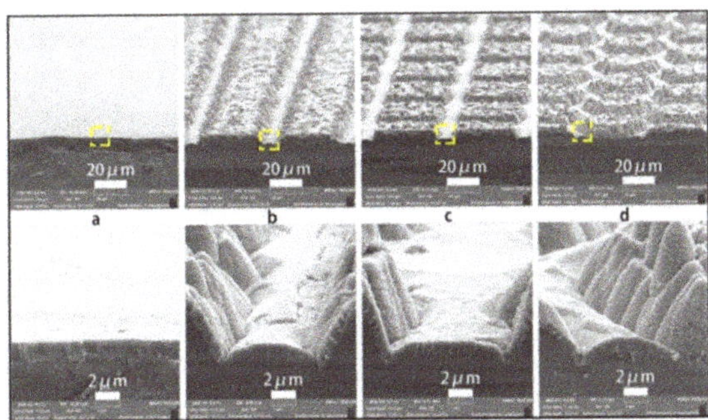

Figure 8. SEM images of (**a**) planar TiO$_2$ film; (**b**) grating-structured TiO$_2$ film; (**c**) square-structured TiO$_2$ film; (**d**) hexagon-structured TiO$_2$ film. All figures are reprinted from [89], with permission from Elsevier, 2019.

To evaluate the activity of photocatalysts, scientists conducted an experiment, in which the dye degradation occurred in water as a result of its exposure with UV (254 nm) on methyl orange. As shown in Figure 9, the microstructured TiO$_2$ films exhibit a more active photocatalytic activity than TiO$_2$ films with a flat surface. According to scientists, this is due to the presence of reaction centers on the surface of microstructured films. The highest photocatalytic activity was observed for TiO$_2$ films with a square microstructure. The experimental results showed that the efficiency of photocatalysts is influenced not only by the surface area, but also by the type of their microstructure. It was revealed, that the TiO$_2$ film with a grating structure, despite its low specific surface area, possessed photoactive properties similar to the TiO$_2$ film with a hexagonal structure [89].

A significant role during photocatalysis is played by the surface microstructure, which influences the mass transfer of degradable organic compounds by the diffusion to the surface of the photocatalyst. It is important to take into account the fact that the fluid flowing over the surface with protrusions meets more resistance from the side of the walls, compared to the fluid flowing through the flat surface. As a result, such protrusions can adversely affect the degradation efficiency of organic pollutants.

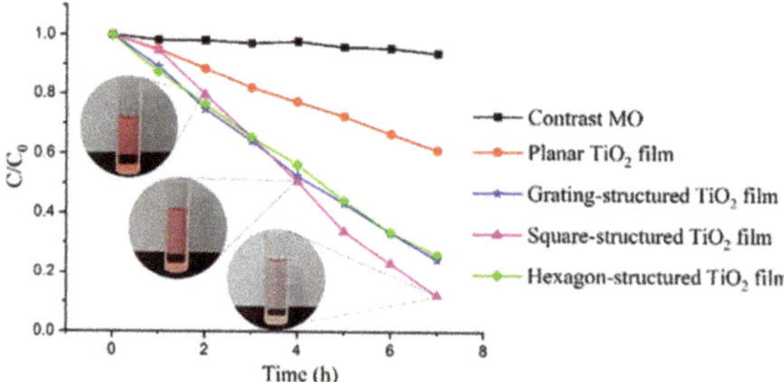

Figure 9. Photocatalytic degradation of MO under ultraviolet light (254 nm) irradiation. C and C_0 represent the real-time and initial concentration of methyl orange solution. This figure is reprinted from [89], with permission from Elsevier, 2019.

TiO_2 doped with Fe_2O_3 is one of the best-known effective photocatalysts. Due to the narrow bandgap of Fe_2O_3 (2.2 eV), its doping leads to a redshift of the light response of the photocatalyst. The phenomenon of decreasing of photoactivity of TiO_2 based photocatalyst doped with non-metals during heating, the high cost of some metals and the availability of Fe_2O_3 in large quantities increases the attractiveness of Fe_2O_3 over other dopants [90]. J.-J. Zhang et al. [91] used Fe_2O_3 nanoparticles, which served as a doping agent for obtaining the TiO_2/graphene aerogel (GA) based photocatalyst with a 3D structure (Figure 10). Due to the narrow bandgap (2.0 eV), Fe_2O_3 can easily generate electron-hole pairs, thereby contributing to the photodegradation of rhodamine B even in visible light. The results showed that the aqueous solution containing rhodamine B (RhB) was purified to 97.7% (Figure 10 b). Fe_3O_4 can also be used as a doping agent for the photocatalysis [92]. For example, F. Soltani-Nezhad et al. [93] presented a method for producing a $GO/Fe_3O_4/TiO_2$-NiO-based photocatalyst, which is able to efficiently degrade imidacloprid (pesticide).

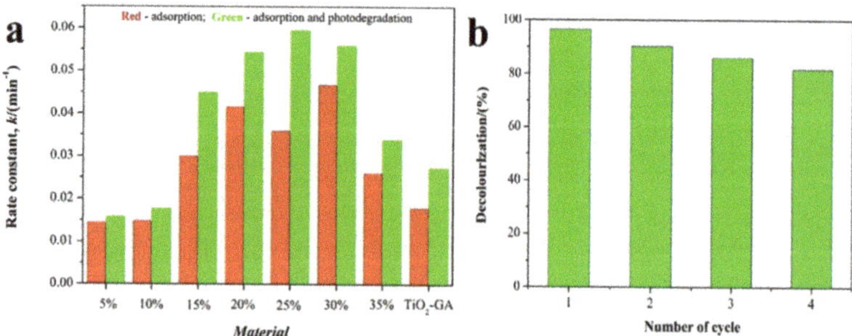

Figure 10. (**a**) The rate constants for the adsorption and the collective removal of rhodamine B (RhB) over TiO_2-GA and the Fe_2O_3-TiO_2-GA composites; (**b**) stability of the Fe_2O_3-TiO_2-GA (25%) composites in the removal of RhB dye. Both figures are reprinted from [91], with permission from Elsevier, 2018.

To increase the efficiency of the process of degradation of organic pollutants by photocatalysis, attempts have been made to combine radiation with ultrasonic cavitation. S. Rajoriya et al. [94] reported the photodegradation of 4-acetamidophenol to 91% using Sm (samarium) and N-doped TiO_2 photocatalysts, in which a combination of UV radiation, hydrodynamic cavitation, and ultrasound was applied. In another work, the use of N and Cu-doped TiO_2@CNTs in sono-photocatalysis for purification of pharmaceutical wastewater is discussed [95]. The results showed that when using a commercial Xe lamp (50 W) and ultrasound for pharmaceutical wastewaters treatment, the photocatalyst removal efficiency within 180 min were 100, 93, and 89% for sulfamethoxazole, chemical oxygen demand, and total organic carbon, respectively.

The great interest in the use of ultrasonic action during photocatalysis is justified by the enhancement of electronic excitations, which leads to an increase in the density of pairs of charged particles. Under the influence of ultrasound, the aggregate photocatalysts are dispersed, contributing to the rapid renewal and expansion of the boundaries of heterogeneous reactions, which improves the mass transfer and the course of chemical reactions [96,97].

4. Conclusions

In this review, information on the main mechanisms of water splitting and decomposition of organic compounds by TiO_2-based photocatalysts was collected and analyzed. The efficiency of TiO_2 doping by various components, which significantly increases the photocatalytic activity, is shown. In addition, other factors, such as the bandgap, charge carrier recombination, the use of sacrificial agents, the size and type of photocatalyst structures, surface morphology, photocatalyst concentration in solution, radiation source power, etc., which can affect the photocatalysis were analyzed. Despite the reports of scientists on the development of effective photocatalysts capable of operating under sunlight, their use is still not economically viable. Therefore, due to the growth of environmental problems and the limited availability of hydrocarbon fuels, investigations in the field of hydrogen production by "artificial photosynthesis" will undoubtedly increase.

Author Contributions: Conceptualization, B.B. and C.D.; writing—original draft preparation, F.S., R.B., A.B., I.C. and B.B.; writing—review and editing, A.U. and A.M.; project administration, R.B. All authors have read and agreed to the published version of the manuscript.

Funding: This study was funded by Ministry of Education and Science of the Republic of Kazakhstan № AP05135273. The name of the focal area of the development of science, on which the application is submitted: Energy and Machine industry. The name of the specialized research area, on which the application is filed, type of study: Alternative energy and technology: renewable energy sources, nuclear and hydrogen energy, other energy sources, applied research.

Acknowledgments: Askhat Bekbaev is grateful to al-Farabi Kazakh National University for financial support in the frame of the postdoctoral fellowship program.

Conflicts of Interest: The authors declare no conflict of interest.

References

1. Samokhvalov, A. Hydrogen by photocatalysis with nitrogen codoped titanium dioxide. *Renew. Sustain. Energy Rev.* **2017**, *72*, 981–1000. [CrossRef]

2. Phoon, B.L.; Lai, C.W.; Juan, J.C.; Show, P.-L.; Pan, G.-T. Recent developments of strontium titanate for photocatalytic water splitting application. *Int. J. Hydrogen Energy* **2019**, *44*, 14316–14340. [CrossRef]

3. Fujishima, A.; Honda, K. Electrochemical Photolysis of Water at a Semiconductor Electrode. *Nature* **1972**, *238*, 37–38. [CrossRef] [PubMed]

4. Xiao, N.; Li, S.; Li, X.; Ge, L.; Gao, Y.; Li, N. The roles and mechanism of cocatalysts in photocatalytic water splitting to produce hydrogen. *Chin. J. Catal.* **2020**, *41*, 642–671. [CrossRef]

5. Katal, R.; Masudy-Panah, S.; Tanhaei, M.; Farahani, M.H.; Jiangyong, H. A review on the synthesis of the various types of anatase TiO_2 facets and their applications for photocatalysis. *Chem. Eng. J.* **2020**, *384*, 123384. [CrossRef]

6. Bingham, M.; Mills, A. Photonic efficiency and selectivity study of M (M = Pt, Pd, Au and Ag)/TiO_2 photocatalysts for methanol reforming in the gas phase. *J. Photochem. Photobiol. A Chem.* **2020**, *389*, 112257. [CrossRef]

7. Paz, Y. Application of TiO_2 photocatalysis for air treatment: Patents' overview. *Appl. Catal. B Environ.* **2010**, *99*, 448–460. [CrossRef]

8. Shayegan, Z.; Lee, C.-S.; Haghighat, F. TiO_2 photocatalyst for removal of volatile organic compounds in gas phase—A review. *Chem. Eng. J.* **2018**, *334*, 2408–2439. [CrossRef]

9. Verbruggen, S.W. TiO_2 photocatalysis for the degradation of pollutants in gas phase: From morphological design to plasmonic enhancement. *J. Photochem. Photobiol. C Photochem. Rev.* **2015**, *24*, 64–82. [CrossRef]

10. Kou, J.; Lu, C.; Wang, J.; Chen, Y.; Xu, Z.; Varma, R.S. Selectivity Enhancement in Heterogeneous Photocatalytic Transformations. *Chem. Rev.* **2017**, *117*, 1445–1514. [CrossRef]

11. Parrino, F.; Bellardita, M.; García-López, E.I.; Marcì, G.; Loddo, V.; Palmisano, L. Heterogeneous Photocatalysis for Selective Formation of High-Value-Added Molecules: Some Chemical and Engineering Aspects. *ACS Catal.* **2018**, *8*, 11191–11225. [CrossRef]

12. Yang, Q.; Dong, L.; Su, R.; Hu, B.; Wang, Z.; Jin, Y.; Wang, Y.; Besenbacher, F.; Dong, M. Nanostructured heterogeneous photo-catalysts for hydrogen production and water splitting: A comprehensive insight. *Appl. Mater. Today* **2019**, *17*, 159–182. [CrossRef]

13. Vajda, K.; Saszet, K.; Kedves, E.Z.; Kása, Z.; Danciu, V.; Baia, L.; Magyari, K.; Hernádi, K.; Kovács, G.; Pap, Z. Shape-controlled agglomeration of TiO_2 nanoparticles. New insights on polycrystallinity vs. single crystals in photocatalysis. *Ceram. Int.* **2016**, *42*, 3077–3087. [CrossRef]

14. Li, D.; Song, H.; Meng, X.; Shen, T.; Sun, J.; Han, W.; Wang, X. Effects of Particle Size on the Structure and Photocatalytic Performance by Alkali-Treated TiO_2. *Nanomaterials* **2020**, *10*, 546. [CrossRef] [PubMed]

15. Nagaraj, G.; Irudayaraj, A.; Josephine, R.L. Tuning the optical band Gap of pure TiO_2 via photon induced method. *Optik* **2019**, *179*, 889–894.

16. Allen, N.S.; Vishnyakov, V.; Kelly, P.J.; Kriek, R.J.; Mahdjoub, N.; Hill, C. Characterisation and photocatalytic assessment of TiO_2 nano-polymorphs: Influence of crystallite size and influence of thermal treatment on paint coatings and dye fading kinetics. *J. Phys. Chem. Solids* **2019**, *126*, 131–142. [CrossRef]

17. Sinhmar, A.; Setia, H.; Kumar, V.; Sobti, A.; Toor, A.P. Enhanced photocatalytic activity of nickel and nitrogen codoped TiO_2 under sunlight. *Environ. Technol. Innov.* **2020**, *18*, 100658. [CrossRef]

18. Liu, X.; Li, Y.; Wei, Z.; Shi, L. A Fundamental DFT Study of Anatase (TiO_2) Doped with 3d Transition Metals for High Photocatalytic Activities. *J. Wuhan Univ. Technol. Mater. Sci. Ed.* **2018**, *33*, 403–408. [CrossRef]

19. Benjwal, P.; De, B.; Kar, K.K. 1-D and 2-D morphology of metal cation co-doped (Zn, Mn) TiO_2 and investigation of their photocatalytic activity. *Appl. Surf. Sci.* **2018**, *427*, 262–272. [CrossRef]

20. Shen, Q.; Wang, Y.; Xue, J.; Gao, G.; Liu, X.; Jia, H.; Li, Q.; Xu, B. The dual effects of RGO films in TiO_2/CdSe heterojunction: Enhancing photocatalytic activity and improving photocorrosion resistance. *Appl. Surf. Sci.* **2019**, *481*, 1515–1523. [CrossRef]

21. Kernazhitsky, L.; Shymanovska, V.; Gavrilko, T.; Naumov, V.; Kshnyakin, V.; Khalyavka, T. A comparative study of optical absorption and photocatalytic properties of nanocrystalline single-phase anatase and rutile TiO_2 doped with transition metal cations. *J. Solid State Chem.* **2013**, *198*, 511–519. [CrossRef]

22. Song, E.; Kim, Y.-T.; Choi, J. Anion additives in rapid breakdown anodization for nonmetal-doped TiO_2 nanotube powders. *Electrochem. Commun.* **2019**, *109*, 106610. [CrossRef]

23. Fadlallah, M.M. Magnetic, electronic, optical, and photocatalytic properties of nonmetal- and halogen-doped anatase TiO_2 nanotubes. *Phys. E Low Dimens. Syst. Nanostruct.* **2017**, *89*, 50–56. [CrossRef]

24. Varnagiris, S.; Medvids, A.; Lelis, M.; Milcius, D.; Antuzevics, A. Black carbon-doped TiO_2 films: Synthesis, characterization and photocatalysis. *J. Photochem. Photobiol. A Chem.* **2019**, *382*, 111941. [CrossRef]

25. Li, G.; Zou, B.; Feng, S.; Shi, H.; Liao, K.; Wang, Y.; Wang, W.; Zhang, G. Synthesis of N-Doped TiO_2 with good photocatalytic property. *Phys. B Condens. Matter* **2020**, *588*, 412184. [CrossRef]

26. Li, H.; Qiu, L.; Bharti, B.; Dai, F.; Zhu, M.; Ouyang, F.; Lin, L. Efficient photocatalytic degradation of acrylonitrile by Sulfur-Bismuth co-doped F-TiO_2/SiO_2 nanopowder. *Chemosphere* **2020**, *249*, 126135. [CrossRef]

27. He, D.; Li, Y.; Wu, J.; Yang, Y.; An, Q. Carbon wrapped and doped TiO_2 mesoporous nanostructure with efficient visible-light photocatalysis for NO removal. *Appl. Surf. Sci.* **2017**, *391*, 318–325. [CrossRef]

28. Boningari, T.; Inturi, S.N.R.; Suidan, M.; Smirniotis, P.G. Novel one-step synthesis of nitrogen-doped TiO_2 by flame aerosol technique for visible-light photocatalysis: Effect of synthesis parameters and secondary nitrogen (N) source. *Chem. Eng. J.* **2018**, *350*, 324–334. [CrossRef]

29. Jiang, J.X.; Zhang, Q.Q.; Li, Y.H.; Li, L. Three-dimensional network graphene aerogel for enhancing adsorption and visible light photocatalysis of nitrogen-doped TiO_2. *Mater. Lett.* **2019**, *234*, 298–301. [CrossRef]

30. Marques, J.; Gomes, T.D.; Forte, M.A.; Silva, R.F.; Tavares, C.J. A new route for the synthesis of highly-active N-doped TiO_2 nanoparticles for visible light photocatalysis using urea as nitrogen precursor. *Catal. Today* **2019**, *326*, 36–45. [CrossRef]

31. Du, M.; Qiu, B.; Zhu, Q.; Xing, M.; Zhang, J. Fluorine doped TiO_2/mesocellular foams with an efficient photocatalytic activity. *Catal. Today* **2019**, *327*, 340–346. [CrossRef]

32. Li, C.; Sun, Z.; Ma, R.; Xue, Y.; Zheng, S. Fluorine doped anatase TiO_2 with exposed reactive (001) facets supported on porous diatomite for enhanced visible-light photocatalytic activity. *Microporous Mesoporous Mater.* **2017**, *243*, 281–290. [CrossRef]

33. Wei, J.Q.; Chen, X.J.; Wang, P.F.; Han, Y.B.; Xu, J.C.; Hong, B.; Jin, H.X.; Jin, D.F.; Peng, X.L.; Li, J.; et al. High surface area TiO_2/SBA-15 nanocomposites: Synthesis, microstructure and adsorption-enhanced photocatalysis. *Chem. Phys.* **2018**, *510*, 47–53. [CrossRef]

34. Hafez, H.S. Synthesis of highly-active single-crystalline TiO_2 nanorods and its application in environmental photocatalysis. *Mater. Lett.* **2009**, *63*, 1471–1474. [CrossRef]

35. Vargas Hernández, J.; Coste, S.; García Murillo, A.; Carrillo Romo, F.; Kassiba, A. Effects of metal doping (Cu, Ag, Eu) on the electronic and optical behavior of nanostructured TiO_2. *J. Alloys Compd.* **2017**, *710*, 355–363. [CrossRef]

36. Belošević-Čavor, J.; Koteski, V.; Umićević, A.; Ivanovski, V. Effect of 5d transition metals doping on the photocatalytic properties of rutile TiO_2. *Comput. Mater. Sci.* **2018**, *151*, 328–337. [CrossRef]

37. Unal, F.A.; Ok, S.; Unal, M.; Topal, S.; Cellat, K.; Şen, F. Synthesis, characterization, and application of transition metals (Ni, Zr, and Fe) doped TiO_2 photoelectrodes for dye-sensitized solar cells. *J. Mol. Liq.* **2020**, *299*, 112177. [CrossRef]

38. Liu, L.; Zhang, X.; Yang, L.; Ren, L.; Wang, D.; Ye, J. Metal nanoparticles induced photocatalysis. *Natl. Sci. Rev.* **2017**, *4*, 761–780. [CrossRef]

39. Yao, G.Y.; Zhao, Z.Y.; Liu, Q.L.; Dong, X.D.; Zhao, Q.M. Theoretical calculations for localized surface plasmon resonance effects of Cu/TiO_2 nanosphere: Generation, modulation, and application in photocatalysis. *Sol. Energy Mater. Sol. Cells* **2020**, *208*, 110385. [CrossRef]

40. Kumaravel, V.; Mathew, S.; Bartlett, J.; Pillai, S.C. Photocatalytic hydrogen production using metal doped TiO_2: A review of recent advances. *Appl. Catal. B Environ.* **2019**, *244*, 1021–1064. [CrossRef]

41. Xiu, Z.; Guo, M.; Zhao, T.; Pan, K.; Xing, Z.; Li, Z.; Zhou, W. Recent advances in Ti^{3+} self-doped nanostructured TiO_2 visible light photocatalysts for environmental and energy applications. *Chem. Eng. J.* **2020**, *382*, 123011. [CrossRef]

42. Pulido Melián, E.; Nereida Suárez, M.; Jardiel, T.; Calatayud, D.G.; Del Campo, A.; Doña-Rodríguez, J.M.; Araña, J.; González Díaz, O.M. Highly photoactive TiO_2 microspheres for photocatalytic production of hydrogen. *Int. J. Hydrogen Energy* **2019**, *44*, 24653–24666. [CrossRef]

43. Huang, G.; Liu, X.; Shi, S.; Li, S.; Xiao, Z.; Zhen, W.; Liu, S.; Wong, P.K. Hydrogen producing water treatment through mesoporous TiO_2 nanofibers with oriented nanocrystals. *Chin. J. Catal.* **2020**, *41*, 50–61. [CrossRef]

44. Chen, W.; Wang, Y.; Liu, S.; Gao, L.; Mao, L.; Fan, Z.; Shangguan, W.; Jiang, Z. Non-noble metal Cu as a cocatalyst on TiO_2 nanorod for highly efficient photocatalytic hydrogen production. *Appl. Surf. Sci.* **2018**, *445*, 527–534. [CrossRef]

45. Gao, H.; Zhang, P.; Zhao, J.; Zhang, Y.; Hu, J.; Shao, G. Plasmon enhancement on photocatalytic hydrogen production over the Z-scheme photosynthetic heterojunction system. *Appl. Catal. B Environ.* **2017**, *210*, 297–305. [CrossRef]

46. El-Bery, H.M.; Matsushita, Y.; Abdel-moneim, A. Fabrication of efficient TiO_2-RGO heterojunction composites for hydrogen generation via water-splitting: Comparison between RGO, Au and Pt reduction sites. *Appl. Surf. Sci.* **2017**, *423*, 185–196. [CrossRef]

47. Wu, L.; Shi, S.; Li, Q.; Zhang, X.; Cui, X. TiO_2 nanoparticles modified with 2D $MoSe_2$ for enhanced photocatalytic activity on hydrogen evolution. *Int. J. Hydrogen Energy* **2019**, *44*, 720–728. [CrossRef]

48. Xing, X.; Zhu, H.; Zhang, M.; Xiao, L.; Li, Q.; Yang, J. Effect of heterojunctions and phase-junctions on visible-light photocatalytic hydrogen evolution in BCN-TiO_2 photocatalysts. *Chem. Phys. Lett.* **2019**, *727*, 11–18. [CrossRef]

49. Li, F.; Xiao, X.; Zhao, C.; Liu, J.; Li, Q.; Guo, C.; Tian, C.; Zhang, L.; Hu, J.; Jiang, B. TiO_2-on-C_3N_4 double-shell microtubes: In-situ fabricated heterostructures toward enhanced photocatalytic hydrogen evolution. *J. Colloid Interface Sci.* **2020**, *572*, 22–30. [CrossRef] [PubMed]

50. Zhang, C.; Zhou, Y.; Bao, J.; Fang, J.; Zhao, S.; Zhang, Y.; Sheng, X.; Chen, W. Structure regulation of ZnS@g-C_3N_4/TiO_2 nanospheres for efficient photocatalytic H_2 production under visible-light irradiation. *Chem. Eng. J.* **2018**, *346*, 226–237. [CrossRef]

51. Melián, E.P.; López, C.R.; Méndez, A.O.; Díaz, O.G.; Suárez, M.N.; Doña Rodríguez, J.M.; Navío, J.A.; Fernández Hevia, D. Hydrogen production using Pt-loaded TiO_2 photocatalysts. *Int. J. Hydrogen Energy* **2013**, *38*, 11737–11748. [CrossRef]

52. Vaiano, V.; Lara, M.A.; Iervolino, G.; Matarangolo, M.; Navio, J.A.; Hidalgo, M.C. Photocatalytic H_2 production from glycerol aqueous solutions over fluorinated Pt-TiO_2 with high {001} facet exposure. *J. Photochem. Photobiol. A Chem.* **2018**, *365*, 52–59. [CrossRef]

53. Yasuda, M.; Matsumoto, T.; Yamashita, T. Sacrificial hydrogen production over TiO_2-based photocatalysts: Polyols, carboxylic acids, and saccharides. *Renew. Sustain. Energy Rev.* **2018**, *81*, 1627–1635. [CrossRef]

54. López, C.R.; Melián, E.P.; Ortega Méndez, J.A.; Santiago, D.E.; Doña Rodríguez, J.M.; González Díaz, O. Comparative study of alcohols as sacrificial agents in H_2 production by heterogeneous photocatalysis using Pt/TiO_2 catalysts. *J. Photochem. Photobiol. A Chem.* **2015**, *312*, 45–54. [CrossRef]

55. Velázquez, J.J.; Fernández-González, R.; Díaz, L.; Pulido Melián, E.; Rodríguez, V.D.; Núñez, P. Effect of reaction temperature and sacrificial agent on the photocatalytic H_2-production of Pt-TiO_2. *J. Alloys Compd.* **2017**, *721*, 405–410. [CrossRef]

56. Kampouri, S.; Stylianou, K.C. Dual-Functional Photocatalysis for Simultaneous Hydrogen Production and Oxidation of Organic Substances. *ACS Catal.* **2019**, *9*, 4247–4270. [CrossRef]

57. Zhang, S.; Wang, L.; Liu, C.; Luo, J.; Crittenden, J.; Liu, X.; Cai, T.; Yuan, J.; Pei, Y.; Liu, Y. Photocatalytic wastewater purification with simultaneous hydrogen production using MoS_2 QD-decorated hierarchical assembly of $ZnIn_2S_4$ on reduced graphene oxide photocatalyst. *Water Res.* **2017**, *121*, 11–19. [CrossRef]

58. Bellardita, M.; García-López, E.I.; Marcì, G.; Nasillo, G.; Palmisano, L. Photocatalytic Solar Light H_2 Production by Aqueous Glucose Reforming: Photocatalytic Solar Light H_2 Production by Aqueous Glucose Reforming. *Eur. J. Inorg. Chem.* **2018**, *2018*, 4522–4532. [CrossRef]

59. Yao, Y.; Gao, X.; Li, Z.; Meng, X. Photocatalytic Reforming for Hydrogen Evolution: A Review. *Catalysts* **2020**, *10*, 335. [CrossRef]

60. Park, Y.K.; Kim, B.J.; Jeong, S.; Jeon, K.J.; Chung, K.H.; Jung, S.C. Characteristics of hydrogen production by photocatalytic water splitting using liquid phase plasma over Ag-doped TiO_2 photocatalysts. *Environ. Res.* **2020**, *188*, 109630. [CrossRef]

61. Wu, G.; Chen, T.; Su, W.; Zhou, G.; Zong, X.; Lei, Z.; Li, C. H2 production with ultra-low CO selectivity via photocatalytic reforming of methanol on Au/TiO_2 catalyst. *Int. J. Hydrogen Energy* **2008**, *33*, 1243–1251. [CrossRef]

62. Udayabhanu; Reddy, N.L.; Shankar, M.V.; Sharma, S.C.; Nagaraju, G. One-pot synthesis of Cu-TiO_2/CuO nanocomposite: Application to photocatalysis for enhanced H_2 production, dye degradation & detoxification of Cr (VI). *Int. J. Hydrogen Energy* **2020**, *45*, 7813–7828.

63. Xing, X.; Tang, S.; Hong, H.; Jin, H. Concentrated solar photocatalysis for hydrogen generation from water by titania-containing gold nanoparticles. *Int. J. Hydrogen Energy* **2020**, *45*, 9612–9623. [CrossRef]

64. Su, E.C.; Huang, B.S.; Lee, J.T.; Wey, M.Y. Excellent dispersion and charge separation of $SrTiO_3$-TiO_2 nanotube derived from a two-step hydrothermal process for facilitating hydrogen evolution under sunlight irradiation. *Sol. Energy* **2018**, *159*, 751–759. [CrossRef]

65. Sultanov, F.; Daulbayev, C.; Bakbolat, B.; Daulbayev, O.; Bigaj, M.; Mansurov, Z.; Kuterbekov, K.; Bekmyrza, K. Aligned composite $SrTiO_3$/PAN fibers as 1D photocatalyst obtained by electrospinning method. *Chem. Phys. Lett.* **2019**, *737*, 136821. [CrossRef]

66. Sultanov, F.; Bakbolat, B.; Mansurov, Z.; Pei, S.-S.; Ebrahim, R.; Daulbayev, C.; Urazgaliyeva, A.; Tulepov, M. Spongy Structures Coated with Carbon Nanomaterialsfor Efficient Oil/Water Separation. *Eurasian Chem. Technol. J.* **2017**, *19*, 127. [CrossRef]

67. Sultanov, F.; Bakbolat, B.; Daulbaev, C.; Urazgalieva, A.; Azizov, Z.; Mansurov, Z.; Tulepov, M.; Pei, S.S. Sorptive Activity and Hydrophobic Behavior of Aerogels Based on Reduced Graphene Oxide and Carbon Nanotubes. *J. Eng. Phys. Thermophys.* **2017**, *90*, 826–830. [CrossRef]

68. Luo, L.; Li, J.; Dai, J.; Xia, L.; Barrow, C.J.; Wang, H.; Jegatheesan, J.; Yang, M. Bisphenol A removal on TiO_2–MoS_2–reduced graphene oxide composite by adsorption and photocatalysis. *Process. Saf. Environ. Prot.* **2017**, *112*, 274–279. [CrossRef]

69. Nguyen, C.H.; Tran, M.L.; Tran, T.T.V.; Juang, R.S. Enhanced removal of various dyes from aqueous solutions by UV and simulated solar photocatalysis over TiO_2/ZnO/rGO composites. *Sep. Purif. Technol.* **2020**, *232*, 115962. [CrossRef]

70. Kumar, K.Y.; Saini, H.; Pandiarajan, D.; Prashanth, M.K.; Parashuram, L.; Raghu, M.S. Controllable synthesis of TiO$_2$ chemically bonded graphene for photocatalytic hydrogen evolution and dye degradation. *Catal. Today* **2020**, *340*, 170–177. [CrossRef]

71. Rajender, G.; Kumar, J.; Giri, P.K. Interfacial charge transfer in oxygen deficient TiO$_2$-graphene quantum dot hybrid and its influence on the enhanced visible light photocatalysis. *Appl. Catal. B Environ.* **2018**, *224*, 960–972. [CrossRef]

72. Ali, M.H.H.; Al-Afify, A.D.; Goher, M.E. Preparation and characterization of graphene—TiO$_2$ nanocomposite for enhanced photodegradation of Rhodamine-B dye. *Egypt. J. Aquat. Res.* **2018**, *44*, 263–270. [CrossRef]

73. Ton, N.N.T.; Dao, A.T.N.; Kato, K.; Ikenaga, T.; Trinh, D.X.; Taniike, T. One-pot synthesis of TiO$_2$/graphene nanocomposites for excellent visible light photocatalysis based on chemical exfoliation method. *Carbon* **2018**, *133*, 109–117. [CrossRef]

74. Lettieri, S.; Gargiulo, V.; Pallotti, D.K.; Vitiello, G.; Maddalena, P.; Alfè, M.; Marotta, R. Evidencing opposite charge-transfer processes at TiO$_2$/graphene-related materials interface through a combined EPR, photoluminescence and photocatalysis assessment. *Catal. Today* **2018**, *315*, 19–30. [CrossRef]

75. Shahbazi, R.; Payan, A.; Fattahi, M. Preparation, evaluations and operating conditions optimization of nano TiO$_2$ over graphene based materials as the photocatalyst for degradation of phenol. *J. Photochem. Photobiol. A Chem.* **2018**, *364*, 564–576. [CrossRef]

76. Khan, H.; Jiang, Z.; Berk, D. Molybdenum doped graphene/TiO$_2$ hybrid photocatalyst for UV/visible photocatalytic applications. *Sol. Energy* **2018**, *162*, 420–430. [CrossRef]

77. Sun, X.; Ji, S.; Wang, M.; Dou, J.; Yang, Z.; Qiu, H.; Kou, S.; Ji, Y.; Wang, H. Fabrication of porous TiO$_2$-RGO hybrid aerogel for high-efficiency, visible-light photodegradation of dyes. *J. Alloys Compd.* **2020**, *819*, 153033. [CrossRef]

78. Sultanov, F.R.; Mansurov, Z.A.; Daulbayev, C.; Urazgaliyeva, A.A.; Bakbolat, B.; Pei, S.-S. Study of Sorption Capacity and Surface Morphology of Carbon Nanomaterials/Chitosan Based Aerogels. *Eur. Chem. Technol. J.* **2016**, *18*, 19. [CrossRef]

79. Sultanov, F.R.; Pei, S.S.S.; Auyelkhankyzy, M.; Smagulova, G.; Lesbayev, B.T.; Mansurov, Z.A. Aerogels Based on Graphene Oxide with Addition of Carbon Nanotubes: Synthesis and Properties. *Eur. Chem. Technol. J.* **2014**, *16*, 265–269. [CrossRef]

80. Ali, M.H.H.; Al-Qahtani, K.M.; El-Sayed, S.M. Enhancing photodegradation of 2,4,6 trichlorophenol and organic pollutants in industrial effluents using nanocomposite of TiO$_2$ doped with reduced graphene oxide. *Egypt. J. Aquat. Res.* **2019**, *45*, 321–328. [CrossRef]

81. Nadimi, M.; Ziarati Saravani, A.; Aroon, M.A.; Ebrahimian Pirbazari, A. Photodegradation of methylene blue by a ternary magnetic TiO$_2$/Fe$_3$O$_4$/graphene oxide nanocomposite under visible light. *Mater. Chem. Phys.* **2019**, *225*, 464–474. [CrossRef]

82. Park, K.; Ali, I.; Kim, J.O. Photodegradation of perfluorooctanoic acid by graphene oxide-deposited TiO$_2$ nanotube arrays in aqueous phase. *J. Environ. Manag.* **2018**, *218*, 333–339. [CrossRef] [PubMed]

83. Al Kausor, M.; Chakrabortty, D. Facile fabrication of N-TiO$_2$/Ag$_3$PO$_4$@GO nanocomposite toward photodegradation of organic dye under visible light. *Inorg. Chem. Commun.* **2020**, *116*, 107907. [CrossRef]

84. Zhou, Q.; Wang, M.; Tong, Y.; Wang, H.; Zhou, X.; Sheng, X.; Sun, Y.; Chen, C. Improved photoelectrocatalytic degradation of tetrabromobisphenol A with silver and reduced graphene oxide-modified TiO$_2$ nanotube arrays under simulated sunlight. *Ecotoxicol. Environ. Saf.* **2019**, *182*, 109472. [CrossRef] [PubMed]

85. Zhao, Y.; Wang, G.; Li, L.; Dong, X.; Zhang, X. Enhanced activation of peroxymonosulfate by nitrogen-doped graphene/TiO$_2$ under photo-assistance for organic pollutants degradation: Insight into N doping mechanism. *Chemosphere* **2020**, *244*, 125526. [CrossRef]

86. Chen, F.; An, W.; Li, Y.; Liang, Y.; Cui, W. Fabricating 3D porous PANI/TiO$_2$-graphene hydrogel for the enhanced UV-light photocatalytic degradation of BPA. *Appl. Surf. Sci.* **2018**, *427*, 123–132. [CrossRef]

87. Ghasemi, A.; Azzouz, R.; Laipple, G.; Kabak, K.E.; Heavey, C. Optimizing capacity allocation in semiconductor manufacturing photolithography area—Case study: Robert Bosch. *J. Manuf. Syst.* **2020**, *54*, 123–137. [CrossRef]

88. Daulbaev, C.B.; Dmitriev, T.P.; Sultanov, F.R.; Mansurov, Z.A.; Aliev, E.T. Obtaining Three-Dimensional Nanosize Objects on a "3D Printer + Electrospinning" Machine. *J. Eng. Phys. Thermophys.* **2017**, *90*, 1115–1118. [CrossRef]

89. Liu, J.; Liu, H.; Zuo, X.; Wen, F.; Jiang, H.; Cao, H.; Pei, Y. Micro-patterned TiO_2 films for photocatalysis. *Mater. Lett.* **2019**, *254*, 448–451. [CrossRef]

90. Wang, Q.; Jin, R.; Zhang, M.; Gao, S. Solvothermal preparation of Fe-doped TiO_2 nanotube arrays for enhancement in visible light induced photoelectrochemical performance. *J. Alloys Compd.* **2017**, *690*, 139–144. [CrossRef]

91. Zhang, J.J.; Qi, P.; Li, J.; Zheng, X.C.; Liu, P.; Guan, X.X.; Zheng, G.P. Three-dimensional Fe_2O_3-TiO_2-graphene aerogel nanocomposites with enhanced adsorption and visible light-driven photocatalytic performance in the removal of RhB dyes. *J. Ind. Eng. Chem.* **2018**, *61*, 407–415. [CrossRef]

92. Lesbayev, A.B.; Smagulova, G.T.; Kim, S.; Prikhod'ko, N.G.; Manakov, S.M.; Guseinov, N.; Mansurov, Z.A. Solution-Combustion Synthesis and Characterization of Fe_3O_4 Nanoparticles. *Int. J. Self Propag. High Temp. Synth.* **2018**, *27*, 195–197. [CrossRef]

93. Soltani-nezhad, F.; Saljooqi, A.; Shamspur, T.; Mostafavi, A. Photocatalytic degradation of imidacloprid using $GO/Fe_3O_4/TiO_2$-NiO under visible radiation: Optimization by response level method. *Polyhedron* **2019**, *165*, 188–196. [CrossRef]

94. Rajoriya, S.; Bargole, S.; George, S.; Saharan, V.K.; Gogate, P.R.; Pandit, A.B. Synthesis and characterization of samarium and nitrogen doped TiO_2 photocatalysts for photo-degradation of 4-acetamidophenol in combination with hydrodynamic and acoustic cavitation. *Sep. Purif. Technol.* **2019**, *209*, 254–269. [CrossRef]

95. Isari, A.A.; Hayati, F.; Kakavandi, B.; Rostami, M.; Motevassel, M.; Dehghanifard, E. N, Cu co-doped TiO_2@functionalized SWCNT photocatalyst coupled with ultrasound and visible-light: An effective sono-photocatalysis process for pharmaceutical wastewaters treatment. *Chem. Eng. J.* **2020**, *392*, 123685. [CrossRef]

96. Wu, D.; Wang, X.; Wang, H.; Wang, F.; Wang, D.; Gao, Z.; Wang, X.; Xu, F.; Jiang, K. Ultrasonic-assisted synthesis of two dimensional $BiOCl/MoS_2$ with tunable band gap and fast charge separation for enhanced photocatalytic performance under visible light. *J. Colloid Interface Sci.* **2019**, *533*, 539–547. [CrossRef]

97. Ding, J.; Liu, Q.; Zhang, Z.; Liu, X.; Zhao, J.; Cheng, S.; Zong, B.; Dai, W.L. Carbon nitride nanosheets decorated with WO_3 nanorods: Ultrasonic-assisted facile synthesis and catalytic application in the green manufacture of dialdehydes. *Appl. Catal. B Environ.* **2015**, *165*, 511–518. [CrossRef]

 *nanomaterials*

Article

Influence of Metal Oxide Particles on Bandgap of 1D Photocatalysts Based on SrTiO₃/PAN Fibers

Fail Sultanov [1,2], **Chingis Daulbayev** [1,2,*], **Seitkhan Azat** [1,2,3,*], **Kairat Kuterbekov** [4,*], **Kenzhebatyr Bekmyrza** [4,*], **Baglan Bakbolat** [1,2,*], **Magdalena Bigaj** [5] and **Zulkhair Mansurov** [1,2]

[1] Faculty of Chemistry and Chemical Technology, al-Farabi Kazakh National University, Almaty 050000, Kazakhstan; sultanov.fail@kaznu.kz (F.S.); zmansurov@kaznu.kz (Z.M.)
[2] Laboratory of Energy-intensive and Nanomaterials, Institute of Combustion Problems, Almaty 050000, Kazakhstan
[3] Institute of Chemical and Biological Technologies, Satbayev University, Almaty 050000, Kazakhstan
[4] Faculty of Physics and Technology, L.N. Gumilev Eurasian National University, Nur-Sultan 010000, Kazakhstan
[5] NanoBioMedical Centre, Adam Mickiewicz University, 61-614 Poznan, Poland; magdalena.bigaj@amu.edu.pl
* Correspondence: chingis.daulbayev@kaznu.kz (C.D.); a.seitkhan@satbayev.university (S.A.); kuterbekov_ka@enu.kz (K.K.); bekmyrza_kzh@enu.kz (K.B.); bakbolat.baglan@icp.kz (B.B.)

Received: 28 July 2020; Accepted: 26 August 2020; Published: 1 September 2020

Abstract: This paper deals with the study of the optical properties of one-dimensional SrTiO₃/PAN-based photocatalysts with the addition of metal oxide particles and the determination of their bandgaps. One-dimensional photocatalysts were obtained by the electrospinning method. Particles of metals such as iron, chromium, and copper were used as additives that are capable of improving the fibers' photocatalytic properties based on SrTiO₃/PAN. The optimal ratios of the solutions for the electrospinning of fibers based on SrTiO₃/PAN with the addition of metal oxide particles were determined. The transmission and reflection of composite photocatalysts with metal oxide particles were measured in a wide range of spectra, from the ultraviolet region (185 nm) to near-infrared radiation (3600 nm), to determine the values of their bandgaps. Thus, the introduction of metal oxide particles resulted in a decrease in the bandgaps of the obtained composite photocatalysts compared to the initial SrTiO₃/PAN (3.57 eV), with the following values: −3.11 eV for SrTiO₃/PAN/Fe₂O₃, −2.84 eV for SrTiO₃/PAN/CuO, and −2.89 eV for SrTiO₃/PAN/Cr₂O₃. The obtained composite photocatalysts were tested for the production of hydrogen by the splitting of water–methanol mixtures under UV irradiation, and the following rates of hydrogen evolution were determined: 344.67 μmol h^{-1} g^{-1} for SrTiO₃/PAN/Fe₂O₃, 398.93 μmol h^{-1} g^{-1} for SrTiO₃/PAN/Cr₂O₃, and 420.82 μmol h^{-1} g^{-1} for SrTiO₃/PAN/CuO.

Keywords: electrospinning; SrTiO₃; fibers; photocatalytic; water splitting; bandgap; hydrogen

---

1. Introduction

Photocatalysis is a well-researched method for renewable energy production in the forms of solar energy and high-purity chemical fuel (H₂, CH₄/CH₂OH) [1,2]. Photocatalytic water splitting occurs upon the solar light irradiation of a semiconductor photocatalyst and results in the formation of hydrogen (H₂) and oxygen (O₂) [3]. An advantage over conventional energy sources, like fossil fuels, is the lack of carbon monoxide production, which, in light of ongoing climate change debates, is a great benefit for the environment [4].

The material used in the photocatalytic system has to be chosen carefully. Strontium titanate (SrTiO₃) is a wide-gap semiconductor that belongs to the perovskite family of ternary oxides with an ABO_3 structure [5]. At room temperature, it exhibits a cubic structure with a lattice parameter $a = 3.9053$ Å [6]. Its attractive properties include strong catalytic activity, high chemical stability, and the long lifetime of electron-hole pairs [5]. Due to the bandgap energy of 3.2 eV, photo-excitation takes place with the use of light with a wavelength λ less than 387 nm (UV light) [2,7], which accounts for about 5% of solar energy [8]. The inherent gap edge positions can be modified by the implementation of elemental doping in the original semiconductor material [5,9]. Doping with metal or non-metal alloys can be applied to extend the activating spectrum, allowing the $SrTiO_3$ photocatalyst to also be used in the visible light region [7,9]. Metal ions implemented in the semiconductor material become electron donors, enhancing the production of hydrogen [3,8,9]. Transition metal ions like Fe, Mn, Cu, Ni, and Cr have been shown to modify the bandgap position of semiconductor materials [10–15] without enhancing the formation of water, which is the case when noble metals (e.g., Rh, Pt) are employed [3,9]. The switch towards the visible light region was also achieved by doping titanate-based materials with Fe [9,16,17], Cr [5,18–20], and Cu [21,22].

The presence of metal in the semiconductor lattice alters electron-hole recombination. Since the transfer of the trapped electron and hole to the semiconductor surface is required for the photocatalytic reaction to occur, it is important that the metal ions are located near the surface of the semiconductor for a more efficient charge transfer [9]. In addition, the effective separation of charges is significantly affected by the specific surface area, which promotes the free diffusion of water [23], as well as the high degree of crystallinity of photocatalysts, leading to a decrease in the number of recombinations of photogenerated charges [24]. Moreover, for the successful use of photocatalysts for the production of hydrogen by water splitting, it is necessary to develop and create inexpensive, efficient, and stable photocatalytic systems, which can result in a decrease in the market price of hydrogen [25].

In our previous work, we reported a low-cost synthesis and thorough characterization of $SrTiO_3$ nanofibers (up to 350 nm in diameter) using the electrospinning method [26]. Here, we show that the modification of this material using oxides of particles of iron (Fe), chromium (Cr), and copper (Cu) results in a narrowing of the bandgap energy. The used metal oxides have different properties for trapping and transferring electrons and holes. An increase in the efficiency of the hydrogen evolution reaction from the water-alcohol mixture under visible light irradiation ($\lambda > 400$ nm), without disrupting the nanofibers, structure, or crystallinity of $SrTiO_3$, was observed for all of the metal oxide particles added to $SrTiO_3$ nanofibers. The results of this work show that metal oxide particles added to $SrTiO_3$ nanofibers create visible-light-responsive photocatalysts, making them potent candidates for the conversion of solar energy to fuel.

2. Materials and Methods

Strontium nitrate ($Sr(NO_3)_2$, 98%), titanium oxide (TiO_2, 99%), and oxalic acid (($COOH)_2 \cdot 2H_2O$, 98%) were purchased from Laborpharma (Almaty, Kazakhstan). Polyacrylonitrile (PAN, average M.W. is 152000), dimethylformamide (99.8%), iron chloride ($FeCl_3$, 45% solution), copper oxide (CuO, 99.995% powder), and chromium sulfate ($Cr_2(SO_4)_3$, 99.99% powder) were purchased from Sigma Aldrich (St. Louis, MO, USA). All chemicals were used without further purification.

2.1. Electrospinning of SrTiO₃/PAN-Based Fibers with the Addition of Metal Oxide Particles

$SrTiO_3$ was obtained as described in [27]. A precursor for the electrospinning of fibers based on $SrTiO_3$ and metal oxide particles was prepared as follows: PAN was used to create the polymer solution by its dissolution in dimethylformamide under constant stirring for 30 min. Then, $SrTiO_3$ powder and $FeCl_3$, $Cr_2(SO_4)_3$, or CuO were added to the polymer solution at different ratios and stirred until the mixture became homogeneous. The obtained suspension was used as a precursor for obtaining fibers based on $SrTiO_3$ with metal oxide particles by pulling under high voltage. Fiber electrospinning was carried out at room temperature with a voltage of 16 kV and a flow rate of 1.5 mL/h. The collector was located at a distance of 15 cm from the needle. Aluminum foil with a 20 cm diameter was used as a collector, which was replaced every 1.5 h throughout the entire process. The obtained fibers were then thermally stabilized at 185 °C for 15 min and calcined at 500 °C for 30 min in an argon medium. The resulting samples were designated as $SrTiO_3/PAN/Fe_2O_3$, $SrTiO_3/PAN/CuO$, and $SrTiO_3/PAN/Cr_2O_3$ depending on the added metal.

2.2. X-ray Diffraction Analysis of Samples

The X-ray diffraction (XRD) analysis was carried out on a Dron-4-type X-ray diffractometer (Omsk, Russian Federation) with a range of rotation angles for diffraction unit detection from −100° to 168°. The minimum step for moving the detection unit is 0.001°. The permissible connection of the detection unit from a given rotation angle is ± 0.015°. The transport rate of the goniometer is 820°/min. The main error in measuring the pulse count of the X-ray measurement was not more than 0.4%.

2.3. Scanning Electron Microscope Characterization of the Surface Morphology of Samples

The surface morphology of the obtained photocatalytic fibers was studied using a Quanta 3D 200i (Waltham, MA, USA) scanning electron microscope (SEM) under an accelerating 15 kV voltage.

2.4. Measurement of the Transmission and Reflection of SrTiO₃/PAN Fibers with the Addition of Metal Oxide Particles in a Wide Spectral Region from Ultraviolet (185 nm) to Near-Infrared Radiation (3600 nm)

The transmission and reflection measurements of photocatalytic fibers with the addition of metal oxide particles in a wide spectral region, from ultraviolet (185 nm) to near-infrared radiation (3600 nm), were carried out on a Shimadzu UV-3600 spectrophotometer (Moscow, Russian Federation) equipped with three detectors: a photoelectron multiplier for operation in the ultraviolet and visible spectral range, a semiconductor InGaAs, and cooled PbS detectors for near-infrared operation.

2.5. Investigation of the Activity of Photocatalysts Based on SrTiO₃/PAN Fibers with the Addition of Metal Oxide Particles

The activity of photocatalysts based on $SrTiO_3/PAN$ fibers with the addition of metal oxide particles was tested by measuring the output of hydrogen during the water–methanol mixture splitting under UV radiation. The mixture, containing a photocatalyst, water, and methanol in different ratios, was loaded into a quartz tube reactor, which was previously purged with inert gas (argon) and exposed to UV irradiation with a wavelength of 320 nm and a power source of 40 W. As a result of the photocatalytic reaction of the water–methanol mixture splitting, the evolved mixture of gases was accumulated in a sealed sampler. The qualitative and quantitative composition of the evolved mixture of gases was analyzed by gas chromatography on a Chromos 1000 chromatograph (Dzershinsk, Russian Federation) with three packed 3 mm columns filled with NAX and PORAPAK Q phases, allowing for the identification of the leading gases: hydrogen, nitrogen, oxygen, carbon monoxide, and carbon dioxide.

3. Results and Discussion

3.1. The Synthesis of Fibers Based on SrTiO₃/PAN with the Addition of Metal Oxide Particles and a Study of Their Physicochemical Properties

To achieve a highly efficient photocatalyst based on $SrTiO_3$ fibers, it is necessary to create its composites with metal oxide particles. The addition of metal oxide particles allows for a narrowing of the bandgap of $SrTiO_3$, leading to the possible use of a wide spectrum of visible light. It also contributes to the improvement of redox reactions that occur during the absorption of light. The experimentally selected optimal ratios of the solution components for obtaining fibers with the required characteristics are 1:9:2:88 $SrTiO_3$:PAN:FeCl$_3$:solvent, 0.5:10:1.5:88 $SrTiO_3$:PAN:Cr$_2$(SO$_4$)$_3$:solvent, and 1.5:8:2.5:88 $SrTiO_3$:PAN:CuO:solvent.

Figure 1 presents SEM images of the obtained polymer fibers based on SrTiO$_3$/PAN with metal oxide particles added. Polymer fibers based on SrTiO$_3$/PAN with the addition of metal oxide particles have a continuous cylindrical shape without defects and are randomly arranged. The samples have the typical structure of fibers obtained by electrospinning, in which they are in contact with each other, forming a three-dimensional polymer network [28]. The average diameter of the obtained fibers is in the range from 200 to 400 nm, which is directly proportional to the viscosity of the solution used for electrospinning and the high voltage applied [29]. According to the SEM images, the presence of metal oxide particles and SrTiO$_3$ does not affect the morphological characteristics of the forming polymer fibers, which is also confirmed by the results obtained in [30], in which the effect of the composition of the electrospinning solution on the diameters of such fibers was studied. For all types of obtained fibers, the size of the agglomerates of SrTiO$_3$ and metal oxides ranges from 1 to 4 μm.

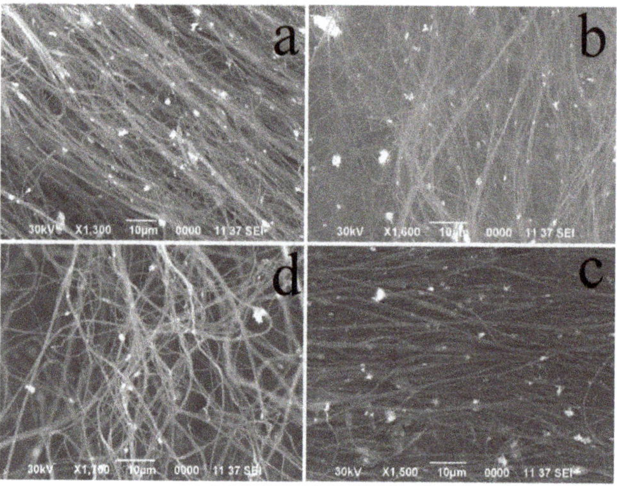

Figure 1. SEM images of non-calcined polymer fibers based on SrTiO$_3$/PAN (**a**) and their composites with metal oxide particles: SrTiO$_3$/PAN/Fe$_2$O$_3$ (**b**), SrTiO$_3$/PAN/Cr$_2$O$_3$ (**c**), and SrTiO$_3$/PAN/CuO (**d**).

The inclusion of metal oxides does not cause a corresponding change in the fiber diameter. On the one hand, the metal particle addition increases the viscosity of the solution, but on the other hand, this effect is balanced by an increase in the electrical conductivity of the initial solution, which contributes to the formation of thinner fibers due to an increase in the charge density on the electrospinning jet, and this, in turn, leads to an elongation of the jet along its axis [31,32].

To confirm the presence of metal oxide particles in the calcined fiber structure based on SrTiO$_3$/PAN, an XRD analysis of the samples was performed. Figure 2 presents the X-ray diffraction patterns of calcined fibers based on SrTiO$_3$/PAN and their composites with metal oxide particles.

Figure 2a shows that for calcined SrTiO$_3$/PAN-based fibers, characteristic peaks are observed at 22.77°, 32.41°, 39.99°, 46.49°, 57.81°, 67.86°, and 77.17°, indicating the presence of perovskite-type SrTiO$_3$ in the structure. These peaks are also present in the X-ray diffraction patterns for samples of SrTiO$_3$/PAN with the addition of metal oxide particles (Figure 2b), indicating the presence of SrTiO$_3$ in each sample (JCPDS:35-0734). At the same time, for samples of SrTiO$_3$/PAN with the addition of metal oxide particles, peaks with a lower intensity are also observed: for Fe$_2$O$_3$, the characteristic peaks are at 33.14°, 35.68°, 49.51°, 52.34°, and 54.07°; for Cr$_2$O$_3$, they are at 22.79°, 24.50°, 33.66°, 36.25°, 41.58°, 63.54°, and 65.24°; and for CuO, they are at 22.74°, 35.46°, 36.46°, 38.66°, and 52.32° (Figure 2b). According to the XRD analysis, for the SrTiO$_3$/PAN/Fe$_2$O$_3$ sample, the content of SrTiO$_3$ in the fibers is 89.6 wt.%, while the content of Fe$_2$O$_3$ is 10.4 wt.% (Figure 2b, black line). The crystal lattice parameter for $SrTiO_3$ is 3.9036 Å (for the standard compound, it is 3.90010 Å), indicating a good crystallinity of the obtained samples of calcined fibers based on $SrTiO_3$/PAN/Fe_2O_3. The crystallite size of the $SrTiO_3$/PAN/Fe_2O_3 composite is 740 Å. Figure 2b (red line) presents the X-ray diffraction pattern of calcined fibers based on $SrTiO_3$/PAN/Cr_2O_3; the content of $SrTiO_3$ in the structure is 54.3 wt.%, while the content of Cr_2O_3 is 45.7 wt.%. The higher content of Cr_2O_3 compared to Fe_2O_3 is associated with the molar masses and densities of $Cr_2(SO_4)_3$ and $FeCl_3$, which were used as additives in the process of electrospinning the composite fibers. The molar volume of $Cr_2(SO_4)_3$ is twice as high as that of $FeCl_3$, which is confirmed by the semi-quantitative analysis of the calcined fibers based on $SrTiO_3$/PAN/Cr_2O_3 and $SrTiO_3$/PAN/Fe_2O_3. The crystallite size for fibers based on $SrTiO_3$/PAN/Cr_2O_3 is 840 Å, and the crystal lattice parameter for $SrTiO_3$ is 3.9036 Å. According to the results of the XRD analysis of the calcined fibers based on $SrTiO_3$/PAN /CuO (Figure 2b, blue line), the content of $SrTiO_3$ is determined to be 77.8 wt.%, and the content of CuO is 21.2 wt.% The crystallite size for the fiber based on $SrTiO_3$/PAN/CuO is 760 Å.

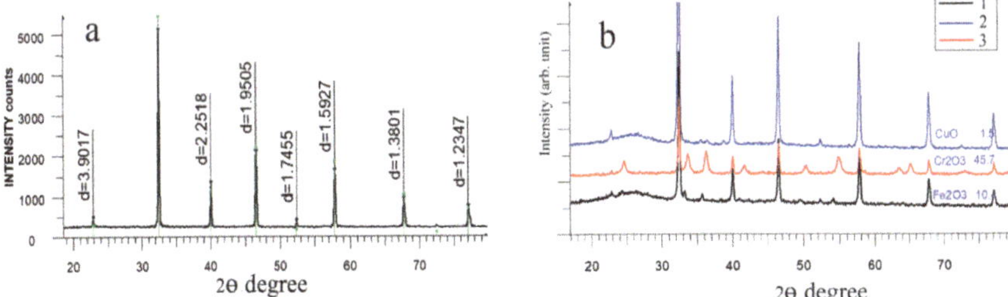

Figure 2. X-ray diffraction patterns of calcined fibers based on $SrTiO_3$/PAN (**a**) and $SrTiO_3$/PAN with the addition of metal oxide particles (**b**): $SrTiO_3$/PAN/Fe_2O_3—1 (black line); $SrTiO_3$/PAN/CuO—2 (blue line); $SrTiO_3$/PAN/Cr_2O_3—3 (red line).

3.2. Investigation of the Transmission and Reflection Spectra of the Obtained Photocatalytic Fibers

The photocatalysis mechanism of $SrTiO_3$ is based on the formation of electron–hole pairs under UV irradiation, where they have sufficiently high energy for the formation of radicals with a high oxidation ability. To study the possible use of a wider spectrum, including visible light, for $SrTiO_3$/PAN-based photocatalysts with the addition of metal oxide particles, their transmission and reflection spectra were determined. Analysis of the transmission spectra allows for the calculation of the bandgap of $SrTiO_3$-based photocatalysts with the addition of metal oxides. For crystalline semiconductors, the following equation is valid for the relationship between the absorption coefficient and the incident photon's energy:

$$\alpha(\vartheta)h\vartheta = B\,(h\vartheta - E_{gap})^m \tag{1}$$

where E_{gap} is the optical bandgap, B is a constant, $h\vartheta$ is incident photon energy, and $\alpha(\upsilon)$ is an absorption coefficient, which is in accordance with the law of Beer–Lambert, equal to

$$\alpha(\vartheta) = 2303Ab(\lambda)/d \tag{2}$$

where d is film thickness and $Ab(\lambda)$ is the film absorption coefficient.

For a more accurate determination of α, a correction that accounts for the reflection spectrum for the absorption coefficient must be made. To calculate the bandgap of $SrTiO_3$/PAN-based photocatalysts with the addition of metal oxide particles, it is necessary to rewrite Equation (1):

$$\alpha(\vartheta) = B(hc)^{(m-1)}\lambda(1/\lambda - 1/\lambda_g)^m \tag{3}$$

where λ_g is the wavelength corresponding to the bandgap, h is the Planck constant, and c is the speed of light.

Using the Beer–Lambert law, Equation (3) can be rewritten as follows:

$$Ab(\lambda) = B(hc)^{(m-1)}d/2303(1/\lambda - 1/\lambda_g)^m + B_1 \tag{4}$$

where B_1 is the constant that takes into account the reflection spectrum.

Using Equation (4), the optical bandgap can be calculated by fitting the absorption spectrum without considering the film thickness. To determine the bandgap (E_{gap}), the absorption coefficient's dependence on the incident radiation energy was plotted and a linear approximation was carried out. Figure 3 presents the absorption and reflection spectra (Figure 3a) and a graph of the absorption coefficient's dependence on the incident radiation energy for SrTiO$_3$/PAN-based photocatalysts with the addition of metal oxide particles (Figure 3b).

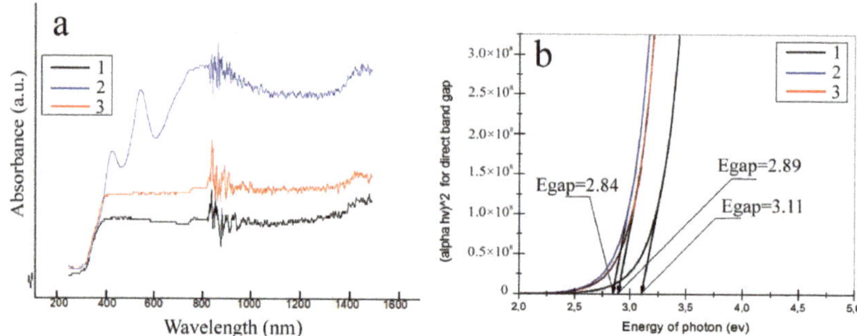

Figure 3. Absorption and reflection spectra (**a**) and values of the bandgap (**b**) of synthesized composite photocatalysts: SrTiO$_3$/PAN/Fe$_2$O$_3$—1 (black line); SrTiO$_3$/PAN/CuO—2 (blue line); and SrTiO$_3$/PAN/Cr$_2$O$_3$—3 (red line).

As a result, the bandgap for SrTiO$_3$/PAN-based photocatalysts with Cr$_2$O$_3$ particles added is determined to be 2.89 eV (Figure 3b, red line). Thus, it is found that the addition of Cr$_2$O particles to SrTiO$_3$-based fibers narrows the bandgap to 2.89 eV, making it possible to use a wide radiation spectrum. The determined value of the bandgap for a SrTiO$_3$/PAN-based photocatalyst with Cr$_2$O$_3$ particles added can be explained by the occupied level of Cr^{3+} cations, which is 1.0 eV above the valence band. The bandgap of the SrTiO$_3$/PAN-based photocatalyst with CuO particles added is 2.84 eV (Figure 3b, blue line). The determined bandgap of the SrTiO$_3$/PAN-based photocatalyst with the addition of Fe$_2$O$_3$ particles is 3.11 eV (Figure 3b, black line). A high bandgap value for a photocatalyst with Fe$_2$O$_3$ particles added is associated with a low Fe$_2$O$_3$ content in calcined fibers, confirmed by XRD (Figure 1b, black line). In turn, the calculated bandgap of the photocatalyst based on the initial SrTiO$_3$/PAN fibers without metal oxide particles added is 3.57 eV.

3.3. Investigation of the Activity of Photocatalysts Based on SrTiO$_3$/PAN Fibers with the Addition of Metal Oxide Particles by the Output of Hydrogen during the Splitting of Water–Methanol Mixture

After determining the bandgaps of photocatalysts based on SrTiO$_3$/PAN fibers with the addition of metal oxide particles, their photocatalytic efficiencies in the splitting of water–methanol mixtures with the production of hydrogen were studied. As seen in Table 1, the composition of the photocatalyst significantly affects the efficiency of hydrogen evolution. As reported in previous work [26], the photocatalytic hydrogen evolution rate from the splitting of the water-methanol mixture using SrTiO$_3$/PAN-based fibers at a 40W UV radiation is 305.96 µmol/h. In turn, the results of the measurement of the average rate of hydrogen evolution during the splitting of the water–methanol mixture at 400W UV irradiation of the composite SrTiO$_3$/PAN-based fibers with the addition of metal oxides are the following:

344.67 μmol/h for the SrTiO$_3$/PAN/Fe$_2$O$_3$ photocatalyst, 398.93 μmol/h for the SrTiO$_3$/PAN/Cr$_2$O$_3$ photocatalyst, and 420.82 μmol/h for the SrTiO$_3$/PAN/CuO photocatalyst. The higher rates of hydrogen evolution for photocatalysts based on SrTiO$_3$/PAN fibers with metal oxides added are explained by the fact that the optical, thermal, and electrochemical properties of metal oxide particles, which also highly depend on their sizes, allow not only for a narrowing of the bandgap of the semiconductors used as a photocatalyst but also for an improvement of their ultraviolet absorption ability.

Table 1. Comparison of the photocatalytic activity of different photocatalysts for producing hydrogen by splitting mixtures of water and alcohol.

Type of Photocatalyst	Parameters of the Process	Composition of the Water Mixture	The Output of Hydrogen, μmol h^{-1} g^{-1}	Reference
SrTiO$_3$/PAN-based fibers	40 W UV lamp, quartz reactor	80% of water and 20% of CH$_3$OH	305.96	[26]
SrTiO$_3$/PAN/Fe$_2$O$_3$ fibers	40 W UV lamp, quartz reactor	80% of water and 20% of CH$_3$OH	344.67	This work
SrTiO$_3$/PAN/Cr$_2$O$_3$ fibers	40 W UV lamp, quartz reactor	80% of water and 20% of CH$_3$OH	398.93	This work
SrTiO$_3$/PAN/CuO fibers	40 W UV lamp, quartz reactor	80% of water and 20% of CH$_3$OH	420.82	This work
MoSe$_2$/TiO$_2$	Xe arc lamp	90% of water and 10% of CH$_3$OH	4.9	[33]
SrTiO$_3$ doped with Cr and N	300 W xenon lamp, quartz reactor	81.5% of water and 18.5% of CH$_3$OH	106.7	[34]
Pt/TiO$_2$	125 W xenon lamp, quartz reactor	70% of water and 30% of CH$_3$OH	523.71	[35]
Pt/ZrO$_2$/TaO$_n$	300 W mercury lamp, quartz reactor	85% of water and 15% of CH$_3$OH	9	[36]

A comparison of the photocatalytic activity of different photocatalysts for hydrogen production shows that the composition of the photocatalyst and the type and intensity of irradiation significantly influence the intensity of hydrogen evolution. The rate of photocatalytic hydrogen evolution from a water–organic alcohol mixture using SrTiO$_3$/PAN-based fibers with the addition of metal oxide particles during UV radiation with 40 W power is several times higher than that of reference analogs [32–36]. Moreover, irradiation with a power of 300 to 400 W was used in these works, which is not economically profitable, in contrast to lamps with a power of 40 W. In [34], TiO$_2$ doped with platinum particles was used as a photocatalyst. The power of ultraviolet radiation was 125 W, and the rate of photocatalytic hydrogen evolution using this photocatalyst with platinum was 523.71 μmol/h, which is still comparable to the hydrogen evolution rate for the SrTiO$_3$/PAN/CuO-based photocatalyst, while the cost and applied irradiation power required to conduct photocatalysis are much lower.

4. Conclusions

One-dimensional photocatalysts based on SrTiO$_3$/PAN fibers with the addition of metal oxide particles were obtained by the electrospinning method. A study of the transmission and reflection spectra showed that the addition of Cr$_2$O$_3$, CuO, and Fe$_2$O$_3$ particles led to decreases in the bandgaps of SrTiO$_3$/PAN-based photocatalysts to 2.89, 2.84, and 3.11 eV, respectively. As a result of the addition of metal oxide particles to the initial SrTiO$_3$/PAN-based photocatalyst, the rate of hydrogen evolution in the photocatalytic splitting of a water–methanol mixture increased to 344.67 μmol h^{-1} g^{-1} for the photocatalyst based on SrTiO$_3$/PAN/Fe$_2$O$_3$, 398.93 μmol h^{-1} g^{-1} for the photocatalyst based on SrTiO$_3$/PAN/Cr$_2$O$_3$, and 420.82 μmol h^{-1} g^{-1} for the photocatalyst based on SrTiO$_3$/PAN/CuO. We believe that this article's proposed approach for increasing the efficiency of photocatalysts by the addition of non-expensive materials, allowing for a reduction in their bandgap, is a promising method for the further development of technologies for efficient solar hydrogen production.

Author Contributions: Conceptualization: F.S. and S.A.; validation and formal analysis: K.B.; investigation: C.D. and B.B.; writing—original draft preparation: F.S. and C.D.; writing—review and editing: M.B. and Z.M.; project administration: K.K. All authors have read and agreed to the published version of the manuscript.

Funding: This research was supported by the Ministry of Education and Science of the Republic of Kazakhstan in the framework of the scientific and technology Program BR05236795 "Development of Hydrogen Energy Technologies in the Republic of Kazakhstan".

Acknowledgments: Seitkhan Azat is grateful to al-Farabi Kazakh National University for financial support in the frame of the «Best University teacher—2019 grant MES RK».

Conflicts of Interest: The authors declare no conflict of interest.

References

1. Shtarev, D.S.; Shtareva, A.V.; Kevorkyants, R.; Rudakova, A.V.; Molokeev, M.S.; Bakiev, T.V.; Bulanin, K.M.; Ryabchuk, V.K.; Serpone, N. Materials synthesis, characterization and DFT calculations of the visible-light-active perovskite-like barium bismuthate $Ba_{1.264(4)}Bi_{1.971(4)}O_4$ photocatalyst. *J. Mater. Chem. C* **2020**, *8*, 3509–3519. [CrossRef]

2. Zhang, X.; Fu, A.; Chen, X.; Liu, L.; Ren, L.; Tong, L.; Ye, J. Highly efficient Cu induced photocatalysis for visible-light hydrogen evolution. *Catal. Today* **2019**, *335*, 166–172. [CrossRef]

3. Ismail, A.A.; Bahnemann, D.W. Photochemical splitting of water for hydrogen production by photocatalysis: A review. *Sol. Energy Mater. Sol. Cells* **2014**, *128*, 85–101. [CrossRef]

4. Sultanov, F.R.; Daulbayev, C.; Bakbolat, B.; Mansurov, Z.A.; Urazgaliyeva, A.A.; Ebrahim, R.; Pei, S.S.; Huang, K.-P. Microwave-enhanced chemical vapor deposition graphene nanoplatelets-derived 3D porous materials for oil/water separation. *Carbon Lett.* **2020**, *30*, 81–92. [CrossRef]

5. Phoon, B.L.; Lai, C.W.; Juan, J.C.; Show, P.-L.; Pan, G.-T. Recent developments of strontium titanate for photocatalytic water splitting application. *Int. J. Hydrogen Energy* **2019**, *44*, 14316–14340. [CrossRef]

6. Collignon, C.; Lin, X.; Rischau, C.W.; Fauqué, B.; Behnia, K. Metallicity and superconductivity in doped strontium titanate. *Annu. Rev. Condens. Matter Phys.* **2019**, *10*, 25–44. [CrossRef]

7. Xu, B.; Ahmed, M.B.; Zhou, J.L.; Altaee, A. Visible and UV photocatalysis of aqueous perfluorooctanoic acid by TiO_2 and peroxymonosulfate: Process kinetics and mechanistic insights. *Chemosphere* **2020**, *243*, 125366. [CrossRef]

8. Sinhmar, A.; Setia, H.; Kumar, V.; Sobti, A.; Toor, A.P. Enhanced photocatalytic activity of nickel and nitrogen codoped TiO_2 under sunlight. *Environ. Technol. Innovation* **2020**, *18*, 100658. [CrossRef]

9. Gonçalves, N.P.F.; Paganini, M.C.; Armillotta, P.; Cerrato, E.; Calza, P. The effect of cobalt doping on the efficiency of semiconductor oxides in the photocatalytic water remediation. *J. Environ. Chem. Eng.* **2019**, *7*, 103475. [CrossRef]

10. Wang, R.; Ni, S.; Liu, G.; Xu, X. Hollow $CaTiO_3$ cubes modified by La/Cr co-doping for efficient photocatalytic hydrogen production. *Appl. Catal. B* **2018**, *225*, 139–147. [CrossRef]

11. Jiang, L.; Ni, S.; Liu, G.; Xu, X. Photocatalytic hydrogen production over Aurivillius compound Bi_3TiNbO_9 and its modifications by Cr/Nb co-doping. *Appl. Catal. B* **2017**, *217*, 342–352. [CrossRef]

12. Chen, P.-W.; Li, K.; Yu, Y.-X.; Zhang, W.-D. Cobalt-doped graphitic carbon nitride photocatalysts with high activity for hydrogen evolution. *Appl. Surf. Sci.* **2017**, *392*, 608–615. [CrossRef]

13. Regmi, C.; Kshetri, Y.K.; Pandey, R.P.; Kim, T.-H.; Gyawali, G.; Lee, S.W. Understanding the multifunctionality in Cu-doped $BiVO_4$ semiconductor photocatalyst. *J. Environ. Sci.* **2019**, *75*, 84–97. [CrossRef]

14. Habba, Y.; Capochichi-Gnambodoe, M.; Leprince-Wang, Y. Enhanced Photocatalytic Activity of Iron-Doped ZnO Nanowires for Water Purification. *Appl. Sci.* **2017**, *7*, 1185. [CrossRef]

15. Sultanov, F.R.; Daulbayev, C.; Bakbolat, B.; Zhurintayeva, A.; Daulbayev, O.; Mansurov, Z.A. Comparison of oil/water separating efficiency of oleophobic membranes based on fluorine containing and fluorine non-containing coatings. *RJC* **2019**, *12*, 1091–1097. [CrossRef]

16. Cheng, G.; Liu, X.; Song, X.; Chen, X.; Dai, W.; Yuan, R.; Fu, X. Visible-light-driven deep oxidation of NO over Fe doped TiO_2 catalyst: Synergic effect of Fe and oxygen vacancies. *Appl. Catal. B* **2020**, *277*, 119196. [CrossRef]

17. Ismael, M. Enhanced photocatalytic hydrogen production and degradation of organic pollutants from Fe(III) doped TiO_2 nanoparticles. *J. Environ. Chem. Eng.* **2020**, *8*, 103676. [CrossRef]

18. Zhang, H.; Chen, G.; Li, X. Synthesis and visible light photocatalysis water splitting property of chromium-doped $Bi_4Ti_3O_{12}$. *Solid State Ionics* **2009**, *180*, 1599–1603. [CrossRef]

19. Chen, Z.; Jiang, X.; Zhu, C.; Shi, C. Chromium-modified $Bi_4Ti_3O_{12}$ photocatalyst: Application for hydrogen evolution and pollutant degradation. *Appl. Catal. B* **2016**, *199*, 241–251. [CrossRef]

20. Zhu, H.; Fang, M.; Huang, Z.; Liu, Y.; Chen, K.; Guan, M.; Tang, C.; Zhang, L.; Wang, M. Novel chromium doped perovskites A_2ZnTiO_6 (A = Pr, Gd): Synthesis, crystal structure and photocatalytic activity under simulated solar light irradiation. *Appl. Surf. Sci.* **2017**, *393*, 348–356. [CrossRef]

21. Liu, J.; Hodes, G.; Yan, J.; Liu, S.F. Metal-doped Mo_2C (metal = Fe, Co, Ni, Cu) as catalysts on TiO_2 for photocatalytic hydrogen evolution in neutral solution. *Chin. J. Catal.* **2021**, *42*, 205–216. [CrossRef]

22. Nikhila, M.P.; John, D.; Pai, M.R.; Renuka, N.K. Cu and Ag modified mesoporous TiO_2 nanocuboids for visible light driven photocatalysis. *Nanostruct. Nano-Objects.* **2020**, *21*, 100420. [CrossRef]

23. Onwubiko, I.; Khan, W.S.; Subeshan, B.; Asmatulu, R. Investigating the effects of carbon-based counter electrode layers on the efficiency of hole-transporter-free perovskite solar cells. *Energ. Ecol. Environ.* **2020**, *5*, 141–152. [CrossRef]

24. Alharbi, A.R.; Alarifi, I.M.; Khan, W.S.; Swindle, A.; Asmatulu, R. Synthesis and characterization of electrospun polyacrylonitrile/graphene nanofibers embedded with $SrTiO_3/NiO$ nanoparticles for water splitting. *J. Nanosci. Nanotechnol.* **2017**, *17*, 5294–5302. [CrossRef]

25. Asmatulu, R.; Shinde, M.A.; Alharbi, A.R.; Alarifi, I.M. Integrating graphene and C_{60} into TiO_2 nanofibers via electrospinning process for enhanced conversion efficiencies of DSSCs. *Macromol. Symp.* **2016**, *365*, 128–139. [CrossRef]

26. Sultanov, F.; Daulbayev, C.; Bakbolat, B.; Daulbayev, O.; Bigaj, M.; Mansurov, Z.; Kuterbekov, K.; Bekmyrza, K. Aligned composite $SrTiO_3$/PAN fibers as 1D photocatalyst obtained by electrospinning method. *Chem. Phys. Lett.* **2019**, *737*, 136821. [CrossRef]

27. Roy, P.K.; Bera, J. Formation of $SrTiO_3$ from Sr-oxalate and TiO_2. *Mater. Res. Bull.* **2005**, *40*, 599–604. [CrossRef]

28. Havlíček, K.; Svobodová, L.; Bakalova, T.; Lederer, T. Influence of electrospinning methods on characteristics of polyvinyl butyral and polyurethane nanofibres essential for biological applications. *Mater. Des.* **2020**, *194*, 108898. [CrossRef]

29. Huang, W.; Tong, Z.; Wang, R.; Liao, Z.; Bi, Y.; Chen, Y.; Ma, M.; Lyu, P.; Ma, Y. A review on electrospinning nanofibers in the field of microwave absorption. *Ceram. Int.* **2020**, in press. [CrossRef]

30. Guo, L.; Xie, N.; Wang, C.; Kou, X.; Ding, M.; Zhang, H.; Sun, Y.; Song, H.; Wang, Y.; Lu, G. Enhanced hydrogen sulfide sensing properties of Pt-functionalized α-Fe_2O_3 nanowires prepared by one-step electrospinning. *Sens. Actuators B* **2018**, *255*, 1015–1023. [CrossRef]

31. Zhao, J.; Lu, Q.; Wei, M.; Wang, C. Synthesis of one-dimensional α-Fe_2O_3/Bi_2MoO_6 heterostructures by electrospinning process with enhanced photocatalytic activity. *J. Alloys Compd.* **2015**, *646*, 417–424. [CrossRef]

32. Koo, B.-R.; Park, I.-K.; Ahn, H.-J. Fe-doped In_2O_3/α-Fe_2O_3 core/shell nanofibers fabricated by using a co-electrospinning method and its magnetic properties. *J. Alloys Compd.* **2014**, *603*, 52–56. [CrossRef]

33. Wu, L.; Shi, S.; Li, Q.; Zhang, X.; Cui, X. TiO_2 nanoparticles modified with 2D $MoSe_2$ for enhanced photocatalytic activity on hydrogen evolution. *Int. J. Hydrogen Energy* **2019**, *44*, 720–728. [CrossRef]

34. Yu, H.; Yan, S.; Li, Z.; Yu, T.; Zou, Z. Efficient visible-light-driven photocatalytic H_2 production over Cr/N-codoped $SrTiO_3$. *Int. J. Hydrogen Energy* **2012**, *37*, 12120–12127. [CrossRef]

35. Li, F.; Gu, Q.; Niu, Y.; Wang, R.; Tong, Y.; Zhu, S.; Zhang, H.; Zhang, Z.; Wang, X. Hydrogen evolution from aqueous-phase photocatalytic reforming of ethylene glycol over Pt/TiO_2 catalysts: Role of Pt and product distribution. *Appl. Surf. Sci.* **2017**, *391*, 251–258. [CrossRef]

36. López-Tenllado, F.J.; Hidalgo-Carrillo, J.; Montes, V.; Marinas, A.; Urbano, F.J.; Marinas, J.M.; Ilieva, L.; Tabakova, T.; Reid, F. A comparative study of hydrogen photocatalytic production from glycerol and propan-2-ol on M/TiO_2 systems (M = Au, Pt, Pd). *Catal. Today* **2017**, *280*, 58–64. [CrossRef]

Article

Kinetics of Hydrogen Generation from Oxidation of Hydrogenated Silicon Nanocrystals in Aqueous Solutions

Gauhar Mussabek [1,2,*], Sergei A. Alekseev [3], Anton I. Manilov [4], Sergii Tutashkonko [5], Tetyana Nychyporuk [5], Yerkin Shabdan [1], Gulshat Amirkhanova [2], Sergei V. Litvinenko [4], Valeriy A. Skryshevsky [4] and Vladimir Lysenko [6]

[1] Faculty of Physics and Technology, Al-Farabi Kazakh National University, 71, Al-Farabi Ave., Almaty 050040, Kazakhstan; nanotechkz2012@gmail.com

[2] Institute of Information and Computational Technologies, 125, Pushkin Str., Almaty 050000, Kazakhstan; gulshat.aa@gmail.com

[3] Chemistry Department, Taras Shevchenko National University of Kyiv, Volodymyrska Street, 64, 01601 Kyiv, Ukraine; alekseev@univ.kiev.ua

[4] Institute of High Technologies, Taras Shevchenko National University of Kyiv, Volodymyrska Street, 64, 01601 Kyiv, Ukraine; anmanilov@univ.kiev.ua (A.I.M.); litvin608pv@gmail.com (S.V.L.); skrysh@univ.kiev.ua (V.A.S.)

[5] Nanotechnology Institute of Lyon (INL), UMR CNRS 5270, INSA de Lyon, University of Lyon, 69621 Villeurbanne, France; s.tutashkonko@gmail.com (S.T.); tetyana.nychyporuk@insa-lyon.fr (T.N.)

[6] Light Matter Institute, UMR-5306, Claude Bernard University of Lyon, 2 rue Victor Grignard, 69622 Villeurbanne, France; vladimir.lysenko@univ-lyon1.fr

* Correspondence: Gauhar.musabek@kaznu.kz; Tel.: +7-727-377-3412

Received: 26 June 2020; Accepted: 15 July 2020; Published: 20 July 2020

Abstract: Hydrogen generation rate is one of the most important parameters which must be considered for the development of engineering solutions in the field of hydrogen energy applications. In this paper, the kinetics of hydrogen generation from oxidation of hydrogenated porous silicon nanopowders in water are analyzed in detail. The splitting of the Si-H bonds of the nanopowders and water molecules during the oxidation reaction results in powerful hydrogen generation. The described technology is shown to be perfectly tunable and allows us to manage the kinetics by: (i) varying size distribution and porosity of silicon nanoparticles; (ii) chemical composition of oxidizing solutions; (iii) ambient temperature. In particular, hydrogen release below 0 °C is one of the significant advantages of such a technological way of performing hydrogen generation.

Keywords: hydrogen generation rate; porous silicon nanopowder; nanosilicon oxidation; engineering of silicon nanoparticles

---

1. Introduction

Hydrogen and fuel cell technologies have strong potential to play a significant role in new energy systems [1]. In view of energy storage purpose, hydrogen can be easily converted into electricity and heat. The amount of energy produced during hydrogen combustion is higher in comparison with any other fuel of the same mass [2]. However, hydrogen cannot be considered as an alternative fuel but rather an energy carrier requesting the consumption of a primary energy for its production. Due to this fact, the number of power-to-gas pilot plants producing hydrogen from fluctuating renewable power sources increases all over the world [3]. Besides, for the energy vector for electricity, mobility and heat, hydrogen can be also used as a raw material for the chemical industry or for the synthesis of various hydrocarbon fuels. Ongoing research is underway to develop environmentally friendly and economical hydrogen production technologies that are essential for the hydrogen economy. It ensures the gradual transition from the actual energy economy to a cleaner and sustainable energy future [4].

Water is an important source of hydrogen within the concept of sustainable development. The splitting of water molecules into hydrogen and oxygen can be performed by means of "green" technologies. For example, water electrolysis constitutes a suitable pathway to hydrogen production [1,5]. There are different technologies for electrolysis, including alkaline, proton-exchange membrane and high-temperature solid oxide processes. They form a basis of the power-to-gas systems converting excess electricity into hydrogen [3,5]. Water electrolysis is convenient to be used in combination with photovoltaics and wind energy as well as in direct relation with the availability of electricity [6–8].

The splitting of water molecules can also be provided by heat obtained from a nuclear reactor or a concentrating solar system, such as a solar tower, solar trough, linear Fresnel system or dish system [1]. The direct decomposition of water driven by heat can be released using thermocatalytic processes [9]. The most promising and useful approach for large-scale hydrogen production is based on thermochemical cycles—redox, sulfur-iodine, copper-chlorine and others [10]. However, the thermal technologies require complicated industrial equipment and are not useful for small-scale and portable applications.

Photocatalytic water splitting under visible light irradiation allows for obtaining hydrogen from water with the use of sunlight in the presence of a suitable catalyst reducing the high activation energy of water decomposition [11,12]. Photocatalysts are typically made of metal oxides, metal sulfides, oxysulfides, oxynitrides and composites thereof. Titanium dioxide, cadmium sulfide and graphitic carbon nitride are the most studied [13–15]. Metal ion implantation and dye sensitization are very effective methods to extend the activating spectrum to the visible range [16]. In recent years, many nanoparticles are synthesized by clean technologies utilizing plants and microbes. Water splitting, based on these nanoparticles, provides a blueprint to sustain the clean energy demands [17]. However, the number of visible light-driven photocatalysts is still limited, and there is no photocatalyst which could satisfy all necessary requirements [1,11].

Water splitting can be also released in a biological way. The absorption of light by microalgae or cyanobacteria results in the generation of hydrogen, known as biophotolysis [18]. Additionally, hydrogen can be produced from CO and H_2O, implementing the biological water gas shift reaction catalyzed by photosynthetic and fermentative bacteria [19]. A high yield of hydrogen and a high production rate can be reached using biomass or biomass-derived chemicals as raw materials. The conversion of biomass is achieved through biological or thermochemical routes, such as fermentation, reforming, gasification or pyrolysis [20,21].

An inexpensive and simple way of hydrogen production is also implemented by hydrolytic systems. In particular, the reaction of water with metals is characterized by a redox potential below the level corresponding to $H^+ \rightarrow H2$ transformation [22]. During the chemical reaction, hydrogen is released, and a metal hydroxide or oxide is formed. Among various available reactions, aluminum-water reactions in either alkaline or neutral solutions are most commonly used [23–25]. However, the existence of a solid oxide on the surface of aluminum particles prevents air and moisture from accessing the underlying metal. The use of strong bases, the application of high temperature or the activation of the aluminum metal [25–27] resolve this problem.

Metal hydrides are also efficient materials for the chemical decomposition of water. Indeed, in this case, not only the hydrogen liberated by oxidation of the corresponding metal is released, but also the hydrogen bonded in the hydride molecules. In particular, borohydrides are very promising materials, due to their very high theoretical capacities of hydrogen content. For example, sodium borohydride (NaBH4), lithium borohydride ($LiBH_4$) and the ammonia–borane complex (NH_3BH_3) were widely investigated [1,22,28–30].

Recent reports devoted to the hydrolysis of magnesium composites (MgH_2, Mg-oxide, Mg_2Si, Mg-graphite, $NdNiMg_{15}$) [21–35] and sodium metal [36] have demonstrated considerable hydrogen generation capabilities. However, the successful utilization of these materials in hydrogen-based energy systems is still quite expensive.

Silicon has also been investigated for hydrogen generation purposes [37]. Indeed, hydrogen can be produced, for example, from the reaction between ferrosilicon, sodium hydroxide and water. Further investigations showed the high potential and economic efficiency of the silicon-based hydrogen fuel [38,39]. In strong contrast to oil and molecular hydrogen, the transport and storage of silicon are free from potential hazards and require a simple infrastructure similar to that requested for coal. Whereas the latter material is converted to carbon dioxide, silicon involved in hydrogen production is transformed in sand. Among a number of the known metalloids, silicon shows the best efficiency for the chemical splitting of water [40,41]. Moreover, photoelectrochemical applications of crystalline silicon have been investigated for solar-induced splitting of water [42,43].

Silicon nanoparticles can be used to generate hydrogen much more rapidly than bulk silicon because of their high specific surface area. Chemical or anodic etching [44–46], the beads milling method [47–49], or laser pyrolysis [50] can produce the nano-powders used for hydrogen generation from oxidation reactions of Si in water. Since the reaction rate strongly depends on pH, the addition of alkalis or ammonia is often used in order to increase the rate of hydrogen production.

The fabrication of silicon nanostructures by electrochemical etching has a significant advantage over the other approaches cited above. Indeed, it results in the formation of nano-porous morphologies covered with an abundant number of silane groups (SiH_X) [51–53]. The maximum specific surface of nanoporous silicon can reach the value of 800 m^2/g, while the content of hydrogen chemically bound on the surface can be as high as 60 mmol of atomic H per gram of porous Si, corresponding to the H/Si ratio ~1.8 or to the 6% mass of H [44]. The presence of this surface hydrogen can increase the output volume of H_2, which releases in the reaction of porous silicon nanoparticles with water by 1.5 times [54]. General hydrogen output and release rate strongly depend on the porosity and sizes of the particles. Moreover, sufficient reaction rates are achieved even at room temperature, without additional heating or mixing [55,56]. A working prototype of cartridge generating hydrogen from PSi nanopowder was designed and coupled with a portable fuel cell [57].

Silicon nanostructures have also been considered as good candidates for photoelectrochemical and photocatalytic water splitting to produce hydrogen. Photo-electrodes based on arrays of silicon nanowires coupled with different catalysts [58–60], as well as porous silicon structures [61–63], were used for the generation of hydrogen. The surface modification of silicon nanostructures plays a considerable role in designing materials for solar-driven catalysis. The applications of Si nanostructures can be moved from photoelectrical to photochemical conversion by taking catalytic sites into account [58]. However, the chemical reactivity of silicon surface with water complicates the implementation of the photocatalytic reaction. This is especially evident for the porous nanostructures, where a higher oxidation degree of nanosilicon can result in the blocking of the nanopores with silicon oxide [64].

2. Hydrogen Generation from Oxidation of Porous Silicon Nanopowders in Water

Research work described in the current paper is devoted to the investigation of influence of various physico-chemical factors on the kinetics of hydrogen generation resulting from chemical reactions of hydrogenated porous silicon (PSi) nanopowders with water. In particular, the impact of ambient temperature, chemical nature and the concentration of used alkalis, grinding degree, and the porosity of PSi nanopowders are examined. In general, this reaction occurs as follows:

$$SiH_X + 3H_2O \rightarrow SiO_2 \bullet H_2O + \left(2 + \frac{X}{2}\right)H_2 \uparrow + Q \tag{1}$$

where Q is equal to 361 kJ/M and to 634 kJ/M for $x = 0$ and $x = 2$, respectively.

Reaction (1) is schematically illustrated in Figure 1.

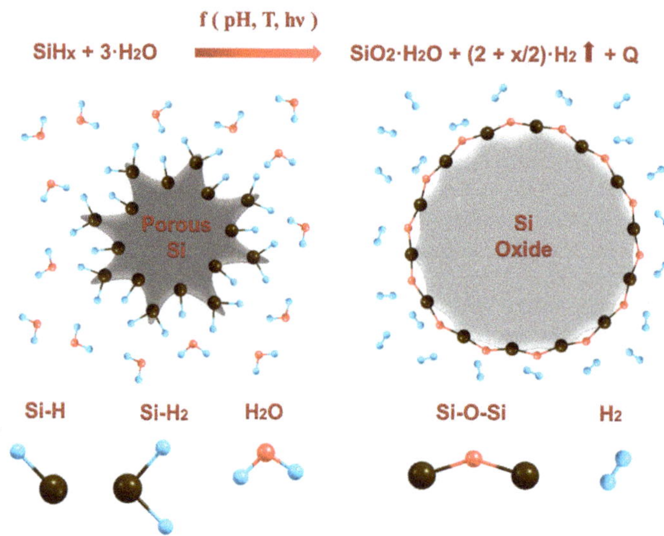

Figure 1. Transformation of partially hydrogenated PSi nanoparticles in silicon oxide in water solution.

As one can see, partially hydrogenated PSi nanoparticles are almost completely transformed in silicon oxide nanoparticles, as shown in Figure S1 in the Supplementary Materials. This transformation is modulated by the pH and temperature of the surrounding aqueous solution, as well as by the intensity of external light results in the abundant production of molecular hydrogen [44,45].

Si-H bonds covering specific surface of PSi nanoparticles are known to be the most chemically active ones. These bonds can react, for example, with the bases, as shown by reaction (a) in Figure 2, substituting surface hydrogen by –OH groups. Si–Si bonds located on the PSi surface can also react with oxidizers, according to the reactions (b) and (c) in Figure 2, leading to the formation of O_3Si–H and Si–OH surface fragments. As a result, a silicon oxide film starts to appear on the PSi surface. The O_3Si–H fragments are rather passive and stable. Nevertheless, they can react with strong bases, as shown by reaction (d) in Figure 2 [44,54].

Figure 2. Schematic representation of gradual transformation of initially hydrogenated PSi nanopowder in silicon oxide during its oxidation in water.

Reactivity of the chemical bonds of PSi surface with water and other oxidants increases in the following order: $O_3Si–H < Si_3Si–H < Si–Si$. In the other words, the oxidation of the Si surface proceeds more preferably via (c)–(d), than via (a)–(b), as shown in Figure 2 [45,65,66]. When a sufficiently thick layer of hydrated silicon oxide is formed on the PSi surface, it efficiently preserves the inner parts of Si NPs from further oxidation, and the reaction stops [40]. The presence of bases in aqueous solution results in the significant acceleration of the reaction between Si and water due to an increase in the primary reaction rate of the surface bonds with water as well as due to the partial dissolution of the surface oxide layer and penetration of water molecules and OH^- ions to the inner parts of the PSi nanoparticles.

3. Materials and Methods

3.1. Formation of PSi Nanopowders

The PSi nanopowders used in this study were obtained by mechanical grinding of porous silicon (PSi) layers. The initial PSi layers were fabricated according to a standard procedure based on electrochemical etching [67] of monocrystalline (100)-oriented boron-doped Si wafers (1–10 Ω cm and 0.01 Ω cm for nano- and meso-PSi structures, respectively) at current densities of 2–340 mA/cm^2. The etching solutions of 9:1, 3:1 and 1:1 mixtures (by volume) of concentrated aqueous hydrofluoric acid (48%) and ethanol were used for nano-PSi formation and an HF/ethanol solution of 1:1 volume mixture for meso-PSi formation. The experimental conditions of the electrochemical etching mentioned above allowed for the formation of nano-PSi and meso-PSi nanopowders with various porosities, as shown in see Figures S2–S4 in Supplementary Materials, and sizes of nanoparticles constituting the porous powders, as shown in see Figure S5 in Supplementary Materials.

3.2. Structural Characterization of PSi Nanopowders

Structural properties and morphologies of the PSi nanopowders were investigated by means of atomic force microscopy (AFM), scanning electron microscopy (SEM) and transmission electron microscopy (TEM). Atomic force microscopy (AFM) characterization was performed on the Digital Instruments 3100 instrument using ultra-sharp silicon cantilevers (Nanosensors™ SSS-NCH, Switzerland) with a typical curvature radius of 10 nm and nominal spring constant of 42 N/m. AFM images were acquired in a tapping mode at room temperature under ambient conditions for the particles deposited onto an atomically flat surface of electronic grade silicon wafers. Scanning electron microscope (SEM) images were taken using the ultra-high resolution Tescan MIRA 3 scanning electron microscope.

3.3. Temperature-Programmed Desorption (TPD)

The TPD studies were performed in a quartz reactor, in the range of 40–1000 °C by means of a Thermoquest TPD/R/O 1100 analyzer. A quadruple mass analyzer (Quadstar 422, Baltzer, Germany) coupled to the TPD apparatus allows for the identification of desorbed species. To obtain quantitative measurements on the molecular hydrogen desorbed from the sample surface, calibration was performed by injecting known amounts of hydrogen into the argon flow and evaluating the area of the detected TPD signal. Figure S6 in the Supplementary Materials shows typical TPD signals obtained on nano- and meso-PSi powders. Comparison between the TPD curves and mass spectrum of the desorbed species indicates that the desorbed phase corresponds perfectly to H_2. These results prove the presence of hydrogen in the initial as-prepared PSi nanopowders before their reaction with oxidizing aqueous solutions. The integration of the shown TPD signals allows us to estimate hydrogen quantity in the PSi nanopowders. In particular, nano-PSi powder can contain hydrogen with quantities more than one order of magnitude compared to the meso-PSi samples. It can be explained by the much more important specific surface area of the nano-PSi powder, which can exceed 1000 m^2 g^{-1} [67].

3.4. Measurement of Hydrogen Generation Kinetics

All of the experiments, aimed at the recording of hydrogen generation kinetics during the reaction (1), were carried out on a working system, shown in Figure S7 in Supplementary Materials. A Plexiglas container was used as the reaction chamber. Samples of the PSi nanopowders were put on a rotating sample holder fixed in the upper part of the Plexiglas container. The weight of the used PSi samples varied from 50 to 100 mg. The volume of the oxidizing solutions was 20 mL. The PSi nanopowders were immersed into the oxidizing solution by rotation of the sample holder. Hydrogen generation started immediately after the beginning of reaction (1). Generated hydrogen was accumulated in the upper part of a graduated glass tube allowing for the measurement of the hydrogen volume. Systematic errors, caused by excess pressure of the superseded water, evaporation of the solution, and solubility of hydrogen in water, are estimated to be inessential (<10%) and were not taken into account. The experiments were carried out at low ambient temperature (−40 °C) and at room temperature (+23 °C) for 160 h and 3 h, respectively.

4. Results and Discussion

Typical view of an as-prepared hydrogenated highly porous cracked silicon film and its SEM top image (in inset) are shown in Figure 3a. The porous film is easily auto-destroyed during its drying in ambient air after formation by electrochemical etching. To obtain porous Si (PSi) nanopowders, the porous films were additionally mechanically grinded, as shown in Figure 3b. The AFM picture shown in Figure 3c confirms the nano-scale size range of the nanoparticles (white spots and points), constituting the PSi nanopowder. The TEM image of the porous nanoparticles, presented in Figure 3d, allows us to estimate their sizes (50–60 nm) as well as clearly reveals their porous morphology (see inset illustrating a single PSi nanoparticle). Porosity values of the PSi nanoparticles used in our experiments were in the range of 30–95%. Their pore sizes varied from several up to 20 nm and specific surface area was between 100–800 m^2 g^{-1} [68].

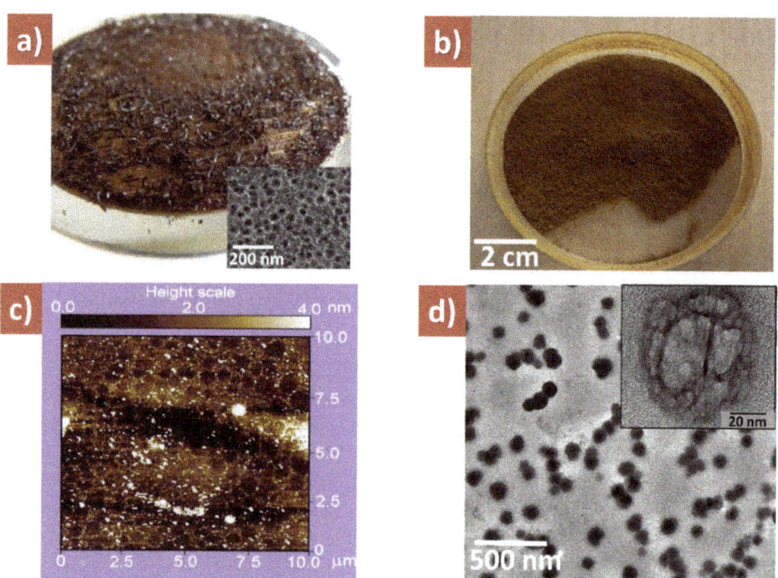

Figure 3. (**a**) As-prepared hydrogenated highly porous silicon film, inset: SEM top view image of the nanoporous Si film; (**b**) PSi nanopowder obtained by mechanical grinding of the PSi layer; (**c**) AFM image of the PSi nanoparticles constituting the PSi nanopowder; (**d**) TEM images of the PSi nanoparticles, inset: single PSi nanoparticle.

Figure 4 shows the kinetics of hydrogen generation due to the oxidation of meso-porous silicon powders in water:ethanol = 3:2 solutions at two ambient temperatures: +23 °C and −40 °C. Ethanol was added to avoid the crystallization of water at negative temperatures as well as for the wetting of the PSi nanoparticles surface, because fresh as-prepared PSi nanopowder is hydrophobic. In both cases, the overall reaction time exceeded 100 h. It is worth remarking that hydrogen generation occurs even at relatively low temperatures (<0 °C) and this fact has a huge practical importance if one uses this reaction for hydrogen generation at negative temperatures. Since the chemical oxidation of silicon in aqueous solutions is thermally activated, a significant increase in H_2 generation rate (factor of three and final global quantity (more than 2-times) of the generated hydrogen can be observed at the higher temperature.

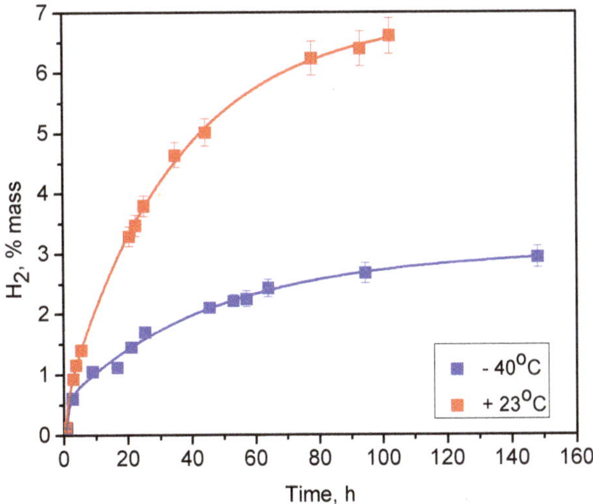

Figure 4. Kinetics of hydrogen generation due to oxidation reaction of meso-PSi nanopowders in water:ethanol solutions at different temperatures. Discrete points correspond to experimentally measured values, while solid lines are the fitting functions given in Supplementary Materials. The values on y-axis correspond to the percentage mass of H_2 related to the mass of the reacted PSi nanopowder and water, which is necessary for the reaction.

The observed difference between the presented curves describing the behavior of the hydrogen generation kinetics at −40 °C and +23 °C can be completely explained by taking into account the temperature-dependent diffusion coefficient of the oxidizing water molecules through the silicon oxide progressively formed at the nanoparticle surface. Indeed, at the lowest temperature (−40 °C), a relatively low value of the diffusion coefficient limits the oxidation reaction rate and, consequently, hydrogen generation kinetics. In addition, the water molecules cannot wade through the formed oxide sheet of a certain thickness and it leads only to a partial oxidation of the silicon nanoparticles and limits the overall quantity of the generated hydrogen at a level of 2.8% mass. The water diffusion coefficient increases at the higher temperature (+23 °C) and it automatically leads to the enhanced hydrogen generation kinetics, as well as to the much higher quantity of hydrogen (6.7% mass) generated at the end of the reaction between the silicon nanopowder and water/ethanol mixture.

Since alkali is known to accelerate the dissolution of silicon, hydrogen generation resulting from the oxidation of meso-PSi powders was studied in water:ethanol solutions containing alkali. Figure 5 allows us to compare the hydrogen generation kinetics, depending on chemical nature of different alkalis at same concentration of 0.21 M/L. As one can see, the presence of any type of alkali accelerates significantly (by several orders of magnitude) the hydrogen generation rate. Among the used alkalis, the relatively weak NH_3 base demonstrates the slowest H_2 generation. The higher H_2 generation rate for NaOH in comparison with KOH is probably due to the better permeability of the surface oxide layer to smaller Na^+ ions than for larger K^+ ions and, as a result, the better access of active OH– ions to the SiO_2/Si interface. At the same time, the final quantity (6% mass) of the generated hydrogen is similar for all the used oxidizing solutions and it is very close to theoretical value (6.25%). White color and the absence of visible photoluminescence of all the products formed in the reaction of PSi nanopowders with aqueous alkaline solutions at room temperature, confirm the total oxidation of the Si phase. Indeed, it can be easily understood, considering that hydrogen generation stops when silicon nanoparticles are completely oxidized.

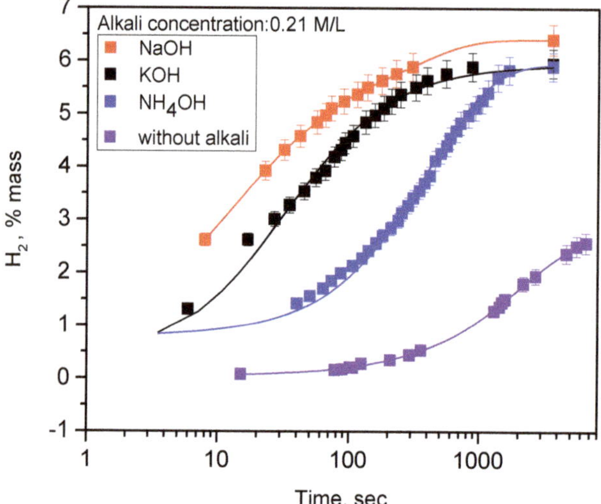

Figure 5. Kinetics of hydrogen generation due to oxidation reaction of meso-PSi nanopowders in water:ethanol:alkali solutions at room temperatures. Concentration of the used alkalis was 0.21 M/L. Discrete points correspond to experimentally measured values, while solid lines are the fitting functions given in Supplementary Materials.

Since NaOH was found to be the most efficient alkali accelerating hydrogen generation from the oxidation of PSi nanopowders in aqueous solution, it was used for further additional studies. In particular, the impact of NaOH concentration on hydrogen generation kinetics was studied and is shown in Figure 6. As one can see, the higher the alkali concentration is, the higher the hydrogen generation rate is, as shown in Figure S8 in Supplementary Materials. Once again, the overall hydrogen generation quantity is independent on the alkali concentration. Finally, as shown in Figure S8 in Supplementary Materials, the same kinetics of hydrogen generation can be achieved at smaller concentrations of NaOH compared to KOH.

Kinetics of hydrogen generation depending on the grinding degree of meso-PSi nanopowder are shown in Figure 7. Two different nanopowder morphologies—as prepared meso-PSi and grinded meso-PSi—shown in Figure 3a,b, respectively, were used. It can be seen that oxidation of the grinded meso-PSi accompanied by formation of hydrogen goes two-times faster in comparison with the as-prepared meso-PSi (analytical expressions of fitting functions can be found in Supplementary Materials. Such behavior of the hydrogen generation kinetics can be explained by much smaller sizes of nanoparticles (50–200 nm) of the grinded meso-PSi compared to the as-prepared powder constituted by porous particles from 100 μm–1 mm range.

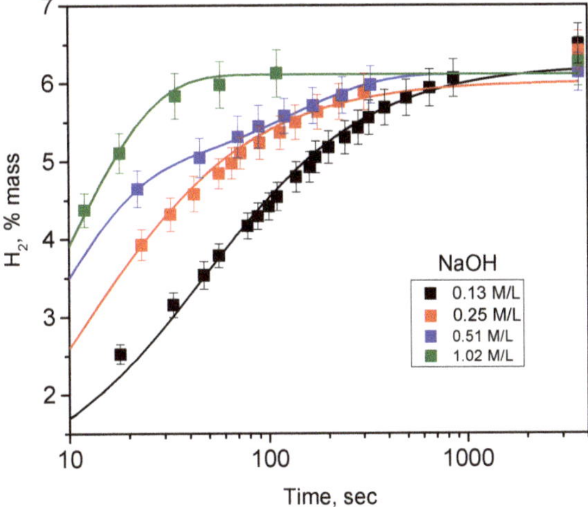

Figure 6. Kinetics of hydrogen generation due to oxidation reaction of meso-PSi nanopowders in water:ethanol:NaOH solutions at different concentrations of the alkali. Discrete points correspond to experimentally measured values, while solid lines are the fitting functions given in Supplementary Materials.

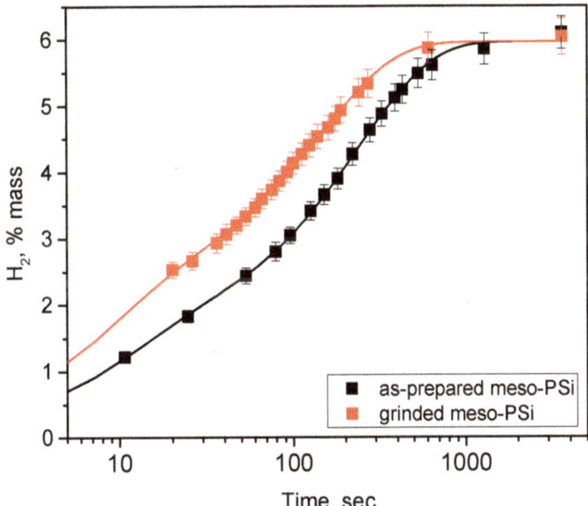

Figure 7. Kinetics of hydrogen generation depending on grinding degree of meso-PSi nanopowders. The oxidation reaction takes place in solutions of water:ethanol = 3:2 containing NH_4OH in a concentration of 0.77 M/L. Discrete points correspond to experimentally measured values, while solid lines are the fitting functions given in Supplementary Materials.

Hydrogen generation kinetics characterizing chemical reaction between aqueous alkaline solution and grinded nano-PSi powders obtained from c-Si substrates with low ($\rho = 1$–10 $\Omega \cdot$cm) and high ($\rho = 1$ $\Omega \cdot$cm) resistivities are compared in Figure 8. As one can see, the resistivity of the initial silicon substrate does not affect the kinetics of hydrogen release. Thus, only the nanoscale morphology of the nano-PSi powder (similar for the both resistivity levels), shown in Figures S2 and S4 in Supplementary Materials, determines the hydrogen generation kinetics.

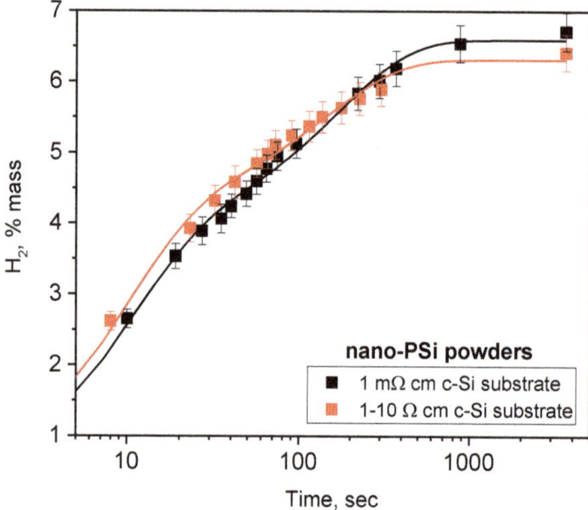

Figure 8. Kinetics of hydrogen generation depending on resistivity of silicon substrates used for fabrication of nano-PSi powder. The oxidation reaction occurs in solutions of water:ethanol = 3:2 containing NaOH in concentration of 0.21 M/L. Discrete points correspond to experimentally measured values, while solid lines are the fitting functions given in Supplementary Materials.

Figure 9 shows dependence of the global amount of generated hydrogen on porosity of the used meso-PSi and nano-PSi powders. In the case of nano-PSi powder, the quantity of the generated hydrogen is porosity-independent, while a significant decrease in hydrogen quantity was observed for higher porosity values of meso-PSi powder. This difference can be explained by smaller nanoparticle size and larger pores for nano-PSi in comparison to meso-PSi, as shown in Figure S5 in Supplementary Materials. Indeed, the hydrogen generation efficiency will be enhanced if the oxidizing solution easily in penetrates nanopores and completely oxidizes the smaller nanoparticles. In addition, the general behavior of the dependence of hydrogen amount on porosity for both the morphologies correlate perfectly with evolution of nanocrystals of the corresponding powders, as shown in Figure S5 in Supplementary Materials.

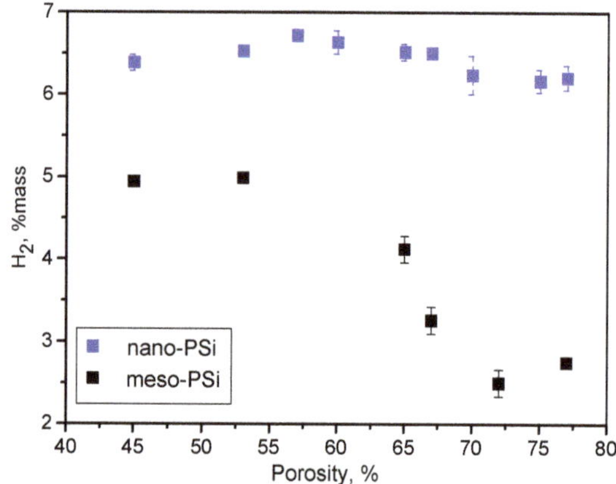

Figure 9. Porosity dependent hydrogen generation amount due to oxidation reaction of meso- and nano-PSi powders in H₂O:Ethanol:NaOH solutions at room temperature. Alkali concentration was 0.5 M/L.

5. Conclusions

Hydrogen generation rate is one of the most important parameters which must be considered for the design of engineering solutions for hydrogen energy applications. Indeed, sometimes a slow and long-term production of hydrogen is requested, while for other kinds of applications the instantaneous generation of a very large amount of hydrogen must be provided. For the given technological approach, various parameters should allow for the control of the hydrogen generation rates in as large a range as possible. In our paper, precise experimental conditions, allowing the tuning of the hydrogen generation process based on the oxidation reaction of porous silicon nanoparticles in aqueous solutions, are reported. The hydrogen generation rate is dependent on chemical composition and the concentration of alkali added in the oxidizing solutions. In particular, the higher the alkali concentration is, the higher the hydrogen generation rate is, while the global amount of the released hydrogen remains constant. The size distribution of the porous silicon nanopowder also affects the generation rate values of produced hydrogen. For example, the smaller the nanoparticle sizes are, the more intense the oxidation reaction and, consequently, the higher the hydrogen generation rates are. Finally, hydrogen release below 0 °C is one of the significant advantages of such a technological way of hydrogen generation in comparison with numerous other developed technologies reported earlier. The reported experimental results confirm a huge technological potential of the hydrogen generation based on the splitting of Si–H bonds of porous silicon nanopowders and water molecules during the oxidation reaction.

Supplementary Materials: The following are available online at http://www.mdpi.com/2079-4991/10/7/1413/s1, Figure S1: Photos of initial partially hydrogenated PSi nanopowder (before its reaction with water) and silicon oxide nanopowder obtained after hydrogen generation from oxidation of the PSi nanopowder in aqueous solutions, Figure S2: Nano-PSi: porosity as a function of the current density for three different HF concentrations (18% (HF:Ethanol = 1:1.66), 24% (1:1), 36% (3:1)), thickness of the porous layer is 7 µm, initial substrate was p-type Si (ρ~1–10 Ω·cm), Figure S3: Meso-PSi: porosity as a function of the current density for three different HF concentrations (18% (HF:Ethanol = 1:1.66), 24% (1:1), 36% (3:1)), thickness of the porous layer is 12 µm, initial substrate was p+- type Si (ρ~10-50 Ω·cm), Figure S4: Nano-PSi: porosity as a function of the current density for three different HF concentrations (18% (HF:Ethanol = 1:1.66), 24% (1:1), 36% (3:1)), thickness of the porous layer is 12 µm, initial substrate was p++- type Si (ρ~1–3 mΩ·cm), Figure S5: Dimentions of silicon nanocrystalls as function of porosity for the meso- and nano-PSi powders, Figure S6: Effusion curves for H2 desorption from the nano- and meso-PSi powders, Figure S7: Photo of the working system used for measurements of hydrogen generation kinetics, Figure S8: Hydrogen generation rates deduced from experimental the curves shown in Figures 5 and 6, Figure S9: Hydrogen amount in terms of %mass (%m) versus the ratio between mass of the oxidizing solution (msol) and mass of meso-PSi nanopowder (mPS). Table S1: Comparison of the hydrogen generation methods, Table S2: Data summarizing the chemical reactions used to produce H2.

Author Contributions: Conceptualization, A.I.M., V.A.S. and V.L.; methodology, S.T., T.N. and S.V.L.; validation, G.M., S.A.A., S.T., T.N. and S.V.L.; formal analysis, all authors; investigation, G.M., S.A.A., S.T., T.N., S.V.L. and V.L.; resources, G.M., G.A., Y.S.; data curation, G.M., S.A.A., T.N. and V.L.; writing—original draft preparation, G.M., S.A.A., A.I.M. and V.L.; writing—review and editing, all authors; visualization, G.M., A.I.M. and T.N.; supervision, Y.S., G.A., S.V.L., V.A.S. and V.L.; project administration, Y.S., G.A. and V.A.S.; funding acquisition, G.A., V.A.S. and V.L. All authors have read and agreed to the published version of the manuscript.

Funding: This research was partially funded by the Committee of Science of the Ministry of Education and Science of the Republic of Kazakhstan, grant number AP05133366, and Ministry of Education and Science of Ukraine, grant number 0119U100326.

Acknowledgments: G.M. is grateful to Al-Farabi Kazakh National University for financial support in the frame of the postdoctoral fellowship program.

Conflicts of Interest: The authors declare no conflict of interest.

References

1. Gandia, L.M.; Arzamendi, G.; Dieguez, P.M. *Renewable Hydrogen Technologies: Production, Purification, Storage, Applications and Safety*; Elsevier: Amsterdam, The Netherlands, 2013; pp. 1–455.
2. Marban, G.; Valdes-Solis, T. Towards the hydrogen economy? *Int. J. Hydrogen Energy* **2007**, *32*, 1625–1637. [CrossRef]

3. Gahleitner, G. Hydrogen from renewable electricity: An international review of power-to-gas pilot plants for stationary applications. *Int. J. Hydrogen Energy* **2013**, *38*, 2039–2061. [CrossRef]

4. Parra, D.; Valverde, L.; Pino, F.J.; Patel, M.K. A review on the role, cost and value of hydrogen energy systems for deep decarbonisation. *Renew. Sustain. Energy Rev.* **2019**, *101*, 279–294. [CrossRef]

5. Buttler, A.; Spliethoff, H. Current status of water electrolysis for energy storage, grid balancing and sector coupling via power-to-gas and power-to-liquids: A review. *Renew. Sustain. Energy Rev.* **2018**, *82*, 2440–2454. [CrossRef]

6. Godula-Jopek, A.; Jehle, W.; Wellnitz, J. *Hydrogen Storage Technologies: New Materials, Transport, and Infrastructure*, 1st ed.; Wiley-VCH Verlag & Co.: Weinheim, Germany, 2012; pp. 1–264.

7. Ahmad, G.E.; El Shenawy, E.T. Optimized photovoltaic system for hydrogen production. *Renew. Energy* **2006**, *31*, 1043–1054. [CrossRef]

8. Kovac, A.; Marciusa, D.; Budin, L. Solar hydrogen production via alkaline water electrolysis. *Int. J. Hydrogen Energy* **2019**, *44*, 9841–9848. [CrossRef]

9. Nakayama, K.; Nishi, M.J.; Taniguchi, T.; Mizusaki, S.; Nagata, Y.; Ozawa, T.C.; Noro, Y.; Samata, H. Catalysts for hydrogen generation from water vapor. *Sci. Technol. Adv. Mater.* **2006**, *7*, 52–55. [CrossRef]

10. Funk, J.E. Thermochemical hydrogen production: Past and present. *Int. J. Hydrogen Energy* **2001**, *26*, 185–190. [CrossRef]

11. Qi, J.; Zhang, W.; Cao, R. Solar-to-hydrogen energy conversion based on water splitting. *Adv. Energy Mater.* **2018**, *8*, 1701620. [CrossRef]

12. Hisatomi, T.; Domen, K. Reaction systems for solar hydrogen production via water splitting with particulate semiconductor photocatalysts. *Nat. Catal.* **2019**, *2*, 387–399. [CrossRef]

13. Matsuoka, M.; Kitano, M.; Takeuchi, M.; Tsujimaru, K.; Anpo, M.; Thomas, J.M. Photocatalysis for new energy production: Recent advances in photocatalytic water splitting reactions for hydrogen production. *Catal. Today* **2007**, *122*, 51–61. [CrossRef]

14. Zhu, J.; Zäch, M. Nanostructured materials for photocatalytic hydrogen production. *Curr. Opin. Colloid Interface Sci.* **2009**, *14*, 260–269. [CrossRef]

15. Gholipour, M.R.; Dinh, C.-T.; Béland, F.; Do, T.-O. Nanocomposite heterojunctions as sunlight-driven photocatalysts for hydrogen production from water splitting. *Nanoscale* **2015**, *7*, 8187–8208. [CrossRef] [PubMed]

16. Ni, M.; Leung, M.K.H.; Leung, D.Y.C.; Sumathy, K. A review and recent developments in photocatalytic water-splitting using TiO_2 for hydrogen production. *Renew. Sustain. Energy Rev.* **2007**, *11*, 401–425. [CrossRef]

17. Basheer, A.A.; Ali, I. Water photo splitting for green hydrogen energy by green nanoparticles. *Int. J. Hydrogen Energy* **2019**, *44*, 11564–11573. [CrossRef]

18. Nagarajan, D.; Lee, D.-J.; Kondo, A.; Chang, J.-S. Recent insights into biohydrogen production by microalgae—From biophotolysis to dark fermentation. *Bioresour. Technol.* **2017**, *227*, 373–387. [CrossRef] [PubMed]

19. Zhao, Y.; Cimpoia, R.; Liu, Z.; Guiot, S.R. Orthogonal optimization of carboxydothermus hydrogenoformans culture medium for hydrogen production from carbon monoxide by biological water-gas shift reaction. *Int. J. Hydrogen Energy* **2011**, *36*, 10655–10665. [CrossRef]

20. Wang, J.; Yin, Y. Fermentative hydrogen production using various biomass-based materials as feedstock. *Renew. Sustain. Energy Rev.* **2018**, *92*, 284–306. [CrossRef]

21. Dou, B.; Zhang, H.; Song, Y.; Zhao, L.; Jiang, B.; He, M.; Ruan, C.; Chen, H.; Xu, Y. Hydrogen production from the thermochemical conversion of biomass: Issues and challenges. *Sustain. Energy Fuels* **2019**, *3*, 314–342. [CrossRef]

22. Eberle, U.; Felderhoff, M.; Schüth, F. Chemical and physical solutions for hydrogen storage. *Angew. Chem. Int. Ed.* **2009**, *48*, 6608–6630. [CrossRef]

23. Wang, H.; Leung, D.C.Y.; Leung, M.K.H.; Ni, M. A review on hydrogen production using aluminum and aluminum alloys. *Renew. Sustain. Energy Rev.* **2009**, *13*, 845–853. [CrossRef]

24. Mahmoodi, K.; Alinejad, B. Enhancement of hydrogen generation rate in reaction of aluminum with water. *Int. J. Hydrogen Energy* **2010**, *35*, 5227–5232. [CrossRef]

25. Fan, M.-Q.; Sun, L.-X.; Xu, F. Experiment assessment of hydrogen production from activated aluminum alloys in portable generator for fuel cell applications. *Energy* **2010**, *35*, 2922–2926. [CrossRef]

26. Bunker, C.E.; Smith, M.J.; Shiral Fernando, K.A.; Harruff, B.A.; Lewis, W.K.; Gord, J.R.; Guliants, E.A.; Phelps, D.K. Spontaneous hydrogen generation from organic-capped Al nanoparticles and water. *ACS Appl. Mater. Interfaces* **2010**, *2*, 11–14. [CrossRef] [PubMed]

27. Xiao, F.; Guo, Y.; Li, J.; Yang, R. Hydrogen generation from hydrolysis of activated aluminum composites in tap water. *Energy* **2018**, *157*, 608–614. [CrossRef]

28. Santos, D.M.F.; Sequeira, C.A.C. Sodium borohydride as a fuel for the future. *Renew. Sustain. Energy Rev.* **2011**, *15*, 3980–4001. [CrossRef]

29. Çakanyıldırım, Ç.; Gürü, M. Production of NaBH$_4$ and hydrogen release with catalyst. *Renew. Energy* **2009**, *34*, 2362–2365. [CrossRef]

30. Gil-San-Millan, R.; Grau-Atienz, A.; Johnson, D.T.; Rico-Francés, S.; Serrano, E.; Linares, N.; García-Martínez, J. Improving hydrogen production from the hydrolysis of ammonia borane by using multifunctional catalysts. *Int. J. Hydrogen Energy* **2018**, *43*, 17100–17111. [CrossRef]

31. Huang, M.; Ouyang, L.; Wang, H.; Liu, J.; Zhu, M. Hydrogen generation by hydrolysis of MgH$_2$ and enhanced kinetics performance of ammonium chloride introducing. *Int. J. Hydrogen Energy* **2015**, *40*, 6145–6150. [CrossRef]

32. Huang, M.; Ouyang, L.; Chen, Z.; Peng, C.; Zhu, X.; Zhu, M. Hydrogen production via hydrolysis of Mg-oxide composites. *Int. J. Hydrogen Energy* **2017**, *42*, 22305–22311. [CrossRef]

33. Tan, Z.H.; Ouyang, L.Z.; Huang, J.M.; Liu, J.W.; Wang, H.; Shao, H.Y.; Zhu, M. Hydrogen generation via hydrolysis of Mg$_2$Si. *J. Alloys Compd.* **2019**, *770*, 108–115. [CrossRef]

34. Ma, M.; Yang, L.; Ouyang, L.; Shao, H.; Zhu, M. Promoting hydrogen generation from the hydrolysis of Mg-Graphite composites by plasma-assisted milling. *Energy* **2019**, *167*, 1205–1211. [CrossRef]

35. Alasmar, E.; Aubert, I.; Durand, A.; Nakhl, M.; Zakhour, M.; Gaudin, E.; Bobet, J.-L. Hydrogen generation from Mg-NdNiMg$_{15}$ composites by hydrolysis reaction. *Int. J. Hydrogen Energy* **2019**, *44*, 523–530. [CrossRef]

36. Stern, A.G. A new sustainable hydrogen clean energy paradigm. *Int. J. Hydrogen Energy* **2018**, *43*, 4244–4255. [CrossRef]

37. Weaver, E.R.; Berry, W.M.; Bohnson, V.L.; Gordon, B.D. *The Ferrosilicon Process for the Generation of Hydrogen*; NACA Technical Report 40; National Bureau of Standards: Washington, DC, USA, 1920; pp. 429–468. Available online: https://ntrs.nasa.gov/search.jsp?R=19930091069 (accessed on 16 July 2020).

38. Auner, N. Silicon as an intermediary between renewable energy and hydrogen. *Dtsch. Bank Res.* **2004**, *11*, 1–11. Available online: https://www.econstor.eu/handle/10419/21871 (accessed on 16 July 2020).

39. Tichapondwa, S.M.; Focke, W.W.; Del Fabbro, O.; Mkhize, S.; Muller, E. Suppressing H$_2$ evolution by silicon powder dispersions. *J. Energetic Mater.* **2011**, *29*, 326–343. [CrossRef]

40. Gevorgyan, A.; Mkrtchyan, S.; Grigoryan, T.; Iaroshenko, V.O. Application of silicon-initiated water splitting for the reduction of organic substrates. *ChemPlusChem* **2018**, 375–382. [CrossRef]

41. Ma, W.; Li, J.; Sun, H.; Chen, J.; Wang, D.; Mao, Z. Robust hydrogen generation over layered crystalline silicon materials via integrated H$_2$ evolution routes. *Int. J. Hydrogen Energy* **2020**. [CrossRef]

42. Zhang, D.; Shi, J.; Zi, W.; Wang, P.; Liu, S. Recent advances in photoelectrochemical applications of silicon materials for solar-to-chemicals conversion. *ChemSusChem* **2017**, *10*, C4324–C4341. [CrossRef]

43. Chandrasekaran, S.; Kaeffer, N.; Cagnon, L.; Aldakov, D.; Fize, J.; Nonglaton, G.; Baleras, F.; Mailley, P.; Artero, V. A robust ALD-protected silicon-based hybrid photoelectrode for hydrogen evolution under aqueous conditions. *Chem. Sci.* **2019**, *10*, 4469–4475. [CrossRef]

44. Lysenko, V.; Bidault, F.; Alekseev, S.; Zaitsev, V.; Barbier, D.; Turpin, C.; Geobaldo, F.; Rivolo, P.; Garrone, E. Study of porous silicon nanostructures as hydrogen reservoirs. *J. Phys. Chem. B* **2005**, *109*, 19711–19718. [CrossRef] [PubMed]

45. Litvinenko, S.; Alekseev, S.; Lysenko, V.; Venturello, A.; Geobaldo, F.; Gulina, L.; Kuznetsov, G.; Tolstoy, V.; Skryshevsky, V.; Garrone, E.; et al. Hydrogen production from nano-porous Si powder formed by stain etching. *Int. J. Hydrogen Energy* **2010**, *35*, 6773–6778. [CrossRef]

46. Goller, B.; Kovalev, D.; Sreseli, O. Nanosilicon in water as a source of hydrogen: Size and pH matter. *Nanotechnology* **2011**, *22*, 305402. [CrossRef]

47. Imamura, K.; Kimura, K.; Fujie, S.; Kobayashi, H. Hydrogen generation from water using Si nanopowder fabricated from swarf. *J. Nanopart. Res.* **2016**, *18*, 116. [CrossRef]

48. Imamura, K.; Kobayashi, Y.; Matsuda, S.; Akai, T.; Kobayashi, H. Reaction of Si nanopowder with water investigated by FT-IR and XPS. *AIP Adv.* **2017**, *7*, 085310. [CrossRef]

49. Kobayashi, Y.; Matsuda, S.; Imamura, K.; Kobayashi, H. Hydrogen generation by reaction of Si nanopowder with neutral water. *J. Nanopart. Res.* **2017**, *19*, 176. [CrossRef]
50. Erogbogbo, F.; Lin, T.; Tucciarone, P.M.; LaJoie, K.M.; Lai, L.; Patki, G.D.; Prasad, P.N.; Swihart, M.T. On-demand hydrogen generation using nanosilicon: Splitting water without light, heat, or electricity. *Nano Lett.* **2013**, *13*, 451–456. [CrossRef]
51. Allongue, P.; Henry de Villeneuve, C.; Bernard, M.C.; Péou, J.E.; Boutry-Forveille, A.; Lévy-Clément, C. Relationship between porous silicon formation and hydrogen incorporation. *Thin Solid Films* **1997**, *297*, 1–4. [CrossRef]
52. Martín, P.; Fernández, J.F.; Sánchez, C. Hydrogen surface coverage of as-prepared nanocrystalline porous silicon. *Mater. Sci. Eng. B* **2004**, *108*, 166–170. [CrossRef]
53. Kale, P.; Gangal, A.C.; Edla, R.; Sharma, P. Investigation of hydrogen storage behavior of silicon nanoparticles. *Int. J. Hydrogen Energy* **2012**, *37*, 3741–3747. [CrossRef]
54. Manilov, A.I.; Litvinenko, S.V.; Alekseev, S.A.; Kuznetsov, G.V.; Skryshevsky, V.A. Use of powders and composites based on porous and crystalline silicon in the hydrogen power industry. *Ukr. J. Phys.* **2010**, *55*, 928–935.
55. Manilov, A.I.; Alekseev, S.A.; Skryshevsky, V.A.; Litvinenko, S.V.; Kuznetsov, G.V.; Lysenko, V. Influence of palladium particles impregnation on hydrogen behavior in meso-porous silicon. *J. Alloys Compd.* **2010**, *492*, 466–472. [CrossRef]
56. Zhan, C.; Chu, P.K.; Ren, D.; Xin, Y.; Huo, K.; Zou, Y.; Huang, N.K. Release of hydrogen during transformation from porous silicon to silicon oxide at normal temperature. *Int. J. Hydrogen Energy* **2011**, *36*, 4513–4517. [CrossRef]
57. Pastushenko, A.; Litvinenko, S.; Lysenko, V.; Skryshevsky, V. Self-regulated hydrogen generation with the use of nano-powders: Application for portable fuel cells. *Phys. Sci. Technol.* **2018**, *5*, 50–59. [CrossRef]
58. Dai, F.; Zai, J.; Yi, R.; Gordin, M.L.; Sohn, H.; Chen, S.; Wang, D. Bottom-up synthesis of high surface area mesoporous crystalline silicon and evaluation of its hydrogen evolution performance. *Nat. Commun.* **2014**, *5*, 3605. [CrossRef]
59. Liu, D.; Ma, J.; Long, R.; Gao, C.; Xiong, Y. Silicon nanostructures for solar-driven catalytic applications. *Nano Today* **2017**, *17*, 96–116. [CrossRef]
60. Chandrasekaran, S.; Nann, T.; Voelcker, N.H. Silicon nanowire photocathodes for photoelectrochemical hydrogen production. *Nanomaterials* **2016**, *6*, 144. [CrossRef]
61. Bahruji, H.; Bowker, M.; Davies, P.R. Photoactivated reaction of water with silicon nanoparticles. *Int. J. Hydrogen Energy* **2009**, *34*, 8504–8510. [CrossRef]
62. Mathews, N.R.; Sebastian, P.J.; Mathew, X.; Agarwal, V. Photoelectrochemical characterization of porous Si. *Int. J. Hydrogen Energy* **2003**, *8*, 629–632. [CrossRef]
63. Song, H.; Liu, D.; Yang, J.; Wang, L.; Xu, H.; Xiong, Y. Highly crystalline mesoporous silicon spheres for efficient visible photocatalytic hydrogen evolution. *ChamNanoMat* **2016**, *3*, 22–26. [CrossRef]
64. Ali, M.; Starkov, V.V.; Gosteva, E.A.; Druzhinin, A.V.; Sattar, S. Water splitting using porous silicon photo-electrodes for hydrogen production. *J. Phys. Conf. Ser.* **2017**, *917*, 052008. [CrossRef]
65. Barabash, R.N.; Alekseev, S.A.; Zaitsev, V.N.; Barbier, D. Oxidation resistance of porous silicon and modification of porous silicon by vinylsilanes. *Ukr. Khim. Zh.* **2006**, *72*, 78–84.
66. Manilov, A.I. Problems of application of porous silicon to chemical and photocatalytic productions of hydrogen. *Ukr. J. Phys.* **2016**, *61*, 233–239. [CrossRef]
67. Halimaoui, A. Porous silicon formation by anodisation. In *Properties of Porous Silicon*; Canham, L., Ed.; INSPEC, The IEE: London, UK, 1997; pp. 2–12.
68. Nychyporuk, T.; Lysenko, V.; Barbier, D. Fractal nature of porous silicon nanocrystallites. *Phys. Rev. B* **2005**, *71*, 115402. [CrossRef]

Article

PMMA Thin Film with Embedded Carbon Quantum Dots for Post-Fabrication Improvement of Light Harvesting in Perovskite Solar Cells

Askar A. Maxim [1,†], Shynggys N. Sadyk [2,†], Damir Aidarkhanov [1], Charles Surya [3], Annie Ng [1], Yoon-Hwae Hwang [4], Timur Sh. Atabaev [2,*] and Askhat N. Jumabekov [5,*]

[1] Department of Electrical and Computer Engineering, Nazarbayev University, Nur-Sultan 010000, Kazakhstan; askar.maxim@nu.edu.kz (A.A.M.); aidarkhanov@nu.edu.kz (D.A.); annie.ng@nu.edu.kz (A.N.)
[2] Department of Chemistry, Nazarbayev University, Nur-Sultan 010000, Kazakhstan; shynggys.sadyk@nu.edu.kz
[3] Nazarbayev University, Nur-Sultan 010000, Kazakhstan; charles.surya@nu.edu.kz
[4] Department of Nanoenergy Engineering and BK21 PLUS Nanoconvergence Technology Division, Pusan National University, Busan 46241, Korea; yhwang@pusan.ac.kr
[5] Department of Physics, Nazarbayev University, Nur-Sultan 010000, Kazakhstan
* Correspondence: timur.atabaev@nu.edu.kz (T.S.A.); askhat.jumabekov@nu.edu.kz (A.N.J.)
† These authors contributed equally.

Received: 2 December 2019; Accepted: 4 February 2020; Published: 9 February 2020

Abstract: Perovskite solar cells (PSCs) with a standard sandwich structure suffer from optical transmission losses due to the substrate and its active layers. Developing strategies for compensating for the losses in light harvesting is of significant importance to achieving a further enhancement in device efficiencies. In this work, the down-conversion effect of carbon quantum dots (CQDs) was employed to convert the UV fraction of the incident light into visible light. For this, thin films of poly(methyl methacrylate) with embedded carbon quantum dots (CQD@PMMA) were deposited on the illumination side of PSCs. Analysis of the device performances before and after application of CQD@PMMA photoactive functional film on PSCs revealed that the devices with the coating showed an improved photocurrent and fill factor, resulting in higher device efficiency.

Keywords: Perovskite solar cell; PMMA; carbon quantum dots; down-conversion; light harvesting

---

1. Introduction

Hybrid organic–inorganic lead halide perovskites are considered to be one of the most promising materials for photovoltaic (PV) applications. In just a decade, the power conversion efficiency (PCE) of perovskite solar cells (PSCs) has improved 6-fold, reaching remarkable values of 25.2% in 2019 [1]. This is the result of tremendous efforts spent on optimization of the device fabrication process, careful tailoring of material properties used for device functional layers (perovskite photo-absorber layer, electron- and hole-transporting layers (ETL and HTL, respectively), and device structure (mesoporous, planar *p-i-n*, and planar *n-i-p*)) [2,3]. Theoretical calculations suggest that PSCs have the potential to reach efficiencies as high as ~31% [3]. Approaching the theoretical performance limit would require even further development and optimization in all aspects of device fabrication. Light management for effective utilization of incident light would be one of them.

One of the drawbacks of PSCs in terms of their light harvesting ability is the sandwich structure of the devices, which induces parasitic light absorption by the substrate (fluorine- or indium-doped tin-oxide-coated glass (glass/FTO or glass/ITO, respectively)) and other functional layers, such as ETL and HTL [4]. For instance, most of the UV light is absorbed by the ETL made of wide band gap materials, such as TiO_2 and SnO_2, or by the HTL (e.g., NiO_x in devices with an inverted structure). This, on the one hand, is beneficial for the perovskite photo-absorber layer, since it prevents degradation of the perovskite material under UV illumination. However, on the other hand, absorption of UV light by the substrate and charge-transporting layers does not allow for an effective utilization of the incident light. One way to overcome this issue without changing the structural composition of PSCs is to employ fluorescent materials with down-converting properties. These are materials that can absorb photons with higher energy (e.g., in the near-UV range) and re-emit them with lower energy (e.g., in the visible light range). Application of such functional materials to improve the light harvesting ability of the photo-absorber layer in PSCs can afford enhanced charge carrier generation and improved device performance.

Carbon quantum dots (CQDs) with excellent fluorescent properties absorb UV light and re-emit in the visible light range. Such a unique optical property of CQDs combined with their low-cost synthesis, non-toxicity, and chemical inertness make them very attractive for electronics, sensing, and bioimaging applications [5–7]. For example, carbon and graphene quantum dots were utilized recently for light management in silicon-based solar cells [8,9]. In both cases, the authors reported enhancement in the performance of silicon-based solar cells, owing to the carbon/graphene quantum dots coatings that emit additional down-converted photons in the visible spectrum. A recent study reported that CQDs with a down-conversion property introduced into a mesoporous TiO_2 film in PSCs can improve the stability (convert UV radiation to visible light) as well as the PCE of the devices [10]. Therefore, by considering the favorable down-conversion optical effects, we embedded the CQDs in a thin poly(methyl methacrylate) (PMMA) film (CQD@PMMA) and used it as an additional coating layer for PSC devices. This coating will be able to convert the UV part of the incident light to visible light, thus allowing for more light to be transmitted to the perovskite photo-absorber layer (see Figure 1).

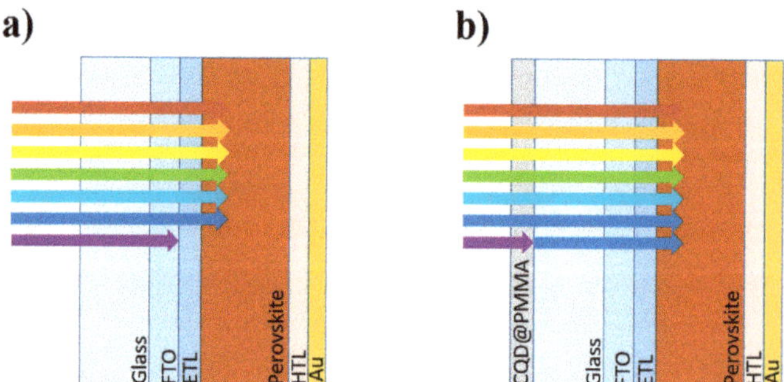

Figure 1. Schematic image of perovskite solar cells (PSCs) (**a**) without and (**b**) with the poly(methyl methacrylate) with embedded carbon quantum dots (CQD@PMMA) coating used for improving light harvesting in a PSC due to the down-conversion effect.

2. Materials and Methods

2.1. Materials

High-purity materials were purchased from Alfa Aesar or Sigma-Aldrich and used as received. The FTO-coated glass substrates (~7 ohm/sq) were obtained from OPV-tech (Yingkou City, China). The organic precursors for synthesis of perovskite thin films were obtained from the GreatCell Solar.

2.2. Synthesis of CQDs

Phosphorus-doped CQDs were prepared according to a previously reported protocol [7]. Briefly, 0.1 g of Na_2HPO_4 and 1 g of dextrose were dissolved in 25 mL of deionized water. Next, the solution was placed into an Erlenmeyer flask (50 mL capacity) with a screw cap, closed, and kept at 200 °C under vigorous stirring for 1 h. The prepared black solution was passed through a 0.1 μm filter, and samples were purified and collected using a lyophilization process.

2.3. Preparation of CQD@PMMA Solution

As-prepared CQDs with sizes around 3–7 nm were mixed with 5 wt% PMMA in chlorobenzene (10 mg of CQDs in 10 mL of 5 wt% PMMA/chlorobenzene). The resulting mixture was sonicated for 30 min in an ultrasonic bath, and then stirred vigorously for 48 hrs at 40 °C to disperse CQDs and obtain the final brownish-colored CQD@PMMA solution.

2.4. Fabrication of PSCs

A glass/FTO/ETL/perovskite/HTL/gold architecture was used to fabricate PSCs. The FTO-coated glass substrates were cleaned by consecutive sonication in the detergent, deionized water (DI), and organic solvents. The substrates were then dried under a nitrogen flow and treated by UV-ozone for 30 min. The ETL was composed of a bilayer structure formed by consecutive depositions of SnO_2 quantum dot (QD-SnO_2) and SnO_2 nanoparticles (NP-SnO_2) as described below. The 0.15 M colloidal SnO_2 quantum dot (QD-SnO_2) solution was prepared by using $SnCl_2 \cdot 2H_2O$ as the precursor dissolved in 30 mL of DI water, and left overnight under vigorous stirring in an open beaker. The prepared solution was spin coated at 3000 rpm for 30 s and thermally annealed at 150 °C for 10 min and 200 °C for 60 min. SnO_2 nanoparticles (NP-SnO_2) were prepared by using SnO_2 colloidal dispersion solution. The prepared solution was spin coated at 3000 rpm for 30 s, and the samples were annealed at 150 °C for 30 min. The perovskite precursor solution was prepared by forming 1.1 M PbI_2, 1 M formamidinium iodide (CH_5IN_2), 0.2 M $PbBr$, and 0.2 M methylammonium bromide (CH_3NH_3Br) in the mixture of dimethyl formamide (DMF) and dimethyl sulfoxide (DMSO) (4:1 volume ratio). Then, CsI (1.5 M stock solution) was added into the precursor solution in a volume ratio of 8:92 to form 1 mL of final perovskite precursor solution. The perovskite thin film was prepared in a nitrogen-filled glove box by a four-step process: (i) spin coating of the precursor solution at 2000 rpm for 30 s; (ii) cryogenic treatment; (iii) blow-drying under a nitrogen gas flow for 60 s; and (iv) thermal annealing at 105 °C for 30 min on a hotplate. The precursor solution for the hole transport layer (HTL) was prepared by dissolving 80 mg of 2,2′,7,7′-Tetrakis [N,N-di(4-methoxyphenyl)amino]-9,9′-spirobifluorene (Spiro-OMeTAD), 954 μL of chlorobenzene, 29 μL of 4-tert-Butylpyridine, and 17.5 μL of Bis(trifluoromethane)sulfonimide lithium salt in acetonitrile with a concentration of 520 mg/mL. The prepared solution was spin coated at 3500 rpm for 30 s. Gold electrodes (70 nm) were thermally evaporated through a shadow mask at a vacuum pressure of 10^{-7} Torr to complete fabrication of PSCs (Glass/FTO/SnO_2/$Cs_{0.05}(MA_{0.17}FA_{0.83})_{0.95}Pb(I_{0.84}Br_{0.16})_3$/spiroMeOTAD/Au) [11].

2.5. Deposition of CQD@PMMA Coating on PSCs

In order to deposit CQD@PMMA coating on PSCs, freshly prepared CQD@PMMA solution was spin coated (30 s under various spinning speed conditions) on the glass slide of already fabricated PSC devices at room temperature. For each deposition speed, at least 15 devices were tested. The estimated number of CQD particles per 1×1 cm^2 geometric area is ~1.48×10^{12}.

2.6. Characterization

A transmission electron microscope (TEM) (JEM 2010F, JEOL Ltd., Tokyo, Japan) was used for morphology and size analysis of CQDs. Fluorescent measurements were performed using a Cary Eclipse Fluorescence Spectrophotometer (Agilent Technologies Inc., Santa Clara, CA, USA). Transmittance measurements were performed using a Lambda 1050 spectrophotometer (Perkin Elmer Inc., Waltham, MA, USA) equipped with an integrating sphere. The J–V characteristics of the devices were measured by using a Keithley 2400 sourcemeter (Tektronix Inc., Beaverrton, OR, USA) under an AAA class ORIEL Sol3A solar simulator (Newport Corporation, Irvine, CA, USA) equipped with an Air Mass (AM) 1.5 filter at 100 mW/cm^2. The light intensity was calibrated by a Si reference cell from Newport (Newport Corporation, Irvine, CA, USA).

3. Results and Discussion

The TEM analysis of CQD samples confirmed a narrow size distribution in the range of 3–7 nm (Figure 2a). The down-converting property of the CQD@PMMA films was tested using fluorescence spectroscopy. For this, freshly made CQD@PMMA solution was spin coated on thin microscope glass slides and fluorescence emission spectra of the samples were recorded. The emission spectra of the coatings were tested at the excitation wavelength of 320 nm. As shown in Figure 2b, the emission spectra of a CQD@PMMA film shows a distinct peak centered at 430–440 nm, which is attributed to a typical emission range of CQDs [7]. The emission spectra of a neat PMMA film deposited on a thin microscope glass slide were also recorded for comparison. The absence of an emission peak in the neat PMMA film indicates that the emission peak in the CQD@PMMA film originates from the CQDs embedded in the PMMA matrix.

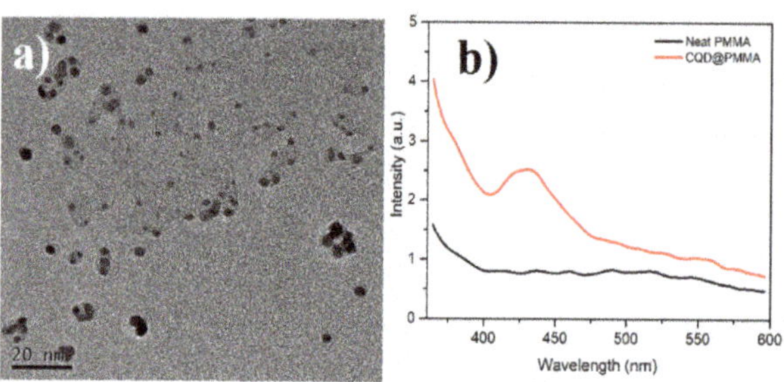

Figure 2. (**a**) TEM image of prepared carbon quantum dots (CQDs), and (**b**) fluorescence emission of neat poly(methyl methacrylate) (PMMA) and CQD@PMMA coatings deposited on thin microscope glass slides.

Figure 3 shows the transmittance spectrum of neat PMMA and CQD@PMMA films deposited on thin microscope glass slides for the measurement range of 300–800 nm. The CQD@PMMA film has lower transmittance in the near-UV region (<350 nm) compared to the neat PMMA film, which suggests absorption of UV light by the CQDs embedded in the PMMA matrix.

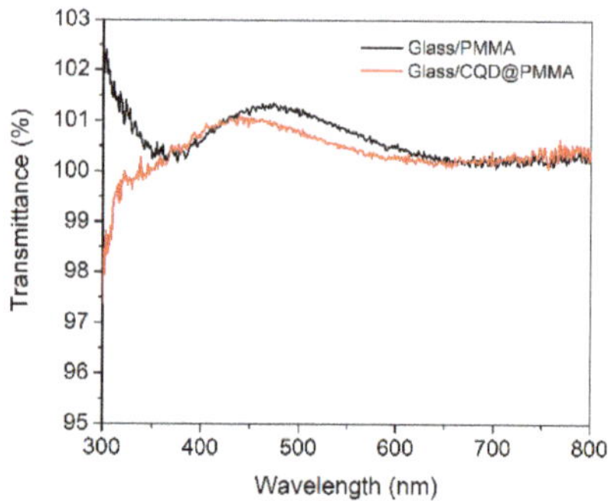

Figure 3. Transmittance of neat PMMA and CQD@PMMA coatings on thin microscope glass slides.

In the next stage, CQD@PMMA coatings were applied to PSCs in order to evaluate their ability to enhance the light harvesting efficiency of the devices. For this, a PMMA/chlorobenzene mixture with dispersed CQDs was spin-coated (30 s at 3000 rpm) on the illumination side of prefabricated PSCs. The thickness of the CQD@PMMA coating was around 480 nm (Figure 4a). To evaluate the effect of CQD@PMMA coating on device performance, J–V characteristics of PSCs before and after their application were measured. Figure 4b shows a comparison of J–V curves, both forward and reverse, for devices before and after application of a CQD@PMMA coating. The active area of the device was

0.075 cm^2. The *J–V* curves show that the device with the CQD@PMMA coating resulted in a higher photocurrent (J_{SC}) compared to the device without the CQD@PMMA coating. The relative increase for J_{SC}, which was around 3%, was consistent with the transmittance measurements presented in Figure 3. The *J–V* curves also show that there was a notable enhancement (around a 6% relative increase) in fill factor (*FF*), and some minor improvements (around a 1% relative increase) in open-circuit voltage (V_{OC}) after the application of the CQD@PMMA layer on the device (see Table 1).

Control experiments were also performed, in which the effect of neat PMMA coating on the light harvesting ability of PSCs was investigated. For this, we obtained the *J–V* characteristics of PSCs before and after the application of a neat PMMA coating (deposited via spin-coating for 30 s at 3000 rpm). This was done in order to separate and estimate the effect of the PMMA layer alone on the performance of the devices. Comparison of the *J–V* curves for a device before and after application of the neat PMMA coating shows that the PMMA coating alone had a negligible or no effect on the performance of the device (Figure 4c). This indicates that the device performance enhancement observed in the case of CQD@PMMA coating indeed originates from the CQDs embedded in the PMMA matrix—i.e., due to the down-conversion effect of CQDs, effectively converting UV to visible light, affording more photons to pass into the perovskite photo-absorber layer, and, thus, generating more charge carriers.

The external quantum efficiency (EQE) measurements were also performed on PSCs to obtain their optical spectral response before and after application of CQD@PMMA coating. Figure 4c shows EQE spectra of a device measured before and after application of a CQD@PMMA coating. The comparison of EQE spectra indicates that after deposition of the CQD@PMMA coating, the EQE response of the device in the blue region has improved [8–10]. The calculated J_{SC} from EQE spectra obtained before and after deposition of the CQD@PMMA coating results in values of 16.31 mA cm^{-2} and 17.14 mA cm^{-2}, respectively, which exhibits the consistent improvement in J_{SC} for the device after deposition of the CQD@PMMA coating. It is noteworthy that an ~26%–28% discrepancy was observed between the calculated J_{SC} from the EQE spectra and the measured J_{SC} from the solar simulator. The underestimation of EQE commonly happens in PSCs. This can be due to the barrier for the photocurrent of PSC becoming dominant under the monochromatic illumination with low light intensity during the EQE measurement, while the barrier will be alleviated by photodoping of the perovskite under the strong illumination. Nevertheless, the enchantment trend, as determined from EQE, is consistent with the trend of the *J–V* characteristics measured under 1 sun illumination.

Finally, we investigated the dependence of the PSC performance on the thickness of the CQD@PMMA coating. For this, we deposited a CQD@PMMA coating on the glass side of PSCs at various spinning speeds. The performance of the devices was measured before and after application of the CQD@PMMA coating, and relative changes in J_{SC}, *FF*, and V_{OC} and the efficiency of the devices were investigated. The relative change for photovoltaic parameters is given as a percentage; i.e., it indicates to what extent a certain photovoltaic parameter is changed (improved or worsened) after the deposition of CQD@PMMA coating.

Figure 5 shows the dependence of relative change in the photovoltaic parameters on the spinning speed used to deposit CQD@PMMA coating on PSCs. In order to verify the reproducibility of the PSCs, at least 15 devices for each spinning speed condition were examined. Similar ranges of deviations in photovoltaic parameters of the devices were observed among all spinning conditions, which were predicted due to the high sensitivity of the device performance to the fabrication ambient. Nevertheless, the trend of the device performance with varying thicknesses of the CQD@PMMA coating can clearly be observed. Figure 5a,b show the device performance parameters extracted from *J–V* curves measured in forward and reverse scan directions, respectively. The curves for device efficiency for forward and reverse scan directions show that there is a gradual increase from a lower spinning speed up to 3000 rpm, and then it drops. This trend, especially for forward scan directions (see Figure 5a), originates from a similar behavior in relative changes of J_{SC} and *FF*. The initial increase up to 3000 rpm and the following decrease in the relative change of J_{SC} in Figure 5a,b can be attributed to the thickness of the CQD@PMMA coating. Thicker CQD@PMMA coating has many CQD particles, which leads to transmission losses, while a CQD@PMMA coating that is too thin is not effective in converting UV into visible light. Based on this assumption, we can cautiously suggest that 3000 rpm for deposition of CQD@PMMA coating, yielding a thickness of around 480 nm, is the optimal value for spinning speed. Meanwhile, a similar dependence of *FF* on spinning speed—i.e., an initial increase up to 3000 rpm and a subsequent decrease at a higher spinning speed—for both scan directions is not as intuitive to explain. However, a similar trend was reported by Jin et al. [10] in related studies. A possible explanation for this phenomenon could be the UV-filtering effect of CQDs, which prevents the perovskite layer from photodegradation. However, a more detailed investigation of this phenomenon is required and will be the subject of future studies. The relative changes in V_{OC} for both scanning directions, however, show very subtle improvements after deposition of CQD@PMMA coating, and slowly decrease with a higher spinning speed, indicating a negligible relative change in V_{OC} for thinner CQD@PMMA layers.

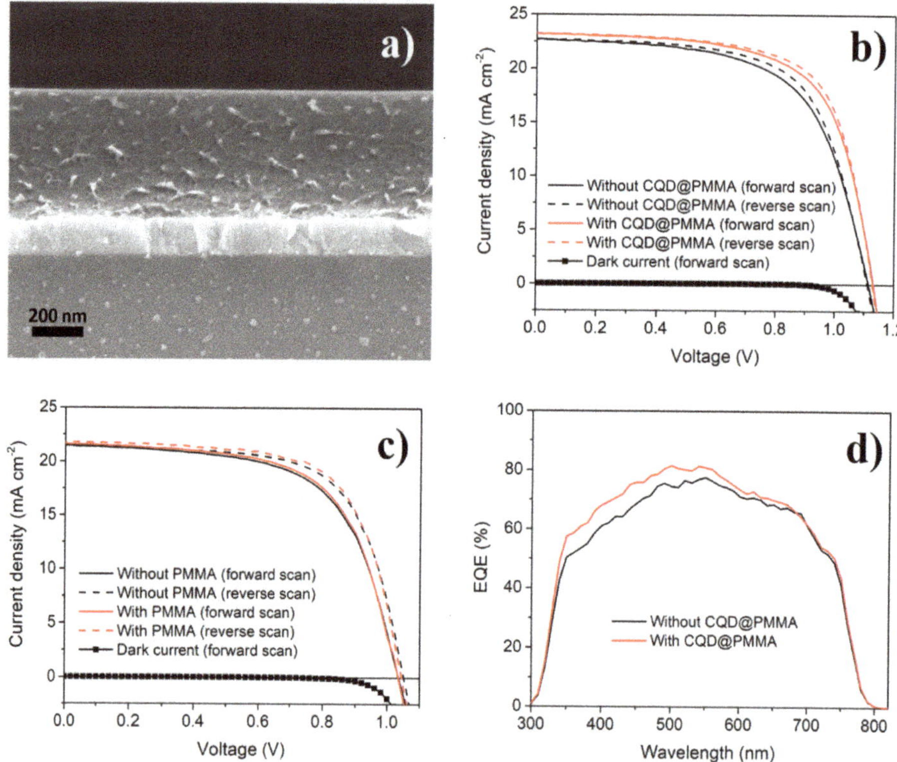

Figure 4. (**a**) Cross-sectional SEM image of a CQD@PMMA coating deposited onto a fluorine-doped tin oxide (FTO) substrate at 3000 rpm. (**b**) *J–V* characteristics (scan rate 0.5 V s^{-1}) of a PSC device before and after application of a CQD@PMMA coating tested under 1 sun illumination and in the dark (blue solid line). (**c**) *J–V* curves of a PSC device before and after application of a neat PMMA layer. (**d**) External quantum efficiency (EQE) spectra of a PSC before and after application of a CQD@PMMA layer.

Table 1. Photovoltaic parameters for a PSC device before and after application of a CQD@PMMA layer tested under 1 sun illumination.

CQD@PMMA Layer	Scan Direction	J_{SC} (mA cm^{-1})	V_{OC} (V)	FF (%)	Efficiency (%)
Without	forward	22.69	1.11	61.98	15.67
Without	reverse	22.63	1.11	65.29	16.4
With	forward	23.25	1.13	65.76	17.29
With	reverse	23.21	1.13	68.19	17.86

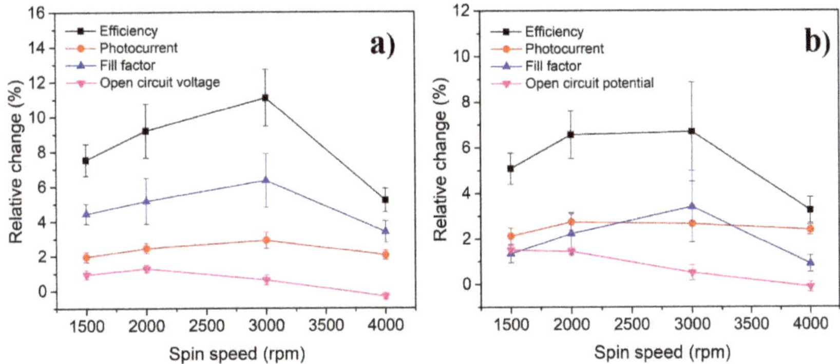

Figure 5. Dependence of relative change in the photovoltaic parameters on the spinning speed used to deposit a CQD@PMMA coating on PSCs. Key photovoltaic parameters extracted from *J–V* curves measured in (**a**) forward and (**b**) reverse scan directions.

4. Conclusions

In conclusion, we developed an easy-to-handle, low-cost, and solution-processable method to fabricate photoactive functional coating for sandwich-structured PSCs to enhance their light harvesting ability and improve device performance. The photoactive functional coating consists of a PMMA matrix embedded with CQDs that possess down-conversion fluorescent properties. The application of this coating on PSC devices showed an improved photocurrent and fill factor, resulting in an overall performance increase. The method is simple and straightforward to apply and can afford further fine-tuning and/or modification to fabricate functional thin films with the desired properties and morphology. In addition, this low-cost and solution-based method offers ease of scale-up of fabrication, which makes it very attractive for future industrial manufacturing.

Author Contributions: Conceptualization, A.N.J. and T.S.A.; methodology, A.N.J., T.S.A. and A.N.; investigation, A.A.M., S.N.S., D.A. and A.N.J.; resources, C.S., A.N., Y.-H.H., T.S.A. and A.N.J.; writing—original draft preparation, A.A.M., S.N.S, and A.N.J.; writing—review and editing, C.S., A.N., Y.-H.H., T.S.A. and A.N.J.; visualization, A.N.J.; supervision, A.N.J.; All authors have read and agreed to the final version of the manuscript.

Funding: This research was funded by Nazarbayev University, grant number 110119FD4512, 090118FD5326, 110119FD4506 and Ministry of Education and Science, Kazakhstan, grant number BR05236524.

Acknowledgments: A.N.J. acknowledges the financial support from the FDCRG (Grant Number: 110119FD4512) and SPG fund (Nazarbayev University). C.S. thanks NU (Grant Number: 090118FD5326) for the support. A.N. acknowledges the grant 110119FD4506 (Nazarbayev University) and the targeted program BR05236524 funded by Ministry of Education and Science, Kazakhstan. The authors thank Rakhima Shamenova from Nazarbayev University for her help with SEM imaging.

Conflicts of Interest: The authors declare no conflict of interest.

References

1. Best Research-Cell Efficiencies. Available online: https://www.nrel.gov/pv/assets/pdfs/best-research-cell-efficiencies.20190802.pdf (accessed on 4 February 2020).
2. Rajagopal, A.; Yao, K.; Jen, A.K. Toward perovskite solar cell commercialization: A perspective and research roadmap based on interfacial engineering. *Adv. Mater.* **2018**, *30*, 1800455. [CrossRef] [PubMed]
3. Correa-Baena, J.P.; Saliba, M.; Buonassisi, T.; Gratzel, M.; Abate, A.; Tress, W.; Hagfeldt, A. Promises and challenges of perovskite solar cells. *Science* **2017**, *358*, 739–744. [CrossRef] [PubMed]
4. Wang, D.L.; Cui, H.J.; Hou, G.J.; Zhu, Z.G.; Yan, Q.B.; Su, G. Highly efficient light management for perovskite solar cells. *Sci. Rep.* **2016**, *6*, 18922. [CrossRef] [PubMed]
5. Semeniuk, M.; Yi, Z.; Poursorkhabi, V.; Tjong, J.; Jaffer, S.; Lu, Z.H.; Sain, M. Future perspectives and review on organic carbon dots in electronic applications. *ACS Nano* **2019**, *13*, 6224–6255. [CrossRef] [PubMed]
6. Molaei, M.J. Carbon quantum dots and their biomedical and therapeutic applications: A review. *RSC Adv.* **2019**, *9*, 6460–6481. [CrossRef]
7. Molkenova, A.; Atabaev, T.S. Phosphorus-doped carbon dots (P-CDs) from dextrose for low-concentration ferric ions sensing in water. *Optik* **2019**, *187*, 70–73. [CrossRef]
8. Dai, D.; Tu, X.; Li, X.; Lv, T.; Han, F. Tuning solar absorption spectra via carbon quantum dots/VAE composite layer and efficiency enhancement for crystalline Si solar module. *Prog. Photovolt. Res. Appl.* **2019**, *27*, 283–289. [CrossRef]
9. Tsai, M.L.; Tu, W.C.; Tang, L.; Wei, T.C.; Wei, W.R.; Lau, S.P.; Chen, L.J.; He, J.H. Efficiency enhancement of silicon heterojunction solar cells via photon management using graphene quantum dots as downconverters. *Nano Lett.* **2016**, *16*, 309–313. [CrossRef] [PubMed]
10. Jin, J.; Chen, C.; Li, H.; Cheng, Y.; Xu, L.; Dong, B.; Song, H.; Dai, Q. Enhanced performance and photostability of perovskite solar cells by introduction of fluorescent carbon dots. *ACS Appl. Mater. Interfaces* **2017**, *9*, 14518–14524. [CrossRef] [PubMed]
11. Ng, A.; Ren, Z.; Hu, H.; Fong, P.W.K.; Shen, Q.; Cheung, S.H.; Qin, P.; Lee, J.W.; Djurišić, A.B.; So, S.K.; et al. A cryogenic process for antisolvent-free high-performance perovskite solar cells. *Adv. Mater.* **2018**, *30*, 1804402. [CrossRef] [PubMed]

 *nanomaterials*

Article

White Light Emission by Simultaneous One Pot Encapsulation of Dyes into One-Dimensional Channelled Aluminophosphate

Rebeca Sola-Llano [1], Ainhoa Oliden-Sánchez [1], Almudena Alfayate [2], Luis Gómez-Hortigüela [2], Joaquín Pérez-Pariente [2], Teresa Arbeloa [1], Johan Hofkens [3], Eduard Fron [3,*] and Virginia Martínez-Martínez [1,*]

[1] Departamento de Química Física, Universidad del País Vasco, UPV/EHU, Apartado 644, 48080 Bilbao, Spain; rebeca.sola@ehu.eus (R.S.-L.); ainhoa.oliden@ehu.eus (A.O.-S.); teresa.arbeloa@ehu.eus (T.A.)

[2] Instituto de Catálisis y Petroleoquímica-CSIC, C/Marie Curie 2, Cantoblanco, 28049 Madrid, Spain; a.alfayatelanza@gmail.com (A.A.); lhortiguela@icp.csic.es (L.G.-H.); jperez@icp.csic.es (J.P.-P.)

[3] Department of Chemistry, Katholieke Universiteit Leuven, Celestijnenlaan 200F, B-3001 Heverlee, Belgium; johan.hofkens@kuleuven.be

* Correspondence: eduard.fron@kuleuven.be (E.F.); virginia.martinez@ehu.eus (V.M.-M.); Tel.: +34-94-601-5969 (V.M.-M.)

Received: 12 May 2020; Accepted: 9 June 2020; Published: 16 June 2020

Abstract: By simultaneous occlusion of rationally chosen dyes, emitting in the blue, green and red region of the electromagnetic spectrum, into the one-dimensional channels of a magnesium-aluminophosphate with AEL-zeolitic type structure, MgAPO-11, a solid-state system with efficient white light emission under UV excitation, was achieved. The dyes herein selected—acridine (AC), pyronin Y (PY), and hemicyanine LDS722—ensure overall a good match between their molecular sizes and the MgAPO-11 channel dimensions. The occlusion was carried out via the crystallization inclusion method, in a suitable proportion of the three dyes to render efficient white fluorescence systems by means of fine-tuned FRET (fluorescence resonance energy transfer) energy transfer processes. The FRET processes are thoroughly examined by the analysis of fluorescence decay traces using the femtosecond fluorescence up-conversion technique.

Keywords: dye-guest encapsulation; zeolite; microporous aluminophosphates; one-pot synthesis; hybrid fluorescent system; white light emitter; FRET

---

1. Introduction

Hybrid organic-inorganic materials are recognized as promising systems for many applications in different fields such as optoelectronics, energy, and biomedicine [1–5]. Generally, the final properties are not a mere sum of the individual contributions of each moiety; new and improved features also emerge as a result of synergetic effects [1,5,6]. In this way, new materials with interesting features for innovative technologies can be produced.

White light emitters are one of those potential materials that are extremely interesting for lightning systems, necessary in technological devices, and present in everyday life. Hitherto, white light emission has been largely pursued to improve existing lightning technology [7–15]. In this regard, several approaches have been postulated. In particular, fluorescence resonance energy transfer, FRET, is a bioinspired process to develop new systems as light-emitting devices [16–18]. The control of energy transfer processes with the combination of proper dyes allows to tune the color of the resultant emitted light. The most common strategy to achieve white light emission consists of combining fluorophores with emission spectra in the three fundamental regions (blue, green, and red), emitting at well-defined intensities into a single host system able to impose a proper spatial distribution regarding intermolecular distances and molecular orientations [8,19–23]. Note that although purely organic compounds can allow a low-cost fabrication of white emitter devices, their major drawback is the aging, usually at different rates for each component [23,24]. However, encapsulated organic dyes within nanocavities, preferably inorganic frameworks, are sheltered from chemical-, photo- and thermo-degradation due to the protection provided by the porous host [25,26]. Microporous zeolitic materials are ideal candidates to be used as inorganic stable hosts, in particular microporous aluminophosphates based on AlO_4 and PO_4 alternate tetrahedral units [26].

Generally, optical applications based on the combination of organic fluorescent dyes with inorganic hosts require optically dense materials. In such cases, molecular aggregation should be completely avoided in order to preserve the optical characteristics of the components. The key to success in the design of these hybrid materials is the matching between dye molecular size and nanochannel dimensions. In this context, the best synthetic approach to reach a tight host-guest fit is via the crystallization inclusion method, in which the dye is embedded while the inorganic framework is being formed [26,27]. Thus, the opening dimension of the pores limits the size of the guest, and since the occlusion process does not involve diffusion, the guest dyes can be accommodated all along the host crystals, avoiding guest entangling at the pore entrance. However, this is not a trivial approach and several aspects should be considered before the encapsulation of organic dyes within inorganic channelled structures. The first condition when choosing an organic dye for this approach is that its chemical structure should not be very different from that of the organic structure-directing agents (SDA) that typically drive through a template effect the crystallization of zeolitic materials, usually amines or quaternary ammonium compounds. In particular, dyes should be able to bear positive charges to maximize interaction with the negatively-charged inorganic framework, enhancing their incorporation. Moreover, the dye must be soluble in an aqueous synthetic gel. For this reason, cationic organic dyes with amine groups are a preferred choice. Besides, a negative net charge should be generated in the zeolitic framework, i.e., by the isomorphic substitution of Al^{3+} by Mg^{2+} in aluminophosphate networks, to facilitate the entry of those cationic dyes.

In previous works, it has been demonstrated that the one-dimensional Mg-containing aluminophosphate MgAPO-11, with AEL structure, is a perfect host for the incorporation of many commercial dyes due to its adequate pore structure having non-intersecting one-dimensional (1D) channels (6.5 × 4 Å) and their special topology [28,29]. By the encapsulation of dyes through the crystallization inclusion method into this framework, not only is molecular aggregation avoided, but preferential alignment along the pores is also induced [27,30,31]. Moreover, the inorganic 1D-host MgAPO-11 offers a very rigid environment limiting molecular motions responsible for the non-radiative deactivation pathways to the ground-state, yielding an astonishing enhancement of fluorescence with respect to the dye in solution, particularly for dyes with flexible molecular structures. As a result, highly fluorescent hybrid materials with an anisotropic response to the linearly polarized light have been obtained using this particular host [31].

In fact, the encapsulation of different commercial fluorescent dyes into the 1D-nanochanneled inorganic host MgAPO-11 (Scheme 1) via a one-pot synthesis has already rendered hybrid materials with diverse optical properties such as (i) an optically switchable system by the simultaneous encapsulation of two chromophores with a complementary response to linearly polarized light [29]; and (ii) a red-emitting hybrid material with Second Harmonic Generation (SHG) properties under IR excitation by the confinement of a nonlinear optics (NLO) dye perfectly aligned within the channels [31].

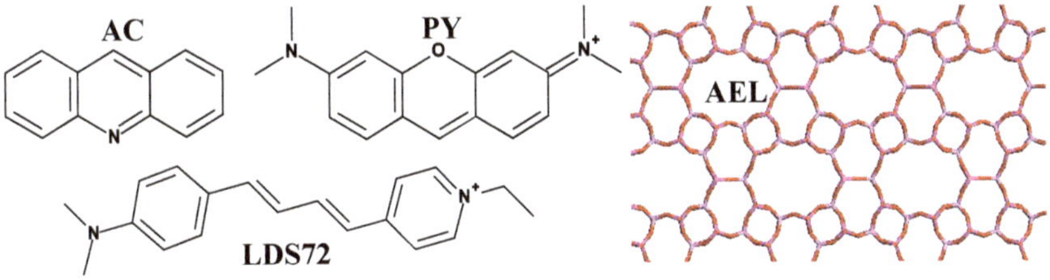

Scheme 1. (**Left**) Molecular structures of dyes: acridine (AC), pyronine Y (PY), and LDS722. (**Right**) Schematic illustrations of AEL-zeolitic type structure, MgAPO-11.

Particularly in this work and as above-mentioned, a critical aspect to attain pure white light after excitation with UV light is to control the efficiency of energy transfer among the different co-adsorbed fluorescence emitters. Taking into account the requirements for an effective Försters Resonance Energy Transfer or FRET [32], acridine (AC), pyronin Y (PY), and hemicyanine LDS722 dyes (Scheme 1) were chosen to achieve a white light hybrid emitter. AC, with bluish-cyan emission, was selected as the primary energy-donor moiety due to its relatively high fluorescent quantum yield and its small dimensions that would provide a high donor-rate into the host. The AC emission spectrum overlaps well with the absorption band of the next energy-acceptor molecule, PY. Indeed, this AC-PY pair was set in a previous work as a suitable donor-acceptor pair for an efficient FRET process [29]. Finally, a hemicyanine-like dye, 4-[4-[4-(dimethylamino)phenyl]-1,3-butadienyl]-1-ethyl-pyridinium (LDS722), characterized by emission in the red region of the electromagnetic spectrum, is chosen as final acceptor dye in the FRET cascade. Although the fluorescence of this dye is rather poor in solution due to the inherent flexibility of its molecular structure, a highly improved red emission efficiency is reached by the great constraint imposed by the AEL nanopores [31].

Here, to achieve a bright emitting system with pure white light, a systematic variation of the relative proportion of the three dyes in the synthesis of the hybrid material is performed. The resultant materials are fully characterized by steady-state and time-resolved spectroscopic techniques. Experimental evidence of FRET processes among the dyes within the pores of the material are also provided via ultrafast spectroscopy experiments where the excited-state dynamics is investigated.

2. Materials and Methods

2.1. Synthesis

Mg-containing aluminophosphates (MgAPO-11) with the three dyes were prepared by hydrothermal crystallization method using phosphoric acid (Aldrich, Madrid, Spain, 85 wt.%), magnesium acetate tetrahydrate (Aldrich, Madrid, Spain, 99 wt.%,), aluminum hydroxide (Aldrich, Madrid, Spain), ethylbutylamine (EBA, Aldrich, Madrid, Spain), and the aforementioned dyes: AC (Aldrich, Spain 97%), PY (Acros Organics, Madrid, Spain, 75%), and LDS722 (Exciton through Lasing S.A, Madrid, Spain, laser grade, purity > 99%). All the compounds were used as received. The synthesis gels have a general molar composition of 0.2 MgO:1 P_2O_5:0.9 Al_2O_3:0.75 EBA:x dye:300 H_2O, where x indicates the total amount of dye added into the gel. In this work, different dye-proportions were added to the gel, keeping constant the total amount of dye at $x = 0.024$ in order to not disrupt the final crystalline phase of the material. For example, a dye-proportion in the gel of 3AC:2PY:1LDS722 means that the molar fraction of AC dye in the synthesis gel is 1.5 times higher than the molar fraction of PY, and three times higher than that of LDS722. Considering that the total amount of dyes in the gel was set at 0.024, the proportion added for AC, PY, and LDS722 is 0.012, 0.008, and 0.004, respectively. The pH of the synthesis gels was between 4 and 5. The gels were heated under autogenous pressure for 24 h at 160 °C. The solid products were recovered by filtration, exhaustively washed with ethanol and water, and dried at room temperature overnight.

2.2. Characterization

X-ray powder diffraction (XRD) was used to determine the crystalline phase obtained; XRD patterns were collected with a Panalytical X'Pro diffractometer (Panalytical, Lelyweg, The Netherlands) using Cu Kα radiation.

The dye content within the solid products was photometrically determined using a double beam Varian spectrophotometer (model Cary 7000, Agilent Technologies, Madrid, Spain) after dissolving the solid material in 5 M hydrochloric acid and comparing the resulting solutions with standard solutions prepared from known concentrations of the dyes at the same pH value of the sample solutions.

The absorption spectra of the dye-containing MgAPO powder samples were recorded with a Varian spectrophotometer (model Cary 7000) detecting the reflected light through an integrating sphere. The respective spectra of the MgAPO powders synthesized under identical conditions but without dyes were recorded and subtracted from the sample signal to eliminate the scattering contribution to the absorption spectra. Emission spectra of the bulk powder were recorded in an Edinburgh Instruments spectrofluorimeter (FLSP920 model, Livingston, U.K.) in front-face configuration (40° and 50° to the excitation and emission beams, respectively) and leaned 30° to the plane formed by the direction of incidence and detection. The fluorescence lifetime decay curves of the bulk powder were measured with the time-correlated single-photon counting technique in the same spectrofluorimeter using a microchannel plate photomultiplier tubes (MCP-PMTs, Hamamatsu R38094-50) with picosecond time-resolution (~150 ps). The fluorescence lifetime (τ) was obtained after deconvolution of the instrumental response signal from the recorded decay curves by means of an iterative method. The goodness of the exponential fit was controlled by statistical parameters (chi-square, χ^2, and analysis of the residuals).

Absolute photoluminescence quantum yields of the dye-containing MgAPO powders were measured in an integrated sphere coupled to this spectrofluorimeter. The absorbance at excitation wavelength was obtained by comparing the scatter signal of the dye-loaded hybrid material and a Teflon disk was used as a reference (with a diffuse reflectance of 100%).

An amplified femtosecond double OPA (optical parametric amplifier) laser system was used to provide excitation pulses of 395 nm. The power of the excitation beam was set to 300 µW (300 nJ/pulse, power density = 1697 mW/cm^2, energy density/pulse = 1.70×10^{-3} Joule/(pulse $\times$ cm^2), number of photons/(pulse $\times$ area) = 3.20×10^{15} photons/(pulse $\times$ cm^2)), and the fluorescence light emitted from the samples was efficiently collected from the same side as the excitation beam using a concave mirror. The fluorescence was then filtered using a 420 nm long-pass filter to suppress the scattered light, directed and overlapped with a gate pulse (800 nm, ca. 10 µJ) derived from the regenerative amplifier onto a lithium triborate (LBO) crystal. By tuning the incident angle of these two beams relative to the crystal plane, the sum frequency of the fluorescence light and the gate pulse was generated. The time-resolved traces were then recorded by detecting this sum-frequency light while changing the relative delay of the gate pulse versus the sample excitation time. Fluorescence gating was done under magic-angle conditions in a time window of 50 ps. Monochromatic detection in heterodyne mode was performed using a PMT (R928, Hamamatsu, Shizuoka, Japan) placed at the second exit of the spectrograph mounted behind a slit. Optical heterodyne detection is a highly sensitive technique that is also used to measure very weak changes in absorption induced by a frequency-modulated pump beam. An additional bandpass filter 260–380 nm was placed in front of the monochromator in order to reject light from the excitation and the second harmonic of the gate pulse. The electrical signal from the photomultiplier tube was gated by a boxcar averager (SR 250, Stanford Research Systems, Sunnyvale, CA, USA) and detected by a lock-in amplifier (SR830, Stanford Research Systems). The prompt response (or instrumental response function, IRF) of this setup (including laser sources) was determined by detection of scattered light of the excitation pulse under identical conditions and found to be approximately 100 fs (FWHM: full width at half maximum). This value was used in the analysis of all measurements for curve fitting using iterative re-convolution of the datasets while assuming a Gaussian shape for the prompt response.

Global analysis of the fluorescence decays obtained at different wavelengths as a sum of exponentials allowed to obtain amplitude-to-wavelength spectra. When interpreting the features of those amplitude-to-wavelength spectra, one should keep in mind that no correction for the wavelength dependency of the sensitivity of the detection setup sensitivity (mixing crystal, filters, PMT) was applied. The samples were prepared in powder form and contained in a quartz cuvette. To improve the signal-to-noise ratio, every measurement was averaged 15 times at 256 delay positions where a delay position is referred to as the time interval between the arrival of the pump and gate pulses at the sample position.

3. Results and Discussion

The three dyes chosen to be simultaneously occluded into the MgAPO-11 inorganic host have already been encapsulated and individually analysed in the same framework. Acridine (AC)-loaded MgAPO-11 material (AC/AEL) has demonstrated a bluish-cyan emission under UV excitation light (absorption band, λ_{abs}: 350–450 nm; emission maximum, λ_{emiss}: 481 nm, and CIE "Commission Internationale de l'Eclairage" coordinates: 0.18,0.33) with a relatively high fluorescence quantum yield of 54% and a long fluorescence lifetime of 27 ns ascribed to the protonated species of AC monomers, ACH^+ [29]. As green-emitting moiety, pyronin Y (PY) dye encapsulated into MgAPO-11 (PY/AEL) has revealed the characteristic green emission of the PY monomers (CIE under blue excitation light: 0.35,0.64), with an improved fluorescence quantum yield and a longer lifetime (ϕ_{fl}: 0.29, τ_{fl}: 4.2 ns) than that of PY in diluted aqueous solution (ϕ_{fl}: 0.21, τ_{fl}: 2 ns) [27]. Finally, LDS722 dye into AEL has shown red emission properties (LDS722/AEL, λ_{fl} = 670 nm, CIE under green excitation light: 0.69, 0.31) with a much higher quantum yield (ϕ_{fl}: 0.55) than that of the corresponding one in aqueous solution ($\phi_{fl} \leq 0.01$) due to the rigidity imposed by the host matrix [31].

As stated before, white light emission in this AEL host containing a mixture of dyes must be assisted by a successive FRET energy transfer process among the fluorophores. Note here that FRET process from AC (donor) to PY (acceptor) into the MgAPO-11 matrix has already been explored, in which a high probability to encounter acceptor molecules in neighbouring channels was demonstrated [29]. In the present case, PY will be the energy acceptor in the first step, and the energy donor with respect to LDS722. Note here that even though LDS722 dye can be directly excited upon UV light through its Locally Excited "LE" state (Figure 1), the most efficient absorption band to obtain red emission is the broad band (450–680 nm) of LDS722, located in the green-visible region and assigned to an Intra Charge Transfer state "ICTS" [31]. This broad absorption band shows a good overlap with the emission band of PY/AEL and enables an efficient FRET process. However, the ICTS absorption band also overlaps with the fluorescence of AC/AEL, opening the possibility that the excitation energy of AC/AEL is directly transferred to the LDS722/AEL system. Therefore, the resultant emission of the present materials is expected to involve competitive FRET channels among the three dyes.

A requirement to obtain a white light-emitting material is optimization of the proportion of dyes occluded into AEL, and consequently, the amount and ratio added to the synthesis gel. It is worth mentioning that predicting the incorporation degree of each dye into the inorganic matrix by the crystallization inclusion method is not an easy task and it is even more difficult when a mixture of dyes is present in the synthesis gel. Apart from typical parameters such as the solubility and competition of the dyes' uptake, which can favour or impede the accommodation of the others, the multiple available energy transfer processes involving the three occluded dyes make finding the proper proportions of each chromophore in the synthesis gel a trial-and-error process, which has been nevertheless based on the previous knowledge gained on the behaviour of the individual dyes when occluded in the aluminophosphate matrix. In this sense, a systematic variation of the relative proportion of the three dyes in the synthesis gel was performed. The most representative samples are listed in Table 1 and Figure 2. The emission color in all samples is confirmed by chromaticity experiments (CIE coordinates) and Color Temperature (CCT).

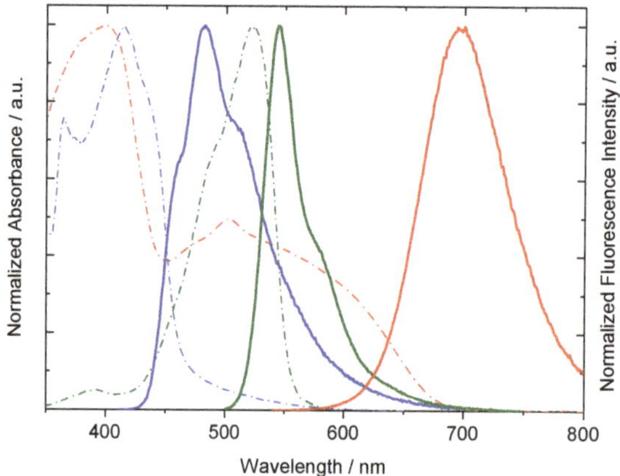

Figure 1. Normalized absorption spectra (dashed lines) and fluorescence spectra (solid lines) of MgAPO-11 samples containing a unique dye: AC/AEL (blue lines), PY/AEL (green lines), and LDS722/AEL (red lines).

As a first attempt, dyes were added to the synthesis gel in equal proportions (Sample 1, $x = 0.008$ of each dye). CIE chromaticity coordinates (Table 1, Figure 2) indicate a resultant yellow emission in this material, showing a deficit mainly of blue components. In fact, the Color Temperature, CCT, obtained from CIE coordinates [33], with a numerical number of 3350 K, is categorized as "warm white" that is characterized by an orange to yellow-white pleasant color suitable for lighting bedrooms, living room, and restaurants.

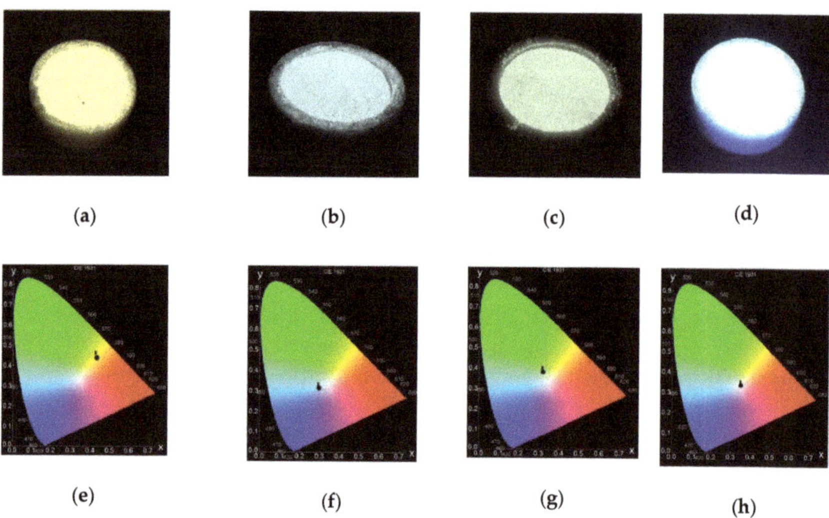

Figure 2. Powder samples under UV illumination, $\lambda_{exc} = 370$ nm, (**a–d**) and the corresponding CIE (Commission Internationale de l'Eclairage) diagram for their fluorescent emission (**e–h**). Sample 1: (1:1:1), warm white (**a,e**); Sample 2: (6:1:1), cool white (**b,f**); Sample 3: (4:2:1), cool white (**c,g**); Sample 4: (3:1:1), neutral white (**d,h**).

Table 1. Relative proportion of each dye in the initial synthesis gel. Final amount of dye occluded into AEL after synthesis, expressed as mmol dye per 100 g powder. Fluorescence quantum yield, Commission Internationale de l'Eclairage (CIE), and Correlated Color Temperature (CCT) number of the overall emission for each sample and the corresponding color temperature are reported.

Sample	Dye Amount in the Gel (in Molar Fraction, x)			Dye Amount into AEL (mmol/100 g)			Φ_{fl}	CIE Coordinates		Correlated Color Temperature CCT [e] (K)
	AC	PY	LDS722	AC	PY	LDS722		x	y	
AC/AEL	0.024	0	0	9.69	-	-	0.54 [a]	0.18	0.33	-
PY/AEL	0	0.024	0	-	0.33	-	0.29 [b]	0.35	0.64	-
LDS722/AEL	0	0	0.008	-	-	1.32	0.55 [c]	0.69	0.31	-
Sample 1 (1:1:1)	0.008	0.008	0.008	3.19	0.14	0.09	0.08 [a]	0.42	0.41	3350
Sample 2 (6:1:1)	0.017	0.003	0.003	7.44	0.02	0.79	0.07 [a]	0.31	0.32	6750
Sample 3 (4:2:1)	0.014	0.007	0.003	6.94	0.08	0.45	0.10 [a]	0.32	0.34	6070
Sample 4 (3:1:1)	0.014	0.005	0.005	8.26	0.06	0.55	0.17 [a]	0.34	0.34	5170
Sample 5 (3:1:2)	0.012	0.004	0.008	9.49	0.06	0.64	0.04 [a]	0.35	0.32	4600

Fluorescence quantum yield values were recorded upon [a] λ_{exc}: 350 nm, [b] λ_{exc}: 490 nm, or [c] λ_{exc}: 530 nm. [d] Sample with 0.024 dye-proportion in the gel was not included since in such synthesis conditions AFI (MgAPO-5) was obtained as a major phase [31]. [e] Correlated Color Temperature derived from CIE coordinates [33] are classified as: "warm white" for 2000–3500 K, "neutral white" from 3500 K to 5300 K and "cool white" > 5300 K.

Then, in the second attempt, the amount of blue-emitting AC dye in Sample 2 (first donor in the FRET cascade of our three-dyes system) was significantly increased in the gel ($x = 0.017$) and the amount of PY and LDS722 was reduced to the same extent (from $x = 0.008$ to $x = 0.003$), resulting in an AC:PY:LDS722 ratio of approximately 6:1:1 (Table 1). In this case, a much higher incorporation of LDS722 with respect to PY and even with respect to Sample 1 was achieved (Table 1). This experimental fact can be ascribed to the strong competition between PY and LDS722 dyes to be embedded into the MgAPO framework, the incorporation of PY being hindered by the presence of LDS722 in the gel. As a result, the fluorescent emission of Sample 2 shows CIE coordinates near the white color under UV excitation, but still with a slightly bluish hue (Figure 2). Accordingly, a relatively high CCT is derived (Table 1), being in the range of the so-called "cool white", which resembles daylight with a bluish-white appearance usually employed when superior brightness is required such as in industrial areas, garages, retails, or art studios. However, neither of these two samples show a high fluorescence quantum yield ($\phi_{fl} \leq 0.10$).

In the next trial, Sample 3 in Table 1 with an AC:PY:LDS722 proportion in the gel of 4:2:1, the relative amount of PY with respect to LDS722 in the initial synthesis gel was increased with a double aim: (i) to provide a slightly higher contribution of green emission compared to Sample 2, and (ii) to improve the energy transfer from this green-emitting dye to the red-emitting entity. Indeed, as stated before, PY is supposed to show a dual behaviour both as an acceptor of the energy coming from AC and as a donor giving its energy to LDS722. Besides, the AC amount was slightly reduced (from $x = 0.017$ to $x = 0.014$) in an attempt to displace the emission-color out of the blue region. However, although not far from the white and with relatively higher efficiency ($\phi_{fl} = 0.10$), in this case, the CCT number of 6070 K indicates a bluish hue of the white light, within the "cool light" range (Table 1, Figure 2).

Taking into account the latest results, in the next try, Sample 4, the amount of AC dye was kept constant ($x = 0.014$), but a slightly lower amount of PY with respect to Sample 3 was used, leading to a proportion of 3:1:1 in the initial gel. Its emission spectrum shows the characteristic bands of each dye, centered in the blue (480 nm), green (540 nm), and red (670 nm) regions of the visible spectrum with very similar intensities (Figure 3). In this way, pure white light emission under UV excitation light was finally achieved (see CIE coordinates in Table 1, Figure 2). According to its corresponding CCT number, a neutral white color is obtained, which is characterized by bright white light useful for the illumination of kitchens, bathrooms, offices, classrooms, or hospitals. Furthermore, this sample shows the highest emission quantum yield (around 17%) among all the samples described in this work (Table 1).

In sum, by (one-pot) adding three dyes to the synthesis gel with the following composition: 0.2 MgO:0.9 Al$_2$O$_3$:1.0 P$_2$O$_5$:0.75 EBA:0.014 AC:0.005 PY:0.005 LDS722:300 H$_2$O, corresponding to a dye molar ratio of 3 AC:1 PY:1 LDS722, a pure white light-emitting material with the highest efficiency was achieved. In this sample, the estimated final amount of dyes loaded was 8.26 mmol of AC, 0.06 mmol of PY, and 0.55 mmol of LDS722 per 100 g of sample powder (Table 1), which corresponds to a relative dye proportion of 138:1:9, respectively. However, it is worth stressing that the quantification of the uptake of all dyes was not a trivial task due to the overlap between the absorption regions of the dyes in solution. Furthermore, the incorporation of each dye into the MgAPO-11 structure does not follow a linear correlation with the initial proportion added to the synthesis gel (Table 1). This fact is a consequence of the strong competition for getting occluded among the dyes, i.e., LDS722 obstructs the PY incorporation, as previously mentioned. The size of PY (perpendicular to the major molecular axis) is larger than that of AC and LDS722, what causes a tighter fit of PY in the AEL nanochannels, and hence a more difficult incorporation.

With the aim of discovering new materials with different dye-ratios to render white light emission of a high variety of white color appearance and compare their emission capacity, a new one, Sample 5, with a higher amount of LDS722 with respect to Sample 4, was prepared (3 AC:1 PY:2 LDS722 in the synthesis gel, Table 1). Note that the incorporation of the dyes at any proportion does not modify the AEL structure (Figure A1). In this sample, neutral white light emission (CCT = 4600 k) was also achieved. However, this sample showed a lower fluorescence quantum yield with respect to Sample 4 (Table 1).

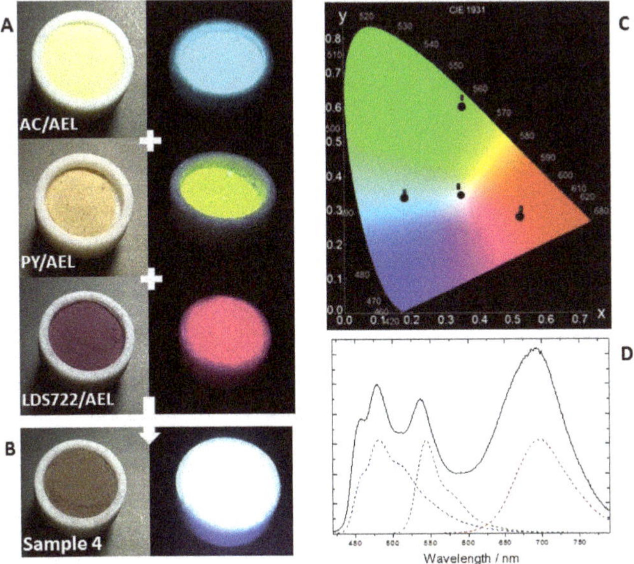

Figure 3. Real images of the samples under ambient light and under UV illumination: (**A**) individual dye (AC, PY, and LDS722) into MgAPO-11 and (**B**) Sample 4 with the optimized mixture of the three dyes (AC-PY-LDS/AEL). (**C**) Commission Internationale de l'Eclairage (CIE) coordinates diagram of the fluorescence emission of the four samples under UV light. (**D**) Emission spectra of Sample 4 (black) upon UV excitation (325–375 nm band pass) recorded in bulk powder, together with the emission spectra of the MgAPO samples containing isolated dyes: AC/AEL (dashed blue), PY/AEL (dashed green), and LDS722/AEL (dashed red).

To gain more insights into the FRET process involved among the three dyes, excited-state dynamics of the most representative samples (Samples 4 and 5) were studied. In this context, FRET process is confirmed by the differences found in the fluorescent lifetimes recorded for each dye in the presence of the other two dyes, with respect to their lifetimes when individually encapsulated into AEL (Table 2). Although it is not easy to discern and assign each lifetime obtained due to the overlapping of the emission bands of the dyes, a general trend was found (Table 2). In the first place, the significant decrease of the long fluorescence lifetime ascribed to the protonated species of AC (donor) from 27 ns in the absence of any acceptor to 18.7 ns and 16.20 ns for Samples 4 and 5, respectively, indicates a deactivation of the AC (donor) emission as a consequence of the energy transfer process. On the other hand, short lifetimes (<1 ns) recorded in both samples at this wavelength (λ_{em}: 480 nm) could be attributed to the FRET process as these short contributions do not show up in the absence of acceptor(s) (vide infra).

Table 2. Fluorescence lifetimes characteristic of each dye, measured in bulk samples upon UV illumination (λ_{exc}: 400 nm), detected in the blue (AC), green (PY), or red (LDS722) emission regions of the visible spectra.

Sample	[a] τ_{fl}/ns 480 nm Detection	[b] τ_{fl}/ns 540 nm Detection	[c] τ_{fl}/ns 690 nm Detection
AC/AEL	27.00	-	-
PY/AEL	-	[1] 4.20	-
LDS722/AEL	-	-	[2] 2.70
Sample 4 (3:1:1)	0.77 (68%) 18.7 (32%)	4.20 (71%) 19.09 (29%) *	2.40 [†] (90%)
Sample 5 (3:1:2)	0.37 (72%) 16.20 (38%)	0.35 (64%) 3.63 (28%) 16.03 (8%) *	2.30 [†] (95%)

Fluorescence lifetime values recorded under 400nm excitation: at [a] λ_{em}: 480 nm, [b] λ_{em}: 540 nm and/or [c] λ_{em}: 690 nm. For samples containing only PY or LDS722, the excitation wavelength was set at [1] 490 nm and [2] 530 nm, respectively. * This contribution is assigned to the emission tail of the AC dye at this wavelength. [†] Only lifetime with major contribution is considered.

Secondly, in Sample 4, the lifetime assigned to PY remains nearly unchanged as when it is individually occluded, likely due to a low participation in the energy transfer, this time from PY (as donor) to LDS722 (as acceptor). The lifetime ascribed to PY in Sample 5 undergoes a slight decrease of its characteristic lifetime in the AEL framework, together with the appearance of a short lifetime (Table 2), indicating that PY acts in this sample by transferring its excitation energy to a large extent to LDS722 as final acceptor dye. Regarding LDS722 lifetime, a slight decrease with respect to the sample with LDS722 alone is also found in both samples, reaching 2.4 ns (Table 2), likely ascribed to a slight change in its molecular conformation.

To better characterize FRET processes in these hybrid samples, fluorescence decay traces were recorded by femtosecond fluorescence up-conversion (100 fs resolution) on a 50 ps time window. Figure 4 displays the fluorescence decays obtained for Sample 4 on a 50-ps time scale detected at 450, 550, and 650 nm corresponding to the emission bands of AC/AEL, PY/AEL, and LDS722/AEL, respectively. The fit analysis retrieved an ultrafast component with a positive amplitude at 550 nm (decay component) and a negative amplitude at 650 nm (rise component). This suggests the presence of a kinetic process that leads to the depopulation of the locally excited state of the PY chromophore and population of the excited state of LDS722. Such a process is characteristic of energy transfer (FRET) and its fast rate indicates a highly efficient excitation energy transfer for the fraction of donor-acceptor involved occurring from PY to LDS722. Since the amplitude of this ultrafast component is roughly one-third of the longer component attributed to the fluorescence decay, we can estimate that for this sample, the fraction of the coupled donor-acceptor systems PY-LDS722 reaches 30%. Interestingly, the overall signal of the fluorescence decay of the AC moiety is rather low, suggesting an energy transfer occurring at a rate faster than the resolution of the setup (100 fs). Nevertheless, the weak fluorescence decays with an ultrafast and a fast component indicates a second energy transfer channel towards the PY or LDS722 moiety.

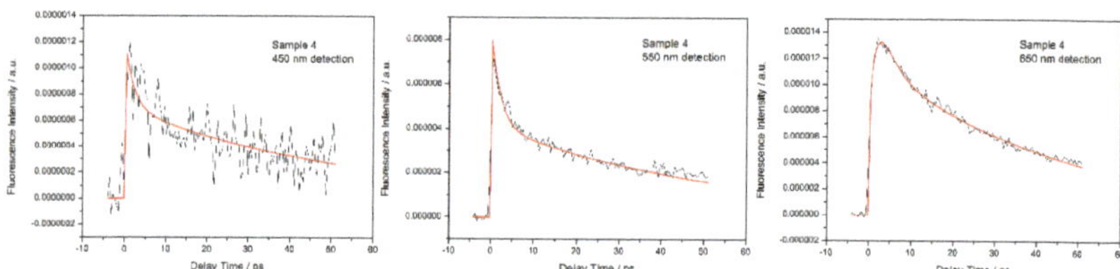

Figure 4. Decay traces recorded by fs fluorescence up-conversion for Sample 4 (3:1:1) at 450, 550 and 650 nm detection wavelength (λ_{exc} = 395 nm).

Figure 5 shows the fluorescence decays for Sample 5 (ratio 3:1:2), plotted on a 50-ps time scale in which the emission bands for AC/AEL, PY/AEL, and LDS722/AEL were recorded at 450, 550, and 650 nm. Similar to that of Sample 4, the fluorescence assigned to PY (550 nm) shows a decay component of 2.5 ps, whereas this appears as a rise component in the emission of LDS722 (650 nm). Such spectroscopic evidence points again to the presence of an energy transfer process from PY to LDS722 as acceptor and donor moieties. Moreover, considering the 60% contribution of this component to the overall decay (see Figure 5, 550 nm detection), we can state that a higher fraction of donor-acceptor systems is present in Sample 5 relative to Sample 4. This is, indeed, in line with the presence of double amount of LDS722 acceptor (ratio 3:1:2) when compared to Sample 4 (where the ratio was 3:1:1). Only a weak emission has been observed in the spectral emission region of AC/AEL (450 nm), indicating a strong quenching process due to FRET.

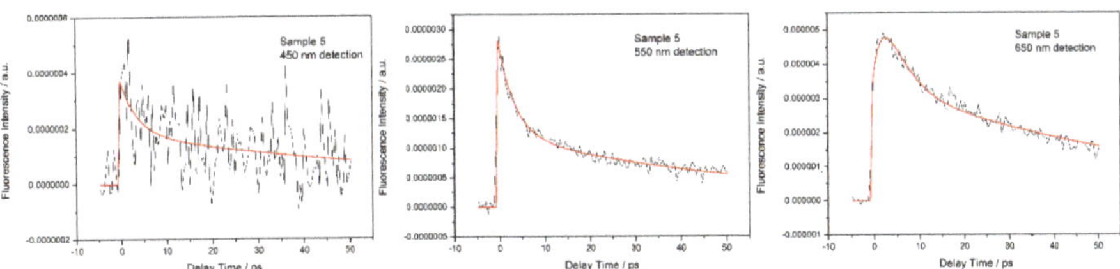

Figure 5. Decay traces recorded by fs fluorescence up-conversion for Sample 5 (3:1:2) at 450, 550, and 650 nm detection wavelength (λ_{exc} = 395 nm).

Figures A2 and A3 show the time-resolved fluorescence emission spectra recorded for Samples 4 and 5, respectively. The spectra show the relative ratio between the emission intensities of the AC/AEL, PY/AEL, and LDS722/AEL constituents in 50 ps time window. It is clear that the fluorescence emission attributed to LDS722 (650–700 nm region) builds up in the first few picoseconds after excitation.

The fluorescence up-conversion experiments clearly indicate, at least for the PY-LDS722/AEL donor-acceptor pair, the presence of an efficient excitation energy transfer process. For the AC dye, two additional fluorescence decay time constants were observed, suggesting the presence of energy transfer channels towards PY and LDS722 within AEL.

The results presented above indicate that it is very difficult to predict the best synthesis conditions in terms of the dye-ratio in the initial gel. However, after optimization of the dye proportion, a solid white light hybrid emitter was achieved, which shows a relatively high fluorescence capacity of nearly 20%.

Finally, a photo-stability experiment was carried out for Sample 4 with the highest brightness (ϕ_{fl} = 0.17) and a CCT in the neutral white range (5170 K). The sample was irradiated with a continuous UV light (λ_{exc} = 355 nm and an irradiance of 2.75 mW/cm^2) for different times. After 6 h of intense UV illumination, the sample lost less than 10% of its efficiency and changed its neutral white to cool white colour (CCT = 6400 K) because the red fluorescent LDS722 dye underwent a faster photobleaching process with respect to the blue acridine and green pyronine Y emissive dyes. The change in colour is depicted in Appendix A Figure A4.

4. Conclusions

A new hybrid material has been designed herein by simultaneous encapsulation in appropriate amounts of three dyes, acridine (AC), pyronin Y (PY), and LDS722, with complementary spectroscopic bands emitting in the blue, green, and red regions of the visible spectrum, into the 1D-Mg-containing aluminophosphate framework with AEL structure, revealing an efficient white light emission under UV irradiation.

The final color is a consequence of successive partial energy transfer processes among the dyes, which can be modulated by the final uptake of each dye in the host matrix. Once the initial dye ratio is optimized in the synthesis gel, this straightforward synthesis through the one-pot occlusion of the chromophores offers an easy and fast approach to render a solid-state system with pure white light emission and an efficiency of around 20%. This type of hybrid material offers versatile white light emitter systems in which their brightness and emission color purity can be easily tuned. More importantly for future commercial applications, this material is easy and inexpensive to manufacture, and the process is not time-consuming since the incorporation of the dyes and synthesis of the hybrid material takes place in a unique step.

Author Contributions: Synthesis of the materials, A.A., L.G.-H., and J.P.-P.; Characterization of the materials, R.S.-L., A.O.-S., T.A., J.H., and E.F.; Conceptualization, R.S.-L. and V.M.-M.; writing—original draft preparation, R.S.-L., V.M.-M., and E.F.; writing—review and editing, V.M.-M. and E.F.; supervision, V.M.-M.; funding acquisition, V.M.-M. All authors have read and agreed to the published version of the manuscript.

Funding: This research was funded by Gobierno Vasco (IT912-16), Ministerio de Economía y Competitividad (MINECO), the Spanish Agencia Estatal de Investigación, and the EU's Fondo Europeo de Desarrollo Regional under projects MAT2017-83856-C3-3-P and MAT2016-77496-R, and C1/C2 KU Leuven, FuEPoNa (3E190382).

Conflicts of Interest: The authors declare no conflict of interest. The funders had no role in the design of the study; in the collection, analyses, or interpretation of data; in the writing of the manuscript, or in the decision to publish the results.

Appendix A

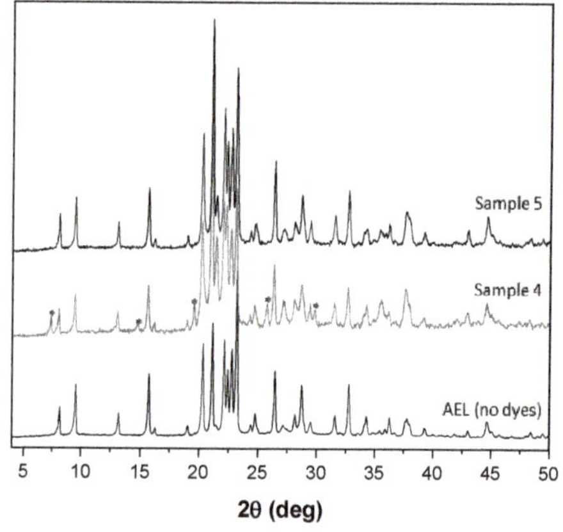

Figure A1. XRD patterns of AEL materials, without dyes (**bottom**), Sample 4 (**middle**), and Sample 5 (**top**); * indicates minor impurities of AFI material.

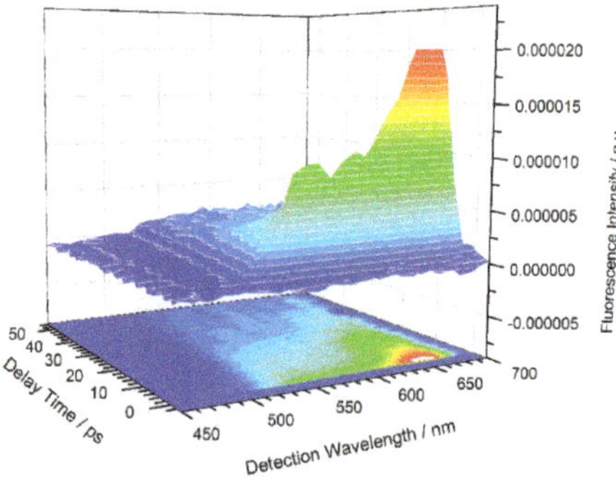

Figure A2. Time-resolved fluorescence emission spectra of Sample 4 recorded by fs fluorescence up-conversion in 50 ps time window (λ_{exc} = 395 nm).

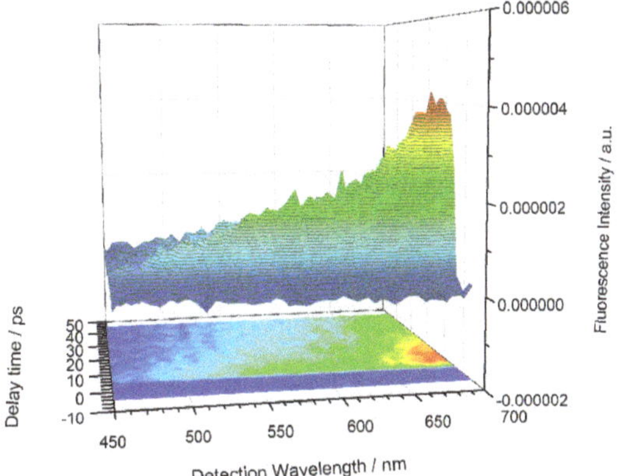

Figure A3. Time-resolved fluorescence emission spectra of Sample 5 recorded by fs fluorescence up-conversion in 50 ps time window (λ_{exc} = 395 nm).

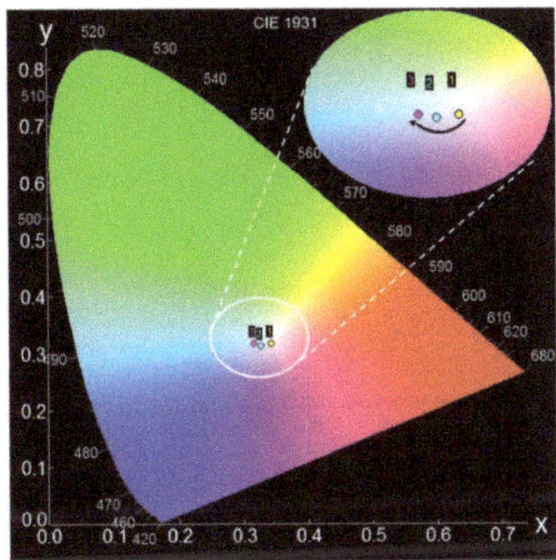

Figure A4. Changes in the CIE coordinates of Sample 4 at different times of intense UV irradiation (λ_{exc} = 355 nm and an irradiance of 2.75 mW/cm^2): t = 0 (point 1); t = 1 h (point 2) and r = 6 h (point 3).

References

1. Nicole, L.; Laberty-Robert, C.; Rozes, L.; Sanchez, C. Hybrid materials science: A promised land for the integrative design of multifunctional materials. *Nanoscale* **2014**, *6*, 6267–6292. [CrossRef]
2. Parola, S.; Julián-López, B.; Carlos, L.D.; Sanchez, C. Optical properties of hybrid organic-inorganic materials and their applications. *Adv. Funct. Mater.* **2016**, *26*, 6506–6544. [CrossRef]
3. Sanchez, C.; Lebeau, B.; Chaput, F.; Boilot, J.-P. Optical properties of functional hybrid organic-inorganic nanocomposites. *Adv. Mater.* **2003**, *15*, 1969. [CrossRef]
4. Park, S.S.; Ha, C.S. Hollow mesoporous functional hybrid materials: Fascinating platforms for advanced applications. *Adv. Funct. Mater.* **2018**, *28*, 1703814. [CrossRef]
5. Sanchez, C.; Julian, B.; Belleville, P.; Popall, M. Applications of hybrid organic-inorganic nanocomposites. *J. Mater. Chem.* **2005**, *15*, 3559–3592. [CrossRef]
6. Kickelbick, G. *Introduction to Hybrid Materials*; Wiley-VCH: Weinheim, Germany, 2007; pp. 1–48.
7. Kido, J.; Hongawa, K.; Okuyama, K.; Nagai, K. White light-emitting organic electroluminescent devices using the poly(*N*-vinylcarbazole) emitter layer doped with three fluorescent dyes. *Appl. Phys. Lett.* **1994**, *64*, 815–817. [CrossRef]

8.   Wen, Y.; Sheng, T.; Zhu, X.; Zhuo, C.; Su, S.; Li, H.; Hu, S.; Zhu, Q.L.; Wu, X. Introduction of red-green-blue fluorescent dyes into a metal–organic framework for tunable white light emission. *Adv. Mater.* **2017**, *29*, 1700778. [CrossRef] [PubMed]

9.   Zhang, Q.; Wang, C.-F.; Ling, L.-T.; Chen, S. Fluorescent nanomaterial-derived white light-emitting diodes: What's going on. *J. Mater. Chem. C* **2014**, *2*, 4358–4373. [CrossRef]

10. Kamtekar, K.T.; Monkman, A.P.; Bryce, M.R. Recent advances in white organic light-emitting materials and devices (WOLEDS). *Adv. Mater.* **2010**, *22*, 572–582. [CrossRef]

11. Wang, Z.; Wang, Z.; Lin, B.; Hu, X.; Wei, Y.; Zhang, C.; An, B.; Wang, C.; Lin, W. Warm-white-light-emitting diode based on a dye-loaded metal-organic framework for fast white-light communication. *ACS Appl. Mater. Interfaces* **2017**, *9*, 35253–35259. [CrossRef]

12. Wang, X.; Tang, J.; Wang, G.; Wang, W.; Ren, J.; Ding, W.; Zhang, X.; Wang, Y.; Shen, W.; Huang, L.; et al. Ln^{3+}-induced diblock copolymeric aggregates for fully flexible tunable white-light materials. *Nanomaterials* **2019**, *9*, 363. [CrossRef] [PubMed]

13. Deng, Y.; Li, P.; Li, H. Single-component warm-white-light materials with high color-rendering index based on Eu^{3+}, Tb^{3+}-Complexes co-doped laponite®under mild reaction conditions. *Opt. Mater. (Amst.)* **2019**, *93*, 6–10. [CrossRef]

14. Zhang, X.; Li, L.; Wang, S.; Liu, X.; Yao, Y.; Peng, Y.; Hong, M.; Luo, J. [(N-AEPz)ZnCl$_4$]Cl: A "Green" metal halide showing highly efficient bluish-white-light emission. *Inorg. Chem.* **2020**, *59*, 3527–3531. [CrossRef] [PubMed]

15. Yao, D.; Xu, S.; Wang, Y.; Li, H. White-emitting phosphors with high color-rendering index based on silver cluster-loaded zeolites and their application to near-UV LED-based white LEDs. *Mater. Chem. Front.* **2019**, *3*, 1080–1084. [CrossRef]

16. Li, J.; Liang, Q.; Hong, J.Y.; Yan, J.; Dolgov, L.; Meng, Y.; Xu, Y.; Shi, J.; Wu, M. White light emission and enhanced color stability in a single-component host. *ACS Appl. Mater. Interfaces* **2018**, *10*, 18066–18072. [CrossRef]

17. Anand, V.; Dhamodharan, R. White light emission from fluorene-EDOT and phenothiazine-hydroquinone based D-π-A conjugated systems in solution, gel and film forms. *New J. Chem.* **2017**, *41*, 9741–9751. [CrossRef]

18. Liu, X.Y.; Xing, K.; Li, Y.; Tsung, C.K.; Li, J. Three models to encapsulate multicomponent dyes into nanocrystal pores: A new strategy for generating high-quality white light. *J. Am. Chem. Soc.* **2019**, *141*, 14807–14813. [CrossRef]

19. Sapsford, K.E.; Berti, L.; Medintz, I.L. Materials for fluorescence resonance energy transfer analysis: Beyond traditional donor-acceptor combinations. *Angew. Chem. Int. Ed. Engl.* **2006**, *45*, 4562–4589. [CrossRef]

20. Maiti, D.K.; Roy, S.; Baral, A.; Banerjee, A. A fluorescent gold-cluster containing a new three-component system for white light emission through a cascade of energy transfer. *J. Mater. Chem. C* **2014**, *2*, 6574–6581. [CrossRef]

21. Wang, T.; Chirmanov, V.; Chiu, W.H.M.; Radovanovic, P.V. Generating tunable white light by resonance energy transfer in transparent dye conjugated metal oxide nanocrystals. *J. Am. Chem. Soc.* **2013**, *135*, 14520–14523. [CrossRef]

22. Vohra, V.; Calzaferri, G.; Destri, S.; Pasini, M.; Porzio, W.; Botta, C. Toward white light emission through efficient two-step energy transfer in hybrid nanofibers. *ACS Nano* **2010**, *4*, 1409–1416. [CrossRef] [PubMed]

23. Li, Y.; Rizzo, A.; Cingolani, R.; Gigli, G. Bright white-light-emitting device from ternary nanocrystal composites. *Adv. Mater.* **2006**, *18*, 2545–2548. [CrossRef]

24. Li, G.; Shinar, J. Combinatorial fabrication and studies of bright white organic light-emitting devices based on emission from rubrene-doped 4,4'-bis(2,2'-diphenylvinyl)-1,1'-biphenyl. *Appl. Phys. Lett.* **2003**, *83*, 5359–5361. [CrossRef]

25. Teepakakorn, A.; Yamaguchi, T.; Ogawa, M. The Improved stability of molecular guests by the confinement into nanospaces. *Chem. Lett.* **2019**, *48*, 398–409. [CrossRef]

26. Laeri, F.; Schüth, F.; Simon, U.; Wark, M. *Host-Guest-Systems Based on Nanoporous Crystals*; Wiley-VCH Verlag GmbH & Co. KGaA: Weinheim, Germany, 2003.

27. Martínez-Martínez, V.; García, R.; Gómez-Hortigüela, L.; Pérez-Pariente, J.; López-Arbeloa, I. Modulating dye aggregation by incorporation into 1D-MgAPO nanochannels. *Chem.-A Eur. J.* **2013**, *19*, 9859–9865. [CrossRef]

28. Baerlocher, C.; McCusker, L.B.; Olson, D.H. *Atlas of Zeolite Framework Types*; Elsevier: Amsterdam, The Netherlands, 2007.
29. Martínez-Martínez, V.; García, R.; Gómez-Hortigüela, L.; Sola Llano, R.; Pérez-Pariente, J.; López-Arbeloa, I. Highly luminescent and optically switchable hybrid material by one-pot encapsulation of dyes into MgAPO-11 unidirectional nanopores. *ACS Photonics* **2014**, *1*, 205–211. [CrossRef]
30. García, R.; Martínez-Martínez, V.; Sola Llano, R.; López-Arbeloa, I.I.; Pérez-Pariente, J. One-dimensional antenna systems by crystallization inclusion of dyes (one-pot synthesis) within zeolitic MgAPO-36 nanochannels. *J. Phys. Chem. C* **2013**, *117*, 24063–24070. [CrossRef]
31. Sola-Llano, R.; Martínez-Martínez, V.; Fujita, Y.; Gómez-Hortigüela, L.; Alfayate, A.; Uji-i, H.; Fron, E.; Pérez-Pariente, J.; López-Arbeloa, I. Formation of a nonlinear optical host–guest hybrid material by tight confinement of LDS722 into aluminophosphate 1D nanochannels. *Chem.-A Eur. J.* **2016**, *22*, 15700–15711. [CrossRef]
32. Lakowicz, J.R. *Principles of Fluorescence Spectroscopy*, 3rd ed.; Springer: New York, NY, USA, 2006.
33. McCamy, C.S. Correlated color temperature as an explicit function of chromaticity coordinates. *Color. Res. Appl.* **1992**, *17*, 142–144. [CrossRef]

 *nanomaterials*

Article

Protonation-Induced Enhanced Optical-Light Photochromic Properties of an Inorganic-Organic Phosphomolybdic Acid/Polyaniline Hybrid Thin Film

Qingrui Zeng [1], Suyue Guo [1], Yuanbo Sun [2], Zhuojuan Li [1] and Wei Feng [1,*]

[1] Key Laboratory of Groundwater Resources and Environment, Ministry of Education, College of New Energy and Environment, Jilin University, Changchun 130021, China; Zengqr36776@163.com (Q.Z.); Guosy36776@163.com (S.G.); lizhuojuan36776@163.com (Z.L.)

[2] College of Earth Science, Jilin University, Changchun 130061, China; Sunyb36776@163.com

* Correspondence: weifeng@jlu.edu.cn

Received: 18 August 2020; Accepted: 12 September 2020; Published: 15 September 2020

Abstract: A phosphomolybdic acid/polyaniline (PMoA/PANI) optical-light photochromic inorganic/organic hybrid thin film was successfully synthesized by protonation between the the multiprotonic acid phosphomolybdic acid ($H_3PO_4 \cdot 12MoO_3$) and the conductive polymer polyaniline. The stable Keggin-type structure of PMoA was maintained throughout the process. Protonation and proton transfer successfully transformed the quinone structure of eigenstate PANI into the benzene structure of single-polarized PANI in the PMoA/PANI hybridized thin film, and proton transfer transformed the benzene structure of single-polarized PANI back to the quinone structure of eigenstate PANI in the PMoA/PANI hybrid thin film, as verified by Fourier transform infrared spectroscopy (FTIR) and X-ray photoelectron spectroscopy (XPS). The average distribution of PMoA/PANI was observed by atom force microscopy (AFM). Interestingly, protonation of PMoA caused PANI to trigger transformation of the quinone structure into the single-polarized benzene structure, which enhanced the electron delocalization ability and vastly enhanced the maximum light absorption of the PMoA/PANI hybrid thin film as confirmed by density functional theory (DFT), electrochemistry, and ultraviolet-visible spectroscopy (UV-Vis) studies. Under optical-light illumination, the pale-yellow PMoA/PANI hybrid thin film gradually turned deep blue, thus demonstrating a photochromic response, and reversible photochromism was also observed in the presence of hydrogen peroxide (H_2O_2) or oxygen (O_2). After 40 min of optical-light illumination, 36% of the Mo^{5+} species in PMoA was photoreduced via a protonation-induced proton transfer mechanism, and this proton transfer resulted in a structural change of PANI, as observed by XPS, generating a dominant structure with high maximum light absorption of 3.46, when compared with the literature reports.

Keywords: phosphomolybdic acid (PMoA); polyaniline (PANI); protonation; photochromism; nanocomposite thin film

---

1. Introduction

Photochromism is a unique physical-chemical phenomenon that is widely applied in various fields, such as information presentation [1], photodriven nanomachines [2], optical switching [3], optical memory devices and sensors [4–6], high-density optical-electron information storage [7], and molecular recognition or examination [8]. Since single-system photochromic materials exhibit poor photochromic properties, such as poor reversibility, low photochromic activity after coloring, narrow spectral response, fatigability, and color monotony, photochromic materials that can easily overcome these limitations have attracted scientists' attention. Koski et al. [9] synthesized a kind of inorganic/inorganic photochromic material by intercalating the zero-valent metals Sn and Co into

MoO_3 nanoribbons, resulting in a change in the of color MoO_3 from transparent white to dark blue and demonstrating the potential of this system in intelligent color windows and convenient color change sensors. Li et al. [10] fabricated a semiconductor/semiconductor-type $MoO_3@TiO_2$ crystalline-core amorphous-shell nanorod photochromic material and the unique structure and heterojunction formed between MoO_3 and TiO_2 enhanced the photochromic properties of MoO_3. Dai et al. [11] synthesized a doped inorganic-type photochromic material by doping Pb^{2+} into $LiEuMo_2O_8$, which expanded and distorted the crystal cell of $LiEuMo_2O_8$, resulting in a 50% enhancement in the red light emission from $LiEuMo_2O_8$. Lu et al. [12] synthesized a new type of inorganic/organic hybrid photochromic material, $As_4Mo_8O_{33}/C_3N_2H_5$, which showed a remarkable photochromic effect, indicating that proton transfer from the imidazole cation to the polyanion plays a key role in the photochromic process.

Polyoxometalates (POMs) are widely applied photochromic chemicals with complex morphological characteristics and unique physical-chemical properties, leading researchers to focus on developing many types of photochromic composites [13]. Posphomolybdic acid ($H_3PMo_{12}O_{40}$, PMoA), a representative Keggin-type polyoxometalate and polyprotic acid, possesses many advantages characteristic of photochromic materials, such as a large molecular volume, high thermal stability, maintenance of the crystal structure regardless of whether the material exists in the solid or liquid state, strong electron and proton storage capacities, and biocompatibility with organisms and it forms "organic-polyacidic compound anions"-type electron donor and electron acceptor hybrid materials. Feng et al. synthesized a series of organic-polyacidic compound anion-type photochromic materials, such as PMoA/polyacrylamide (PAM), PMoA/polyvinyl pyrrolidone (PVP) [14], PMoA/polyvinyl alcohol resin (PVA) [15], and PMoA/polyethylene glycol (PEG) [16], which can easily combine with PMoA to form "organic-polyacidic compound anion" hybrid films for diverse uses. However, the poor photochromic properties of these materials because of their functional groups and poor electron conversion properties limit their application. In addition, Xiao et al. [17] fabricated the Ni/Na inorganic/organic hybrid supermolecule $NiEDTA-PW_{12}$, which exhibited rapid and reversible photochromism and strong antifatigue properties. Li et al. [18] fabricated polyvinyl pyrrolidone (ppy)/hexatungstate (HTA) and applied it to filter paper, realizing rapid and reversible color conversion.

Polyaniline (PANI) is a kind of industrial conductive polymer with unique electrical properties [19], optical properties [20], and magnetic properties [21,22]. Furthermore, as a kind of conductive polymer, PANI also easily undergoes protonation in the presence of a protonic acid, which will change the electrical conductivity and improve its photochromic properties, thus increasing its application value [19,20,23]. In this work, PMoA was combined with PANI to form an inorganic-organic hybrid photochromic PMoA/PANI hybrid film. PMoA, a polyprotic acid, could easily have a protonation effect on PANI, which would significantly change the conductivity of PANI. We explored the protonation effect of PMoA on PANI and the physical-chemical change of the PMoA/PANI hybrid film during the photochromic process through atomic force microscopy (AFM), Fourier transform infrared spectroscopy (FTIR), X-ray photoelectron spectroscopy (XPS), electrochemical studies, ultraviolet-visible (UV-vis) spectroscopy and density functional theory (DFT) calculations, and these techniques were also used to elucidate the photochromic mechanism of the PMoA/PANI hybrid film. This characterization demonstrated that the protonation effect of PMoA on PANI played a significant role in enhancing proton conversion and the high maximum visible light absorbance of this photochromic material.

2. Experimental

2.1. Substrate Modification

We used *piranha* solution, which is a kind of superoxidant, to modify the hydrophilic substrates. The *piranha* solution consisted of a mixture of concentrated sulfuric acid (98%, analytical reagent, Sinopharm Chemical Reagent Co. Ltd., Shanghai, China) and hydrogen peroxide (30%, analytical reagent, Shanghai Macklin Reagent Co. Ltd., Shanghai, China) with a volume ratio of 7:3. Quartz substrates (20.0 mm × 15.0 mm × 1.00 mm) were treated with *piranha* solution at 353 K for 24 h and then rinsed repeatedly with deionized water followed by ethanol (analytical reagent, Shanghai Macklin Reagent Co. Ltd., Shanghai, China); after that, the substrate was stored in acetone (analytical reagent, Shanghai Macklin Reagent Co. Ltd., Shanghai, China). After the acetone is completely volatilized, the quartz substrate can be used. A silicon wafer (20.0 mm × 15.0 mm × 1.00 mm) was treated with the same procedure for AFM and XPS characterization. A KBr pellet (radius of 10.0 mm, thickness of 2.00 mm) was formed from KBr powder (analytical reagent, Shanghai Macklin Reagent Co. Ltd., Shanghai, China) by using a tablet press and was used for the FTIR experiments.

2.2. Preparation of the PMoA/PANI Hybrid Thin Film

The double crystal $H_3PMo_{12}O_{40}$ (PMoA, analytical reagent, Shanghai Reagents Co. Ltd., Shanghai, China) was stored until use. A 0.02 g aliquot of polyaniline (PANI 98%, Shanghai Macklin Biochemical Co. Ltd., Shanghai, China) was mixed with 20 mL of a mixture of *N,N*-dimethylformamide (DMF, analytical reagent, Shanghai Macklin Reagent Co. Ltd., Shanghai, China) and ethanol to form solution No. 1 with a density of 0.1 mg/mL. A 0.04 g aliquot of $H_3PMo_{12}O_{40}$ (PMoA, analytical reagent, Shanghai Reagents Co. Ltd., Shanghai, China) was mixed with 20 mL of a mixture of *N,N*-dimethylformamide (DMF, analytical reagent, Shanghai Reagents Co. Ltd.) and ethanol (analytical reagent, Shanghai Reagents Co. Ltd.) to form solution No. 2 with a density of 0.2 mg/mL. Solution No. 1 and solution No. 2 were mixed to form solution No. 3. A pipetting gun was used to drop this mixed solution onto a quartz substrate to form a PMoA/PANI hybrid thin film with a amount of 100 µL, as shown in Figure 1. To test the thin film thickness, solution No. 3 was dropped on a quartz substrate to form a thin film, than the quartz substrate was put on the test bed of an thin film thickness measurement system (FCT-1030, Changchun Institute of Optics, Fine Mechanics and Physics, LCD Lab, Chinese Academy of Science. Changchun, China). When visible light is vertically illuminated on the measured film, one part of the light is reflected on the surface of the film, the other part penetrates into the film and then reflects at the interface between the film and the bottom layer. The thickness of the hybrid thin films was nearly 1.8 µm, as calculated by the supporting software of the thin film thickness measurement system.

2.3. Characterization

AFM experiments were performed on a 300HV atomic force microscope (Seiko, Tokyo, Japan). FTIR spectroscopy was performed on a 550 Fourier transform infrared spectrometer (Nicolet, Madison, WI, USA) over a wavenumber range of 500–4000 cm^{-1}. XPS analyses were performed on an ESCALAB 250 spectrometer (Thermo, Waltham, MA, USA) with an AlKα (1486.6 eV) and MgKα (1253.6 eV) mono X-ray source. Photochromic property measurements were performed on a V-500 UV-optical spectrophotometer (JASCO, Tokyo, Japan) in the range of 350–900 nm with an optical detection resolution of 1 nm. Electrochemical experiments were performed on a CHI660E electrochemical workstation (Chinstruments, Shanghai, China). DFT calculations of FMOs were carried out by means of the Gaussian 16 Quantum-Chemical calculation software package (Gussian, Inc. Cambridge, UK) with the B3LYP hybridizing functional and the 6–31G+(d,p) basis set.

2.4. Photochromic Experiments

Photochromic properties experiments were performed with a 300 W xenon light source (PLS-SXE, Beijing Perfectlight Technology Co. Ltd., Beijing, China) filtered by a UV filter (pass above 400 nm wavelength) as the optical-light source underneath the circumstance of entire light shield. The straight-line distance between the light and PMoA/PANI hybridizing thin film was 150 mm. The PMoA/PANI hybrid thin film samples were exposed to air during the process of the xenon light source illumination. The radiation was increased over time to generate a series curve of the results. Another sample of PMoA/PANI hybridizing thin film was exposed to air but not exposed to light, and absorption spectrogram curve were concerned regularly to supervise the bleaching process. All experiments were performed at room temperature.

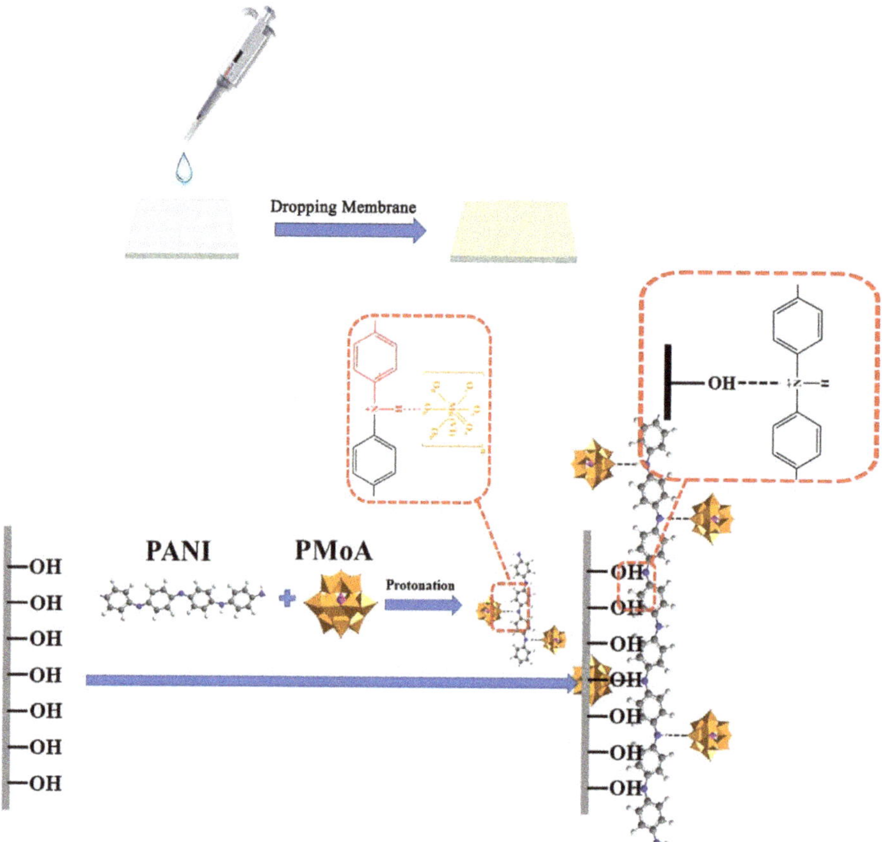

Figure 1. The protonation process to fabricate a PMoA/PANI hybridizing thin film.

3. Results and Discussion

The FTIR spectra of the PMoA, PANI, and the PMoA/PANI hybrid thin film before and after optical-light illumination in the wavenumber range of 500 cm^{-1}–2000 cm^{-1} are shown in Figure 2a. An enlargement of the FTIR spectrogram in the PMoA region is displayed in Figure 2b. The infrared-active vibrations of ν(Mo-Od), ν(P-O), ν(Mo-Oc-Mo), and ν(Mo-Ob-Mo) were observed in the spectrogram of PMoA and the PMoA/PANI hybrid thin film, and the infrared vibration at 1060 cm^{-1} attributed to ν(P-O) did not change in the spectrum of PMoA and the PMoA/PANI hybrid thin film before and after optical-light illumination, which demonstrated that the PMoA Keggin structure was well maintained in the PMoA/PANI hybrid thin film.

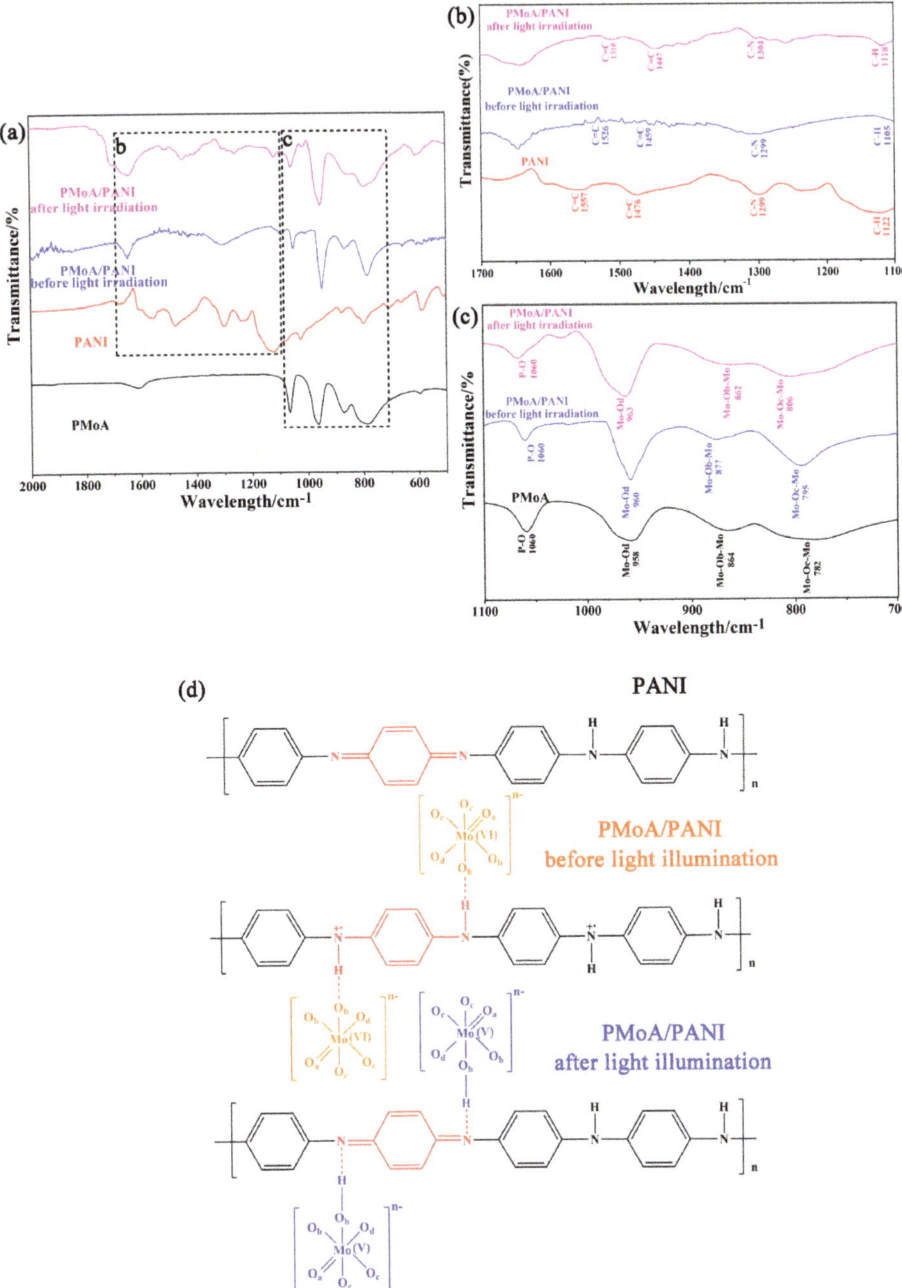

Figure 2. (**a**) FTIR spectrogram of PMoA, PANI, and the PMoA/PANI hybridizing thin film before and after optical-light illumination. (**b**) FTIR spectrogram in the PANI region of PANI and PMoA/PANI before and after optical-light illumination. (**c**) FTIR spectrogram in the PMoA region of PMoA and the PMoA/PANI hybridizing thin film before and after optical-light illumination. (**d**) Chemical structures of PANI and the PMoA/PANI hybridizing thin film before and after optical-light illumination.

The vibrations at 795 cm^{-1} attributed to ν(Mo-Oc-Mo) and at 877 cm^{-1} attributed to ν(Mo-Ob-Mo) in the spectrum of PMoA are attributed to the balance vibration, the intensity of which can be used to analyze the interaction between heteropolyacids and polymers. Compared with the peak positions in the spectrum of pristine PMoA, the stretching vibrations of ν(Mo-Oc-Mo) and ν(Mo-Ob-Mo) peaks were blueshifted by 13 cm^{-1} and the ν(Mo-Od) peak redshifted by 2 cm^{-1} in the spectrum of the PMoA/PANI hybridizing thin film. The FTIR spectrogram characterization of the four samples under different conditions indicated that PMoA maintained the base Keggin structure and interacted interfacially with PANI in the PMoA/PANI thin film during the process of protonation. Because Mo-Ob-Mo and Mo-Od served as of the pathways for charge transfer, the electron density was enhanced in the regions of Mo-Ob-Mo and Mo-Oc-Mo, which led to blueshifts of the stretching vibrations. Mo-Od maintains the stability of PMoA. Because of the protonation effect, PMoA lost protons, and the electron density decreased during the interaction with PANI, which caused a redshift in the spectrum, thus verifying the existence of an interfacial interaction between PMoA and PANI, as shown in the chemical structure of the PMoA/PANI hybridizing thin film before light illumination in Figure 2c. Compared with the spectrum of PMoA/PANI before optical-light illumination, the spectrum of the PMoA/PANI hybridizing thin film after optical-light illumination exhibited a decrease in the peaks at 806 cm^{-1} attributed to ν(Mo-Oc-Mo) and at 862 cm^{-1} attributed to ν(Mo-Ob-Mo), which was attributed to the reduction of heteropoly acid to heteropoly blue and proton (hydrogen ion) transfer to PMoA.

An enlargement of the FTIR spectrum in the PANI region is displayed in Figure 2c. The infrared peaks at 1476 cm^{-1} attributed to ν(C=C) and at 1122 cm^{-1} attributed to ν(C–H) are attributed to the PANI benzene structure in pure PANI [24,25], the infrared peak at 1557 cm^{-1} attributed to ν(C=C) is attributed to the PANI quinone structure in pure PANI [26–28] and the infrared peak at 1299 cm^{-1} attributed to ν(C–N) is attributed to the PANI benzene-quinone-benzene structure in pure PANI, which reflects the protonation doping position and protonation ratio of PANI [29]. Compared with the spectrum of PANI, the spectrum of the PMoA/PANI hybridizing thin film before optical-light illumination exhibited substantial decreases in the infrared peaks and redshifts of 17 cm^{-1} at 1459 cm^{-1} attributed to ν(C=C) and at 1105 cm^{-1} attributed to ν(C–H), redshifts of 21 cm^{-1} of the infrared peaks at 1526 cm^{-1} attributed to ν(C=C), respectively, and the infrared peak at 1299 cm^{-1} attributed to ν(C–N). These results demonstrated that the basic geometrical structure of the benzene ring for PANI was not destroyed, whereas the quinone-type structure of PANI transformed into a benzene-type structure after protonation by PMoA, and the protonation position was the nitrogen atom in the quinone ring. Correspondingly, compared with the spectrum of the PMoA/PANI hybridizing thin film before optical-light illumination, the spectrum of the PMoA/PANI after optical-light illumination exhibited similar peaks at 1188 cm^{-1}, 1447 cm^{-1} and 1510 cm^{-1} attributed to ν(C–H), ν(C=C) and ν(C=C) and a change in the peak at 1304 cm^{-1} attributed to ν(C–N). These results verify that protons (hydrogen ions) transfer to PMoA and that the benzene-type structure converts back to a quinone-type structure after optical-light illumination of the PMoA/PANI hybridizing thin film [24–28], which makes possible the protonation effect generated again in PANI and PMoA, as shown in the chemical structure of the PMoA/PANI hybridizing thin film after light illumination in Figure 2d.

The AFM characterization, which was used to examine the change in surface morphology, is shown in Figure 3. The 2D images of PANI and the PMoA/PANI hybrid thin film before and after optical-light illumination are displayed in Figure 3a1,b1,c1 and show that all three types of thin films are distributed evenly on the treated substrate in the range of $5.0 \times 5.0\ \mu m^2$, resulting in the three images having an even distribution color. The 3D images of PANI and the PMoA/PANI hybrid thin film before and after optical-light illumination in the range of $5.0 \times 5.0\ \mu m^2$ are exhibited in Figure 3a2,b2,c2. The peaks of the three types of thin films all had average surface sizes, which also confirmed that they are distributed evenly on the treated substrate. The root mean square (RMS) roughness of the PMoA/PANI hybridizing thin film before optical-light illumination was 2.34 nm (compared with 0.305 nm for PANI), which confirmed that the protonation effect of PMoA increased the spatial angle of the PANI polymer. Upon optical-light illumination, the root mean square roughness of the PMoA/PANI hybridizing thin film increased from 2.34 to 3.43 nm, which verified that the formation of heteropoly blue (PMo(V)A) and proton transfer increased the spatial angle of the PANI polymer. The change in morphology in PANI and the PMoA/PANI hybridizing thin film before and after optical-light illumination indicated that the addition of PMoA, protonation and proton transfer during photochromism resulted in an increasingly rough morphology, as shown by AFM. Therefore, the AFM characterization demonstrated that PMoA molecules interact with the PANI polymer via ionic bonding upon protonation of PMoA after the PMoA molecules complex with the PANI polymer, which also changes the spatial arrangement of PANI to narrow the peaks in the PMoA/PANI hybridizing thin film before optical-light illumination.

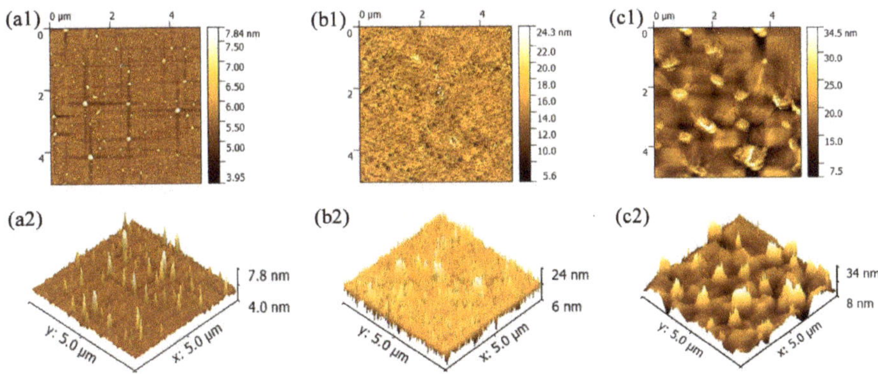

Figure 3. (**a1,a2**) AFM 2D and 3D image of PANI, (**b1,b2**) AFM 2D and 3D image of the PMoA/PANI hybridizing thin film before optical-light illumination, (**c1,c2**) AFM 2D and 3D image of the PMoA/PANI hybridizing thin film after optical-light illumination.

XPS photoelectron spectrograms of pure PANI and the PMoA/PANI hybrid thin film before illumination are shown in Figure 4a. In the full spectrogram (Figure 4a), all the peaks could be attributed to C, N and O (three different elements) in PANI and P, Mo, C, N and O (five different elements) in the PMoA/PANI hybridizing thin film before illumination; no other elements were observed. The XPS survey scan of C and N significantly reflected the structural change of PANI upon protonation. The C 1s XPS survey spectrogram of PANI and the PMoA/PANI hybrid thin film are displayed in Figure 4b. For pure PANI, the peaks fit the binding energies of C 1s at 284.5 eV, attributed to the C–H of the benzene ring and quinone ring, and 285.7 eV, attributed to the C=N of the quinone ring. For the PMoA/PANI hybridizing thin film, the peaks fit the binding energies of C 1s at 284.5 and 285.4 eV, attributed to the C–H and C=N groups, and a binding energy of C 1s at 285.9 eV, attributed to C–N$^+$, was observed upon protonation by PMoA. Furthermore, protonation by PMoA contributed to the redshift of the binding energy of the C=N groups. The N 1s XPS survey spectrogram of PANI and the PMoA/PANI hybridizing thin film are shown in Figure 4c; the peak fit of the binding energies of N 1s at 397.7 and 399.0 eV were attributed to the –N= and –NH– groups of pure PANI, and the bonding energies of N 1s at 397.25 and 398.70 eV were attributed to the –N= and –NH– groups of PMoA/PANI. A new peak observed at 401.5 eV was attributed to the –NH$^+$– group, which was caused by protonation by PMoA. Similarly, protonation by PMoA also contributed to the redshift of the binding energy of the –N= and –NH– groups in the spectrogram of the PMoA/PANI hybrid thin film. Thus, protonation by PMoA caused the binding energy of the C–H and C=N groups and the –N= and –NH– groups to redshift. Upon protonation by PMoA, the group consisting of 43.26% C–H and 8.39% C=N was converted to C–N$^+$, and the group consisting of 4.27% –N= and 12.77% –NH– was converted to –NH$^+$– in PANI of the PMoA/PANI hybridizing thin film. Table 1 shows that the percentages of C–H and C=N groups were 70.80 and 29.20% in pure PANI, and for the PMoA/PANI hybrid thin film, the C–H, C=N, and C–N$^+$ groups contributed 55.36, 22.75 and 23.89%, demonstrating that one part of C–H transferred from the quinone ring to the benzene ring and that a portion of the C=N groups were converted to C–N$^+$. Similarly, the –N= and –NH– groups contributed 27.89 and 72.11% in pure PANI, and for the PMoA/PANI hybrid thin film, the –N=, –NH–, and –NH$^+$– groups contributed 18.21, 64.75 and 17.04%, confirming that a portion of the –N= and –NH– groups are converted to –NH$^+$– upon protonation and unipolar rearrangement. The change in the contributions of the functional groups indicated that the site of protonation by PMoA was on the nitrogen atoms of the quinone structure of PANI and spread through the PANI polymer chain and rearranged via the unipolar effect.

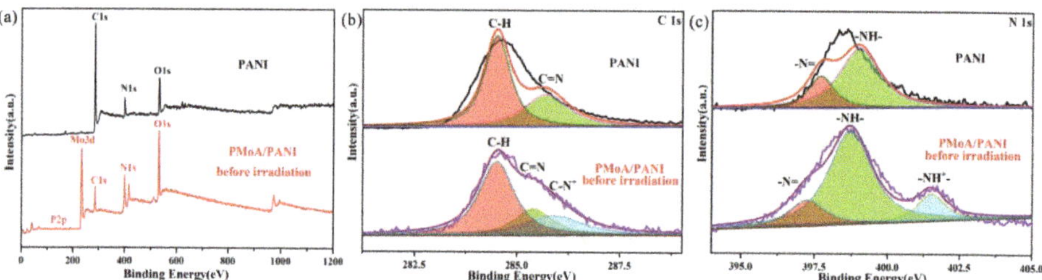

Figure 4. (**a**) XPS photoelectron spectrograms of PANI and PMoA/PANI hybrid thin films before illumination. (**b**) XPS photoelectron spectrogram and Gaussian deconvolution curve fitting for C 1s. (**c**) XPS photoelectron spectrogram and Gaussian deconvolution curve fitting for N 1s.

Table 1. Comparison of C– and N-related functional groups of pure PANI and the. PMoA/PANI hybrid film before irradiation.

Sample	Pure PANI					
Tested element	C			N		
Functional group	C–H	C=N	C–N$^+$	–N=	–NH–	–NH$^+$–
Binding Energy (eV)	284.5	285.7	-	397.7	399.0	-
Contribution	70.80%	29.20%	-	27.89%	72.11%	-
Sample	PMoA/PANI Hybrid Film before Irradiation					
Tested element	C			N		
Functional group	C–H	C=N	C–N$^+$	–N=	–NH–	–NH$^+$–
Binding Energy (eV)	284.5	285.4	285.9	397.25	398.70	401.5
Contribution	53.36%	22.75%	23.89%	18.21%	64.75%	17.04%

The UV-Vis absorption spectrogram of the coloration process of the PMoA/PANI hybrid thin film at wavelengths of 400 nm to 1000 nm for different times are displayed in Figure 5a. Before optical-light illumination, a characteristic absorption by the PMoA/PANI hybridizing thin film in the ultraviolet-optical (UV-Vis) region is observed. Then, after optical-light illumination, the hybrid thin film gradually converted from pale yellow to a deep blue color and the characteristic absorption peak at 732–740 nm was attributed to the intervalence charge transfer (IVCT) [30] transition from Mo^{6+} to Mo^{5+}, which was the characteristic absorption in the spectrogram of PMoA blue. Moreover, as the absorption intensity increased, the characteristic peak site of IVCT underwent a redshift from 732 nm to 740 nm with incremental illumination time. Every 10 min test showed that the coloration process was gradual and was not accomplished in one step. The characteristic absorption peak eventually reached coloration saturation with a maximum absorption intensity of 3.46 in 40 min under optical-light illumination, which verified that the illumination time demanded for the absorbance of the PMoA/PANI hybridizing thin film to reach coloration saturation was long and the coloration rate was low, reflecting that the volume of d-d charge transfer was large.

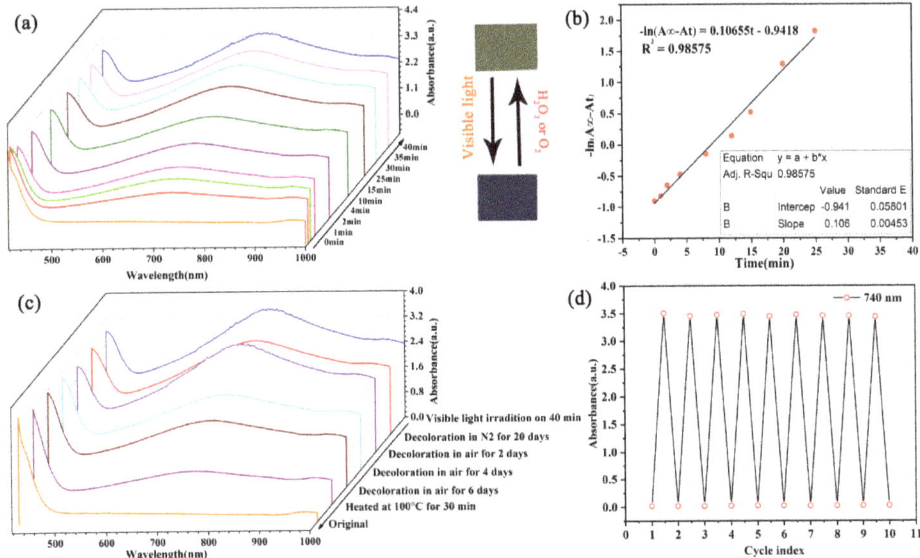

Figure 5. (**a**) UV-Vis spectrogram of the coloration process of the PMoA/PANI hybridizing thin film for different coloration time. (**b**) Kinetic plot of the first-order photochromic process in the PMoA/PANI hybridizing thin film. (**c**) UV-Vis spectrogram of the decoloration process of the PMoA/PANI hybridizing thin film for different decoloration approach. (**d**) The reversibility of the coloration cycle of the PMoA/PANI hybridizing thin film in H_2O_2.

The kinetics of the photochromic coloration process of the PMoA/PANI hybridizing thin film at a wavelength of 740 nm are displayed in Figure 5b. The coloration process can be described by the following first-order kinetic formula:

$$-ln(A\infty - At) = kt + b \tag{1}$$

where $A\infty$ is the absorbance value when the coloration process is complete, At is the absorbance value at each test time during the coloration process, k is the coloration rate constant and b is a constant for the formula. The PMoA/PANI hybrid thin film fits the first-order kinetic formula, and the constant k is 0.08 min^{-1}, which reflects the coloration rate.

The different decoloration processes of the PMoA/PANI hybridizing thin film are exhibited in Figure 5c. The image shows that without any light illumination, the decoloration process proceeds successfully when the PMoA/PANI hybrid thin film is placed in air, which contains a large amount of oxygen, confirming that oxygen performs a key role in the decoloration process. However, when the hybrid thin film is placed in a nitrogen gas environment, the decoloration process does not proceed. When the PMoA/PANI hybridizing thin film was heated at 100 °C for 30 min, the intensity of the absorbance peak decreased to 1.097, and the bleaching process reached 80%.

The invertibility of coloration and decoloration cycle of PMoA/PANI hybridizing thin film at 740 nm is displayed in Figure 5d. The PMoA/PANI hybridizing thin film was irradiated with optical-light during the coloration process, and then 15% H_2O_2 was dropped onto the substrate during the decoloration process. The primary absorbance of every cycle increased as the times of illumination cycles increased, attributing to the intensity of the emission of optical-light. Different absorbance between coloration and decoloration was in the range of 3.59 to 3.42 over 10 cycles, indicating that the PMoA/PANI hybrid thin film presented favorable photochromic properties, with high stability and good invertibility [31].

As Table 2 shows, the PMoA/PANI hybrid thin film exhibited a high maximum absorbance compared with the other representative samples, which indicated that the protonation effect of PMoA toward PANI in the PMoA/PANI hybrid thin film could increase the maximum absorbance and greatly improve the photochromic properties of the PMoA/PANI hybrid thin film.

Table 2. Comparison of the maximum absorbance of photochromic materials.

Photochromic Material	Maximum Light Absorbance	Reference
Fe(III) R-PLG$_{23}$ metal complex	0.12	[32]
PMoA/TiO$_2$	0.13	[31]
PMoA/Na-MMT/PVPd	0.19	[33]
PMoA/PVPd	0.225	[14]
CP/TiO$_2$	0.25	[34]
(NH$_4$)$_{14}$[NaP$_5$W$_{30}$O$_{110}$]	0.28	[35]
Gold Nanoparticle-Molybdenum Trioxide Thin Films	0.75	[36]
CsPbBr$_3$ Quantum Dot Films	0.78	[37]
MoO$_3$ Nanoribbons	0.8	[9]
Two Tri-Lacunary α-Dawson-Type Polyoxotungstates	1.25	[38]
Non-Stoichiometric Monoclinic Structured Tungsten Trioxide (WO$_{3-x}$)	1.27	[39]
Spiropyran-Containing Fluorinated Polyacrylate Hydrophobic Coatings	1.5	[40]
PEG-400 assisted WO$_3$–TiO$_2$–ZnO films	1.6	[41]
WO$_{3-x}$ QDs	2.8	[34]
Rhodamine Joined with Polyurethane	3.2	[42]
Aromatic Sulfonium Octamolybdates	3.35	[43]
PMoA/PANI Hybridizing Thin Film	3.46	**This work**

The XPS spectrogram of the PMoA/PANI hybrid thin film before and after optical-light illumination are displayed in Figure 6a. All the peaks were attributed to P, Mo, C, N and O, and no other elements were observed. Mo XPS spectrogram of the PMoA/PANI hybridizing thin film are shown in Figure 6b. The fitting peaks of the Mo spin-split energy levels correspond to Mo 3d$_{3/2}$ and Mo 3d$_{5/2}$. The binding energies of the PMoA/PANI hybrid thin film before optical-light illumination at 233.06 and 236.21 eV are attributed to Mo^{6+}, while the binding energies at 235.58 and 231.73 eV are attributed to Mo^{5+}. Similarly, the binding energies of the PMoA/PANI hybrid thin film after optical-light illumination at 232.85 and 236.00 eV are attributed to Mo^{6+}, while the binding energies at 234.7 and 231.6 eV are attributed to Mo^{5+}. To further investigate the change in electronic structure during the coloration process of PMoA/PANI, Gaussian deconvolution peak integration was used to calculate the respective proportions of Mo^{6+} and Mo^{5+} in the PMoA/PANI hybridizing thin film before and after optical-light illumination.

The relative proportions of Mo^{6+} and Mo^{5+} before and after optical-light illumination are shown in Table 3. The ligand-to-metal charge transfer (LMCT) triggered a change in the valence of Mo, causing Mo^{6+} to be reduced to Mo^{5+}. Table 2 shows that the proportion of Mo^{5+} in the PMoA/PANI hybrid thin film was 7%, which was attributed to the X-ray illumination of the PMoA/PANI hybrid thin film before optical-light illumination. The proportion of Mo^{5+} in the PMoA/PANI hybrid thin film after optical-light illumination was 36%, almost 5 times larger than that before optical-light illumination. Compared with that of the PMoA/PANI hybrid thin film before optical-light illumination, the Gaussian deconvolution peak integration area for Mo^{5+} of the PMoA/PANI hybrid thin film after optical-light illumination is obviously increased, which confirmed that under the stimulus of photons, electronic transitions from the original energy level occurred on a large scale, such that a large amount of Mo^{6+} gained electrons to be reduced to Mo^{5+}.

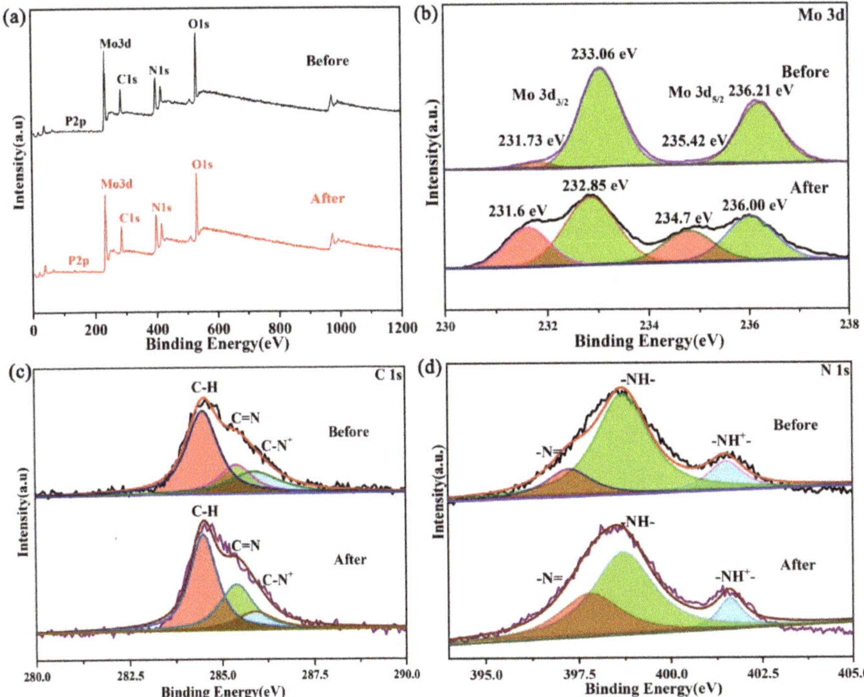

Figure 6. (**a**) XPS survey spectrogram of the PMoA/PANI hybrid thin film before and after optical-light illumination. (**b**) XPS survey spectrogram and Gaussian deconvolution curve fitting for Mo 3d in the PMoA/PANI hybrid thin film. (**c**) XPS survey spectrogram and Gaussian deconvolution curve fitting for C 1s in the PMoA/PANI hybrid thin film. (**d**) XPS survey spectrogram and Gaussian deconvolution curve fitting for N 1s in the PMoA/PANI hybrid thin film.

Table 3. The change in the valence state of Mo in the PMoA/PANI hybrid thin film.

Sample	Mo^{5+}		Mo^{6+}		Mo^{5+}/Mo Ratios
	$3d_{3/2}$	$3d_{5/2}$	$3d_{3/2}$	$3d_{5/2}$	
Before	231.85	235.50	233.05	236.20	0.07
After	231.65	234.80	232.90	236.0	0.36

The C 1s XPS spectrogram of the PMoA/PANI hybrid thin film before and after optical-light illumination is exhibited in Figure 6c. Compared with the spectrum of the PMoA/PANI hybrid thin film before illumination, that of the PMoA/PANI hybrid thin film after illumination exhibited binding energies at 284.5, 285.4, and 285.9 eV that were attributed to C–H, C–N$^+$, and C=N. The relative proportions of C–H, C–N$^+$, and C=N change during the photochromic process because of proton transfer, which results in a valence change of the Mo in PMoA. Compared with the relative proportions in the PMoA/PANI hybrid thin film before illumination, the relative proportion of C–N$^+$ decreased, that of C=N increased, and that of C–H remained constant after illumination, which surely indicated that the PANI in the PMoA/PANI hybrid thin film was converted from the benzene-type structure of single-polarized PANI back to the quinone-type structure of eigenstate PANI because of proton transfer. The N 1s XPS spectrogram of the PMoA/PANI hybrid thin film before and after optical-light illumination is displayed in Figure 6d. Compared with the spectrogram of the PMoA/PANI hybrid thin film before illumination, that of the PMoA/PANI hybrid thin film after illumination exhibited binding energies at 397.76, 398.70, and 401.5 eV that were attributed to –N=, –NH–, and –NH$^+$–. The change in the proportions of –N=, –NH–, and –NH$^+$– during the photochromic process was attributed to proton transfer; the relative proportions of –NH– and –NH$^+$– decreased, whereas that of –N= increased, which also triggered conversion of the PANI in the PMoA/PANI hybrid thin film from the benzene-type structure of single-polarized PANI back to the quinone-type structure of eigenstate PANI upon proton transfer [44,45].

The change in the valence state of Mo also triggered the transformation of PANI. As Table 4 shows, the proportions of the C–H, C=N and C–N$^+$ groups were 53.36, 22.75 and 23.89%, respectively, in the PMoA/PANI hybrid thin film before illumination, and for the PMoA/PANI hybridizing thin film after illumination, the proportions of the C–H, C=N and C–N$^+$ groups were 53.36, 32.44 and 14.20%, respectively. A portion of the C–N$^+$ groups were reconverted to C=N, but the proportion of C–H did not change in the process. Similarly, the proportions of the –N=, –NH–, and –NH$^+$– groups were 18.21, 64.75 and 17.04%, respectively, in PMoA/PANI before illumination, and for the PMoA/PANI hybrid thin film after illumination, the proportions of the –N=, –NH–, and –NH$^+$– groups were 31.89%, 58.30% and 9.81%, respectively. Moreover, a portion of the -NH$^+$- groups were reconverted to –N= and –NH– upon proton transfer. The change in the proportions of these functional groups indicated that site of proton transfer from PMoA was the nitrogen atoms of the quinone structure of PANI and spread through the PANI polymer chain and rearranged via the unipolar effect.

Table 4. Comparison of C– and N—containing functional groups of PMoA/PANI hybrid thin films before and after optical-light illumination.

Sample	PMoA/PANI Hybrid Thin Film before Illumination					
Tested element		C			N	
Functional group	C–H	C=N	C–N$^+$	–N=	–NH–	–NH$^+$–
Binding Energy (eV)	284.5	285.4	285.9	397.2	398.2	401.5
Contribution	53.36%	22.75%	23.89%	18.21%	64.75%	17.04%
Sample	PMoA/PANI Hybrid Thin Film after Illumination					
Tested element		C			N	
Functional group	C–H	C=N	C–N$^+$	–N=	–NH–	–NH$^+$–
Binding Energy (eV)	284.5	285.4	285.9	397.80	398.70	401.5
Contribution	53.36%	32.44%	14.20%	31.89%	58.30%	9.81%

The EIS results of PANI and the PMoA/PANI hybrid thin film are shown in Figure 7a. The electrochemical impedance arc radius of PANI was larger than that of the PMoA/PANI hybrid thin film, which indicated that electrons delocalized more easily in the PMoA/PANI hybrid thin film than in PANI after protonation, providing strong evidence that protonation enhanced the electron transfer efficiency in the PMoA/PANI hybrid thin film.

For further investigating the electron transfer efficiency, the Mott-Schottky curve of PANI and the PMoA/PANI hybrid thin film are given in Figure 7b. After estimation by the Mott-Schottky formula, the flat-band potential of PANI was −0.61 V, and flat-band potential of PMoA was −0.86 V. The flat-band potential of PMoA/PANI was lower than that of PMoA, indicating the PMoA/PANI had more tendency to be donate electrons than PANI and protonation increased the electron transfer properties in the PMoA/PANI hybrid thin film.

The FMOs of eigenstate PANI and unipolar PANI calculated by DFT are displayed in Figure 8. Figure 8a shows that in the eigenstate PANI unit, the electron density in the highest occupied molecular orbital (HOMO) was concentrated on benzene rings 1 and 3 and the nitrogen atoms connected to them. Conversely, the electron density in the lowest unoccupied molecular orbital (LUMO) was concentrated on quinone ring 2 and the nitrogen atoms connected to it. Upon electron transfer from the HOMO to the LUMO, the electron density on quinone was increased in the eigenstate PANI unit. Figure 8b shows that in the unipolar PANI unit arising from protonation of eigenstate PANI, the electron density of the HOMO was distributed evenly across benzene rings 1, 2 and 3 and the nitrogen atoms connected to them. Furthermore, the electron density of the LUMO was distributed evenly across benzene rings 1 to 4 and the nitrogen atoms connected with them. Upon electron transfer from the HOMO to the LUMO, the electron density on the quinone ring was increased in the unipolar PANI unit. In the FMOs of eigenstate PANI and unipolar PANI, all nitrogen atoms exhibit obvious π properties, and those hybridizing atoms evidently contributed to electron delocalization across the four rings of the polymer. Moreover, compared with eigenstate PANI, unipolar PANI has a more homogeneous electron density distribution regardless of whether a benzene ring or a nitrogen-containing ring is present. The HOMO consisting of orbitals on the ring structure exhibits π properties, facilitating electron delocalization. In addition, the LUMO consisting of orbitals on the ring structure exhibits π^* properties. The FMO calculation provides strong evidence that the unipolar PANI possesses strong electron delocalization and transfer abilities via protonation of eigenstate PANI, which was demonstrated by the EIS test and confirmed the hypothesis that the PANI/PMoA hybridizing thin film exhibits improved photochromic properties [46].

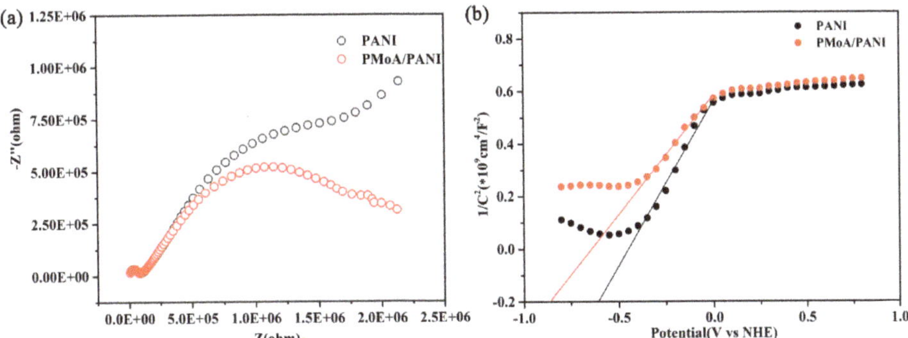

Figure 7. Electrochemistry study of EIS of PANI and the PMoA/PANI hybrid thin film (**a**) EIS of PANI and the PMoA/PANI hybrid thin film. (**b**) Mott-Schottky curve of PANI and the PMoA/PANI hybrid thin film.

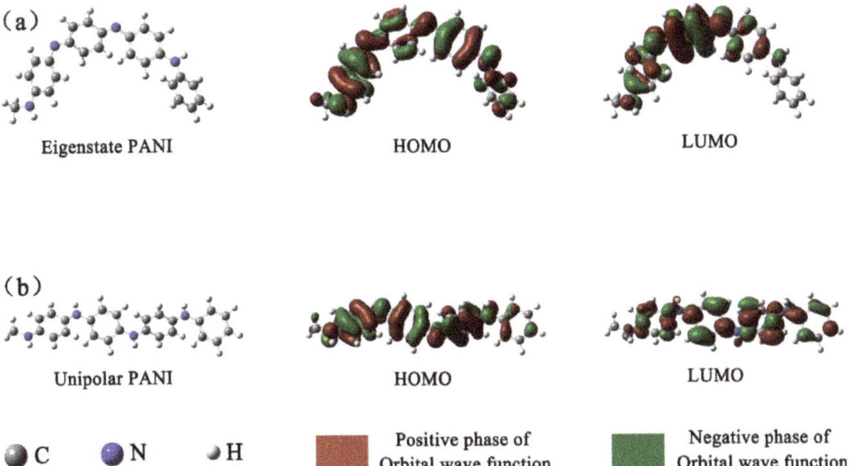

Figure 8. (**a**) DFT calculations of the FMOs of the eigenstate PANI unit. (**b**) DFT calculations of the FMOs of the unipolar PANI unit.

According to the mechanism of photochromism for PMoA and the application of PANI, the mechanism photochromism of the PMoA/PANI hybrid thin film was attributed to IVCT and LMCT [30] of PMoA and the change in electrical conductivity of PANI via protonation. As shown in Figure 9a, the eigenstate PANI consisted of benzene and quinone with a ratio of approximately 3:1 in every polymer unit, resulting in poor electrical conductivity that was approximately equal to that of an insulator before protonation [47]. When PANI was combined with PMoA, the chemical structure of PANI underwent a series of transformations due to protonation by PMoA. Furthermore, the electrons associated with nitrogen atoms in amino groups (–N=) delocalized to the adjacent benzene ring, causing the density of the electron cloud to decrease; such electron delocalization induced a conjugation effect, which caused the electron cloud on the PANI chain to rearrange, as shown in the FTIR results in Figure 2.

This rearrangement of the π electrons triggered a transformation of the quinone-bipolaron state into the benzene monopolaron state. The benzene monopolaron structure exhibited a semi-full conductive band, which resulted in electrons moving easily within the energy band when stimulated, enhancing the electrical conductivity of PANI. Moreover, PMoA interacted with PANI via ionic bonds, as shown in Figure 9b1.

As Figure 9c shows, since PMoA has a special Keggin-type structure [48], the Mo^{6+} ion in oxidized PMoA has a d^0 electronic structure, which causes PMoA to exhibit a single absorption peak in the ultraviolet and visible region that is attributed to the IVCT [30] effect, and then low-energy electrons mainly exist in the 2p orbital of the O atoms of PMoA. Upon visible light irradiation of the PMoA/PANI hybrid film, the electrons in the LUMO, consisting of the 2p orbital of the O atoms, were excited to the HOMO and then captured by electron trapping of the middle energy band formed by metal ions, converting d^0 electrons to d^1 electrons [49,50] by LMCT. Then, the d^1 electron were transferred through the molecule because of different valence states and d-d charges of adjacent metal centers such that Mo^{6+} ions were reduced to Mo^{5+} ions by IVCT, which promoted the absorption of visible light [30]. Moreover, the bridged oxygen (Od) of PMoA has a unique redox absorption point, and protons (H^+) can easily be absorbed by Ob to compensate for the negative charge, resulting in a LUMO with antibonding properties [51]. The terminal oxygen (Od) of the Keggin-type structure, which is linked to Mo by double bonds, also possessed the strongest chemical activity in PMoA.

As Figure 9b1,b2 show, in general, combining the unique electron transfer ability of PMoA and the proton transfer ability of PANI, the photochromic mechanism of the PMoA/PANI hybrid film can be described as follows: PANI is protonated by multiproton PMoA in the liquid phase; then, the eigenstate of the quinone-type structure in every polymer unit of PANI is converted to a single-polarized benzene-type structure, the electrons in the single-polarized benzene structure are easily delocalized, $N-H^+$ groups establish in every protonated PANI polymer unit, and PMoA interacts firmly with PANI by ionic bonds. Upon optical-light illumination of the PMoA/PANI hybrid film, the PMoA undergoes IVCT and LMCT [30] to reduce Mo^{6+} to Mo^{5+}, and the color of the PMoA/PANI hybrid film changes from pale yellow $(PMoA(Mo^{6+}))$ to blue $(PMoA(Mo^{5+}))$. Furthermore, because of the LMCT effect, the electrons in $N-H^+$ are transferred to the LUMO, consisting of the 2p orbitals of O atoms, and H^+ are captured by Ob, protons and transferred such that PANI loses protons. Thus, the benzene-type structure is reduced back to the quinone structure, while PMoA comes into contact with PANI via hydrogen bonding. Because of the strong electron delocalization and transfer properties of unipolar PANI, the d^1 electrons are transferred through the molecule due to the different valence states and d-d charges of adjacent metal centers, which vastly enhances the maximum absorbance of optical-light, significantly improving photochromic properties of the PMoA/PANI hybrid film. The electrons must overcome ionic bonding to be transferred to the LUMO, consisting of O 2p orbitals, and the coloration process becomes slower. During the decoloration process, the PMoA/PANI hybrid films are exposed to air, and $PMoA(Mo^{5+})$ reacts with O_2 and is then reoxidized to $PMoA(Mo^{6+})$. $PMoA(Mo^{6+})$ interacts with PANI again, resulting in a cycle of coloration and decoloration of $PMoA(Mo^{5+})/PMoA(Mo^{6+})$ in the PMoA/PANI hybrid film.

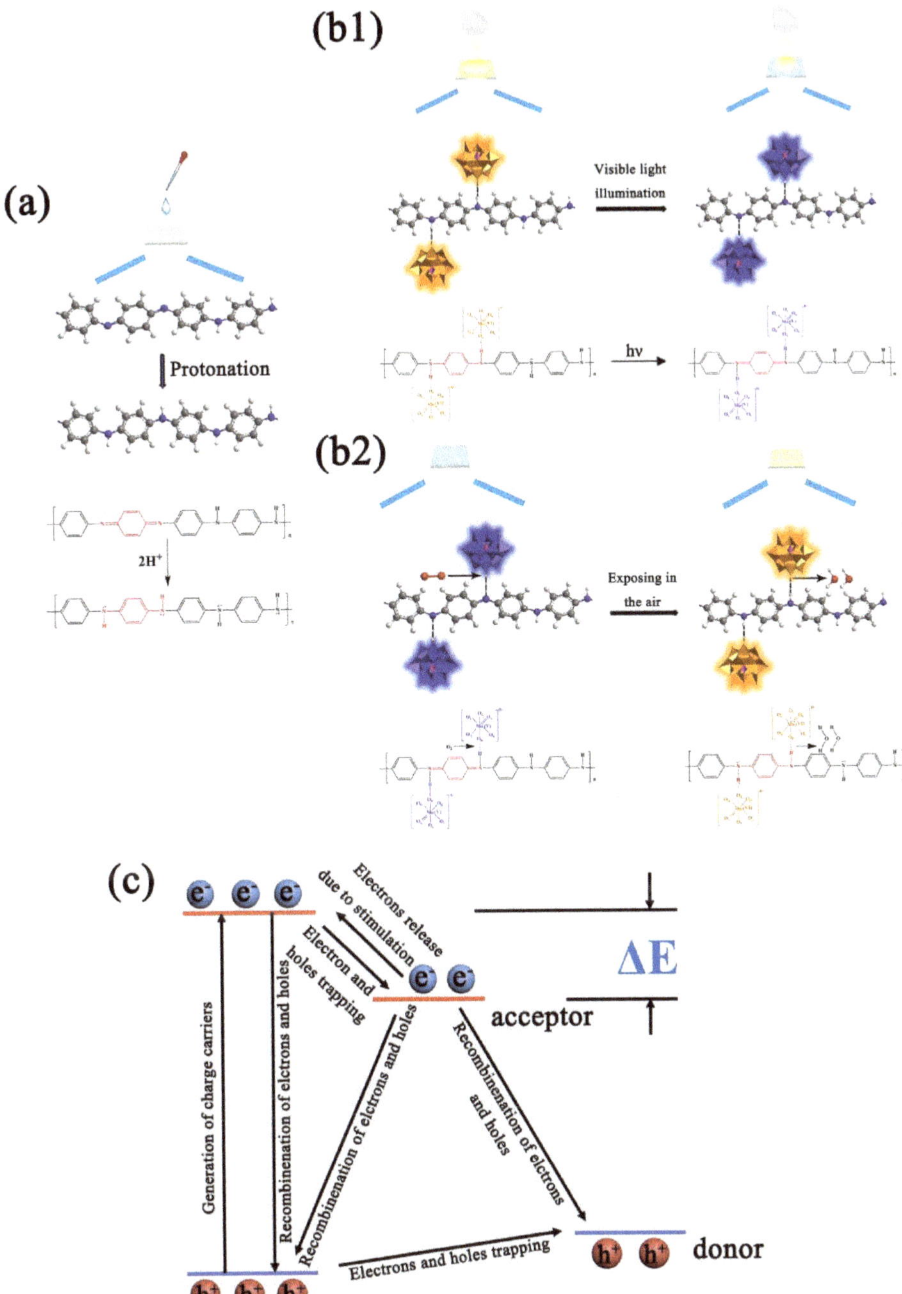

Figure 9. (**a**) The protonation process of PANI. (**b1**) Process of coloration of the PMoA/PANI hybrid thin film. (**b2**) Process of decoloration of the PMoA/PANI hybrid thin film in air. (**c**) Schematic diagram of LMCT.

4. Conclusions

A photochromic organic/inorganic PMoA/PANI hybrid thin film was easily synthesized by the protonation effect and demonstrated great photochromic properties, especially high light absorption of 3.46 in the UV-vis spectrum. AFM presented an even distribution of PMoA/PANI hybrid thin film root mean square roughness with 2.34 nm before illumination and 3.43 nm after illumination. All the characterizations and mechanistic analyses showed that protonation played a key role in the transformations of the quinone structure of eigenstate PANI and the single-polarized benzene structure of PANI observed by FTIR and XPS, which induced absorption by PMoA, enhancing the electrons transfer property verified by electrochemistry and DFT studies and providing electrons for PMoA to alter the valence state in order to achieve exceptional photochromic performance. All these results provided a new possibility for the application of PMoA and PANI, especially in the region of laser ink-free printing and optical control switches.

Author Contributions: Q.Z. contributed to overall organizing of all experiments, and writing-original draft preparation. S.G. and Y.S. contributed to photochromic properties experiments, data analysis and visualization. Z.L. contributed to sample preparation and characterization. W.F. contributed to providing ideas for all the experiments, methodology, and supervision. All authors have read and agreed to the published version of the manuscript.

Funding: This study was funded by the National Natural Science Foundation of China, grant number 61774073, the Open Project of State Key Laboratory of Inorganic Synthesis and Preparative Chemistry, Jilin University, grant number 2016-25), and the Science and Technology Development Program of Jilin Province, grant number 20170101086JC. And the APC was funded by the Science and Technology Development Program of Jilin Province, grant number 20170101086JC.

Conflicts of Interest: The authors declare no conflict of interest.

References

1. Shallcross, R.C.; Körner, P.O.; Maibach, E.; Köhnen, A.; Meerholz, K. A Photochromic Diode with a Continuum of Intermediate States: Towards High Density Multilevel Storage. *Adv. Mater.* **2013**, *25*, 4807–4813. [CrossRef]
2. Feringa, B.L. The Art of Building Small: From Molecular Switches to Motors (Nobel Lecture). *Angew. Chem. (Int. Ed. Engl.)* **2017**, *56*, 11060–11078. [CrossRef] [PubMed]
3. Xu, L.; Mou, F.; Gong, H.; Luo, M.; Guan, J. Light-driven micro/nanomotors: From fundamentals to applications. *Chem. Soc. Rev.* **2017**, *46*, 6905–6926. [CrossRef]
4. Yeh, P.; Yeh, N.; Lee, C.-H.; Ding, T.-J. Applications of LEDs in optical sensors and chemical sensing device for detection of biochemicals, heavy metals, and environmental nutrients. *Renew. Sustain. Energy Rev.* **2017**, *75*, 461–468. [CrossRef]
5. Yao, J.N.; Hashimoto, K.; Fujishima, A. Photochromism induced in an electrolytically pretreated MoO_3 thin film by visible light. *Nature* **1992**, *355*, 624–626. [CrossRef]
6. Liu, H.; Maruyama, H.; Masuda, T.; Arai, F. Vibration-assisted optical injection of a single fluorescent sensor into a target cell. *Sens. Actuators B Chem.* **2015**, *220*, 40–49. [CrossRef]
7. Zhang, Q.; Yue, S.; Sun, H.; Wang, X.; Hao, X.; An, S. Nondestructive up-conversion readout in Er/Yb co-doped $Na_{0.5}Bi_{2.5}Nb_2O_9$-based optical storage materials for optical data storage device applications. *J. Mater. Chem. C* **2017**, *5*, 3838–3847. [CrossRef]
8. Mingzhe, L.; Hiroyuki, A.; Makoto, K. Azobenzene-tethered T7 promoter for efficient photoregulation of transcription. *J. Am. Chem. Soc.* **2006**, *128*, 1009.
9. Mengjing, W.; Koski, K.J. Reversible chemochromic MoO_3 nanoribbons through zerovalent metal intercalation. *ACS Nano* **2015**, *9*, 3226–3233.
10. Li, N.; Li, Y.; Zhou, Y.; Li, W.; Ji, S.; Yao, H.; Cao, X.; Jin, P.; Li, N.; Li, Y. Interfacial-charge-transfer-induced photochromism of $MoO_3@TiO_2$ crystalline-core amorphous-shell nanorods. *Sol. Energy Mater. Sol. Cells* **2017**, *160*, 116–125. [CrossRef]
11. Gurgel, G.M.; Lovisa, L.X.; Pereira, L.M.; Motta, F.V.; Bomio, M.R.D. Photoluminescence properties of (Eu, Tb, Tm) co-doped $PbMoO_4$ obtained by sonochemical synthesis. *J. Alloys Compd.* **2017**, *700*, 130–137. [CrossRef]
12. Lu, J.; Zhang, X.; Ma, P.; Singh, V.; Zhang, C.; Niu, J.; Wang, J. Photochromic behavior of a new polyoxomolybdate/alkylamine composite in solid state. *J. Mater. Sci.* **2018**, *53*, 3078–3086. [CrossRef]

13. Feng, W.; Zhang, T.R.; Liu, Y.; Wei, L.; Lu, R.; Li, T.J.; Zha, Y.Y.; Yao, J.N. Novel Hybrid Inorganic–organic Film Based on the Tungstophosphate Acid–polyacrylamide System: Photochromic Behavior and Mechanism. *J. Mater. Res.* **2002**, *17*, 133–136. [CrossRef]

14. Sun, Y.; Wang, X.; Lu, Y.; Xuan, L.; Xia, S.; Feng, W.; Han, X. Preparation and visible-light photochromism of phosphomolybdic acid/polyvinylpyrrolidone hybrid film. *Chem. Res. Chin. Univ.* **2014**, *30*, 703–708. [CrossRef]

15. Ai, L.-M.; Feng, W.; Chen, J.; Liu, Y.; Cai, W.-M. Evaluation of microstructure and photochromic behavior of polyvinyl alcohol nanocomposite films containing polyoxometalates. *Mater. Chem. Phys.* **2008**, *109*, 131–136. [CrossRef]

16. Chen, J.; Mei Ai, L.; Feng, W.; Xiong, D.Q.; Liu, Y.; Cai, W.M. Preparation and photochromism of nanocomposite thin film based on polyoxometalate and polyethyleneglycol. *Mater. Lett.* **2007**, *61*, 5247–5249. [CrossRef]

17. Parrot, A.; Bernard, A.; Jacquart, A.; Serapian, S.A.; Bo, C.; Derat, E.; Oms, O.; Dolbecq, A.; Proust, A.; Métivier, R. Photochromism and Dual-Color Fluorescence in a Polyoxometalate-Benzospiropyran Molecular Switch. *Angew. Chem. Int. Ed.* **2017**, *129*, 4872. [CrossRef]

18. Li, D.; Wei, J.; Dong, S.; Li, H.; Xia, Y.; Jiao, X.; Wang, T.; Chen, D. Novel PVP/HTA Hybrids for Multifunctional Rewritable Paper. *ACS Appl. Mater. Interfaces* **2018**, *31*, 1701–1706. [CrossRef]

19. Macdiarmid, A.G.; Chiang, J.C.; Richter, A.F.; Epstein, A.J. Polyaniline: A new concept in conducting polymers. *Synth. Met.* **1987**, *18*, 285–290. [CrossRef]

20. Epstein, A.J.; Macdiarmid, A.G. Electronic Phenomena in Polyanilines. *Synth. Met.* **1989**, *29*, 395–400.

21. Kahol, P.K.; Pinto, N.J. Electron paramagnetic resonance investigations of electrospun polyaniline fibers. *Solid State Commun.* **2002**, *124*, 195–197. [CrossRef]

22. Kahol, P.K.; Raghunathan, A.; Mccormick, B.J. A magnetic susceptibility investigation of emeraldine base polyaniline. *Synth. Met.* **2004**, *140*, 261–267. [CrossRef]

23. Diaz, A.F.; Logan, J.A. Electroactive polyaniline films. *J. Electroanal. Chem. Interfacial Electrochem.* **1980**, *111*, 111–114. [CrossRef]

24. Athawale, A.A.; Kulkarni, M.V.; Chabukswar, V.V. Studies on chemically synthesized soluble acrylic acid doped polyaniline. *Mater. Chem. Phys.* **2002**, *73*, 106–110. [CrossRef]

25. Macdiarmid, A.G.; Epstein, A.J. The concept of secondary doping as applied to polyaniline. *Synth. Met.* **1994**, *65*, 103–116. [CrossRef]

26. Zhu, J.; Wei, S.; Lei, Z.; Mao, Y.; Guo, Z. Polyaniline-tungsten oxide metacomposites with tunable electronic properties. *J. Mater. Chem.* **2010**, *21*, 342–348. [CrossRef]

27. Quillard, S.; Louarn, G.; Lefrant, S.; Macdiarmid, A. Vibrational analysis of polyaniline: A comparative study of leucoemeraldine, emeraldine, and pernigraniline bases. *Phys. Rev. B* **1994**, *50*, 12496–12508. [CrossRef]

28. Furukawa, Y.; Ueda, F.; Hyodo, Y.; Harada, I.; Nakajima, T.; Kawagoe, T. Vibrational spectra and structure of polyaniline. *Macromolecules* **1988**, *21*, 1297–1305. [CrossRef]

29. Miao, Y.E.; Fan, W.; Chen, D.; Liu, T. High-performance supercapacitors based on hollow polyaniline nanofibers by electrospinning. *ACS Appl. Mater. Interfaces* **2013**, *5*, 4423–4428. [CrossRef]

30. Yamase, T. Photo- and Electrochromism of Polyoxometalates and Related Materials. *Chem. Rev.* **1998**, *98*, 307–326. [CrossRef]

31. Wei, Y.; Han, B.; Dong, Z.; Feng, W. Phosphomolybdic acid-modified highly organized TiO$_2$ nanotube arrays with rapid photochromic performance. *J. Mater. Sci. Technol.* **2019**, *35*, 1951–1958. [CrossRef]

32. Gao, J.-Y.; Huang, W.-C.; Huang, P.-Y.; Song, C.-Y.; Hong, J.-L. Light-Up of Rhodamine Hydrazide to Generate Emissive Initiator for Polymerization and to Afford Photochromic Polypeptide Metal Complex. *Polymers* **2017**, *9*, 419. [CrossRef]

33. Wang, X.-Y.; Dong, Q.; Meng, Q.-L.; Yang, J.-Y.; Feng, W.; Han, X.-K. Visible-light photochromic nanocomposite thin films based on polyvinylpyrrolidone and polyoxometalates supported on clay minerals. *Appl. Surf. Sci.* **2014**, *316*, 637–642. [CrossRef]

34. Ullah, S.; Acuña, J.J.S.; Pasa, A.A.; Bilmes, S.A.; Vela, M.E.; Benitez, G.; Rodrigues-Filho, U.P. Photoactive layer-by-layer films of cellulose phosphate and titanium dioxide containing phosphotungstic acid. *Appl. Surf. Sci.* **2013**, *277*, 111–120. [CrossRef]

35. Liu, S.; Moehwald, H.; Volkmer, D.; Kurth, D. Polyoxometalate-Based Electro- and Photochromic Dual-Mode Devices. *Langmuir ACS J. Surf. Colloids* **2006**, *22*, 1949–1951. [CrossRef]

36. He, T.; Ma, Y.; Cao, Y.; Jiang, P.; Zhang, X.; Yang, W.; Yao, J. Enhancement Effect of Gold Nanoparticles on the UV-Light Photochromism of Molybdenum Trioxide Thin Films. *Langmuir* **2001**, *17*, 8024–8027. [CrossRef]

37. Qaid, S.M.; Alharbi, F.H.; Bedja, I.; Nazeeruddin, M.K.; Aldwayyan, A.S. Reducing Amplified Spontaneous Emission Threshold in CsPbBr$_3$ Quantum Dot Films by Controlling TiO$_2$ Compact Layer. *Nanomaterials* **2020**, *10*, 1605. [CrossRef]

38. Kato, C.; Kato, D.; Kashiwagi, T.; Nagatani, S. Synthesis, X-ray Crystal Structure, and Photochromism of a Sandwich-Type Mono-Aluminum Complex Composed of Two Tri-Lacunary α-Dawson-Type Polyoxotungstates. *Materials* **2019**, *12*, 2383. [CrossRef]

39. Chen, S.; Xiao, Y.; Xie, W.; Wang, Y.; Hu, Z.; Zhang, W.; Zhao, H. Facile Strategy for Synthesizing Non-Stoichiometric Monoclinic Structured Tungsten Trioxide (WO$_{3-x}$) with Plasma Resonance Absorption and Enhanced Photocatalytic Activity. *Nanomaterials* **2018**, *8*, 553. [CrossRef]

40. Yang, Y.; Zhang, T.; Yan, J.; Fu, L.; Xiang, H.; Cui, Y.; Su, J.; Liu, X. Preparation and Photochromic Behavior of Spiropyran-Containing Fluorinated Polyacrylate Hydrophobic Coatings. *Langmuir* **2018**, *34*, 15812–15819. [CrossRef]

41. Chen, W.; Shen, H.; Zhu, X.; Yao, H.; Wang, W. Preparation and photochromic properties of PEG-400 assisted WO$_3$–TiO$_2$–ZnO composite films. *Ceram. Int.* **2015**, *41 Pt B*, 14008–14012. [CrossRef]

42. Wang, Z.; Ma, Z.; Wang, Y.; Xu, Z.; Luo, Y.; Wei, Y.; Jia, X. A Novel Mechanochromic and Photochromic Polymer Film: When Rhodamine Joins Polyurethane. *Adv. Mater.* **2015**, *27*, 6469–6474. [CrossRef] [PubMed]

43. Kumar, A.; Gupta, A.K.; Devi, M.; Gonsalves, K.E.; Pradeep, C.P. Engineering Multifunctionality in Hybrid Polyoxometalates: Aromatic Sulfonium Octamolybdates as Excellent Photochromic Materials and Self-Separating Catalysts for Epoxidation. *Inorg. Chem.* **2017**, *56*, 10325–10336. [CrossRef] [PubMed]

44. Beard, B.C.; Spellane, P. XPS Evidence of Redox Chemistry between Cold Rolled Steel and Polyaniline. *Chem. Mater.* **1997**, *9*, 1949–1953. [CrossRef]

45. Barthet, C.; Armes, S.P.; Chehimi, M.M.; Bilem, C.; Omastova, M. Surface Characterization of Polyaniline-Coated Polystyrene Latexes. *Langmuir* **1998**, *14*, 5032–5038. [CrossRef]

46. Zhang, Y.; Xi, Q.; Chen, J.; Duan, Y. Theoretical Investigation of the Protonation Mechanism of Doped Polyaniline. *J. Clust. Sci.* **2014**, *25*, 1501–1510. [CrossRef]

47. Li, S.; Macosko, C.; White, H. Electrochemical processing of electrically conductive polymer fibers. *Adv. Mater.* **1993**, *5*, 575–576. [CrossRef]

48. Keggin, J.F.J.N. Structure of the Molecule of 12-Phosphotungstic Acid. *Nature* **1933**, *131*, 908–909. [CrossRef]

49. Gómez-Romero, P.; Lira-Cantú, M. Hybrid organic-inorganic electrodes: The molecular material formed between polypyrrole and the phosphomolybdate anion. *Adv. Mater.* **1997**, *9*, 144–147. [CrossRef]

50. Eguchi, K.; Toyozawa, Y.; Yamazoe, N.; Seiyama, T. An infrared study on the reduction processes of dodecamolybdophosphates. *J. Catal.* **1983**, *83*, 32–41. [CrossRef]

51. Taketa, H.; Katsuki, S.; Eguchi, K.; Seiyama, T.; Yamazoe, N. Electronic structure and redox mechanism of dodecamolybdophosphate. *J. Phys. Chem.* **1986**, *90*, 2959–2962. [CrossRef]

Article

2D Bi$_2$Se$_3$ van der Waals Epitaxy on Mica for Optoelectronics Applications

Shifeng Wang [1,2,*], Yong Li [1,2], Annie Ng [3,*], Qing Hu [4,5], Qianyu Zhou [1], Xin Li [1] and Hao Liu [1]

[1] Department of Physics, Innovation Laboratory of Materials for Energy and Environment Technologies, College of Science, Tibet University, Lhasa 850000, China; xzuliyong@utibet.edu.cn (Y.L.); zhouqianyu@utibet.edu.cn (Q.Z.); lixin@utibet.edu.cn (X.L.); liuhao@utibet.edu.cn (H.L.)

[2] Institute of Oxygen Supply, Center of Tibetan Studies (Everest Research Institute), Tibet University, Lhasa 850000, China

[3] Department of Electrical and Computer Engineering, Nazarbayev University, Nur-Sultan 010000, Kazakhstan

[4] School of Environmental Science and Engineering, Southern University of Science and Technology, Shenzhen 518055, China; huq@sustech.edu.cn

[5] Engineering Innovation Center of Southern University of Science and Technology, Beijing 100083, China

* Correspondence: wsf@utibet.edu.cn (S.W.); annie.ng@nu.edu.kz (A.N.)

Received: 24 July 2020; Accepted: 14 August 2020; Published: 22 August 2020

Abstract: Bi$_2$Se$_3$ possesses a two-dimensional layered rhombohedral crystal structure, where the quintuple layers (QLs) are covalently bonded within the layers but weakly held together by van der Waals forces between the adjacent QLs. It is also pointed out that Bi$_2$Se$_3$ is a topological insulator, making it a promising candidate for a wide range of electronic and optoelectronic applications. In this study, we investigate the growth of high-quality Bi$_2$Se$_3$ thin films on mica by the molecular beam epitaxy technique. The films exhibited a layered structure and highly c-axis-preferred growth orientation with an XRD rocking curve full-width at half-maximum (FWHM) of 0.088°, clearly demonstrating excellent crystallinity for the Bi$_2$Se$_3$ deposited on the mica substrate. The growth mechanism was studied by using an interface model associated with the coincidence site lattice unit (CSLU) developed for van der Waals epitaxies. This high (001) texture favors electron transport in the material. Hall measurements revealed a mobility of 726 cm^2/(Vs) at room temperature and up to 1469 cm^2/(Vs) at 12 K. The results illustrate excellent electron mobility arising from the superior crystallinity of the films with significant implications for applications in conducting electrodes in optoelectronic devices on flexible substrates.

Keywords: van der Waals epitaxy; Bi$_2$Se$_3$; mica; two-dimensional materials; optoelectronics; transparent conductive electrode

---

1. Introduction

Since the discovery of graphene by Geim and Novoselov in 2004 [1], two-dimensional materials have gained renewed interest, because of their exceptional electronic properties, low dangling bond density, and high specific surface areas that are important for optoelectronics, sensing, catalysis, and energy storage applications. Recently, transition metal dichalcogenides (TMDs), such as MoS$_2$, WoS$_2$, and Bi$_2$Se$_3$, have become the research focus. The latter material is of particular interest as a topological insulator and thermoelectric material [2–6]. Layered Bi$_2$Se$_3$ possesses an insulating bulk gap of 0.3–0.35 eV and metallic surface states with a single Dirac cone [7–10], enabling excellent transport properties with a high carrier mobility [11]. In addition, the surface conduction can be significantly boosted and easily tailored in few-layer nanostructures consisting of large surface-to-volume ratios [12–15]. Being analogous to the optoelectronic applications of graphene, a thin layer of Bi_2Se_3 has been proved to be a promising material for broadband and high-performance optoelectronic devices, such as photodetectors, terahertz lasers, and transparent electrodes [3,16]. In particular, transparent conductive electrodes (TCE) play a vitally important role in modern optoelectronic devices, including displays, light-emitting diodes, touch screens, and photovoltaics. Doped metal oxides, such as indium tin oxides (ITO), are predominantly utilized for such applications, owing to their high electrical conductivity and good optical transparency in the visible light region. However, due to the limited reserves of indium, the optoelectronic applications of ITO are severely restricted. Therefore, next-generation transparent electrodes consisting of the emerging 2D materials are highly desirable.

In this work, mica was used as the substrate, owing to its 2D layered structural nature with relatively weak interlayer interaction, which favors the 2D/2D van der Waals epitaxial (vdWE) growth. Furthermore, mica presents an atomically flat, chemically saturated, mechanically flexible, electrically insulating, and optically transparent surface, which is considered an ideal substrate for flexible transparent conductive electrodes [17,18]. Then, molecular beam epitaxial (MBE) growth of high-quality Bi_2Se_3 thin films has been realized on mica substrates, exhibiting a high (001) texture with an XRD rocking curve full-width at half-maximum (FWHM) of 0.088° and a mobility of 726 $cm^2/(Vs)$ at room temperature and up to 1469 $cm^2/(Vs)$ at 12 K. An optoelectronic device consisting of the Bi_2Se_3/mica structure as the 2D flexible TCE and PTB7:PC_{71}BM as the light absorber was fabricated, producing a photocurrent of 2.75 mA/cm^2 and an open-circuit voltage of 697 mV, demonstrating the feasibility of Bi_2Se_3 for optoelectronic applications.

2. Materials and Methods

Bi_2Se_3 was synthesized by the MBE technique on the mica 2D substrate. In addition, 5N purity Bi_2Se_3 pieces (purchased from American Elements, Los Angeles, CA, USA) were used as the evaporation source, loaded in a K-cell and heated up to 470 °C for the film deposition. The substrate temperature was maintained at ~340 °C, and the film growth rate was evaluated to be around 4.5 nm/min by ex-situ measurement of the film thickness with the help of an Alpha Step 500 profiler. The Bi_2Se_3/mica 2D material-based transparent conductive electrode was then covered by a thin layer of V_2O_5 as the hole-transporting layer by thermal evaporation, followed by spin-coating the PTB7:PC_{71}BM absorber. Consequently, a calcium/aluminum electrode was evaporated onto the layers, forming an optoelectronic device.

The microstructure and crystal phase of the prepared Bi_2Se_3 thin films were characterized by high-resolution XRD and recorded on a Rigaku Smartlab 9 kW X-ray diffractometer (Rigaku Corporation, Tokyo, Japan), equipped with a Cu-Kα1 radiation source (λ = 1.5406 Å) and a two-crystal Ge (220) two-bounce hybrid monochromator (Rigaku Corporation, Tokyo, Japan). The film surface morphology was examined by AFM, which was performed on a Bruker NanoScope 8 (Billerica, MA, USA) in tapping mode using a silicon cantilever with a tip radius of less than 10 nm and resonance frequency of 341 kHz. Scanning electron microscopy (SEM) was also employed to investigate the morphology and the microstructure of the films, using a JEOL 6490 (JEOL Ltd., Tokyo, Japan) microscope by applying an accelerating voltage of 20 kV. Transmission electron microscopy (TEM) was employed to characterize the film microstructure, which was recorded through a JEM-2100F (field emission) scanning transmission electron microscope (JEOL Ltd., Tokyo, Japan) equipped with an Oxford INCA x-sight EDS Si(Li) detector (Oxford Instruments, Abingdon, UK)at an acceleration voltage of 200 kV. The film electrical properties were determined by Hall measurements using the four-probe van der Pauw method, performed on a Bio-Rad 5500 Hall system (Hercules, CA, USA) at room temperature using a permanent magnet with a magnetic field of 0.32 T, and conducted in a cryostat equipped with a CTI-Cryogenics 22 Refrigerator (Helix Technology Corporation, Mansfield, MA, USA) from room temperature to ~12 K under an electromagnetic field of 0.32 T. The work function of the Bi_2Se_3 thin film was deduced from ultraviolet photoelectron spectroscopy (UPS) measurements, which were conducted in ultra-high vacuum at 5×10^{-8} mbar, by irradiating with 21.22 eV photons (He I line).

The photoelectric response of the prepared device was recorded on an Agilent B1500A semiconductor device parameter analyzer (Santa Clara, CA, USA). The photocurrent was collected under a simulated Air Mass 1.5 Global (AM 1.5 G) illumination, which was provided by a Newport Oriel Sol3A Solar Simulator (Irvine, CA, USA) with a 100 mW/cm^2 radiance intensity.

3. Results and Discussion

Being a 2D layered material, high-quality Bi_2Se_3 can be deposited on various substrates, especially on 2D substrates, in spite of the large lattice mismatch. Figure 1 shows the XRD patterns of Bi_2Se_3 on the mica substrate. As indicated by the red crosses in Figure 1a, the predominant diffractions originated from the Bi_2Se_3 {003} family planes, suggesting that the Bi_2Se_3 film stacks vertically on the mica substrate along a highly c-axis-oriented direction. The presence of the Bi_2Se_3 {015} diffractions, as labeled by the inverted triangles, indicated that a certain quantity of crystals with the Bi_2Se_3 (015) planes parallel to the mica (001) surface simultaneously emerged. The inset in Figure 1a gives the XRD rocking curve, in which a full-width at half-maximum (FWHM) of 0.088° (~317 arc seconds) was attained, implying an excellent Bi_2Se_3 crystallinity and a perfect crystal alignment of vertical growth. Figure 1b shows the in-plane phi scans conducted on the Bi_2Se_3 (10$\overline{1}$0) plane and the mica (114) plane. Six sharp peaks separated by 60° showed up in the Bi_2Se_3 (10$\overline{1}$0) diffraction pattern, consistent with reports [19–23]. The observed six-fold symmetry, instead of three peaks expected for the trigonal system, was ascribed to the twin domains either stacking in the sequence -[ABCAB]- or -[CABCA]- [20]. Azimuthal diffractions of mica (114) and (1$\overline{1}$4) planes were superimposed on the same plot, from which the mica <100> and <010> directions on the mica (001) surface can be derived, as indicated by the blue dotted lines in Figure 1b [17]. The in-plane crystallographic relationship between Bi_2Se_3 and the mica substrate is then deduced as follows: Bi_2Se_3 (001) // mica (001) and Bi_2Se_3 <100> // mica <100>.

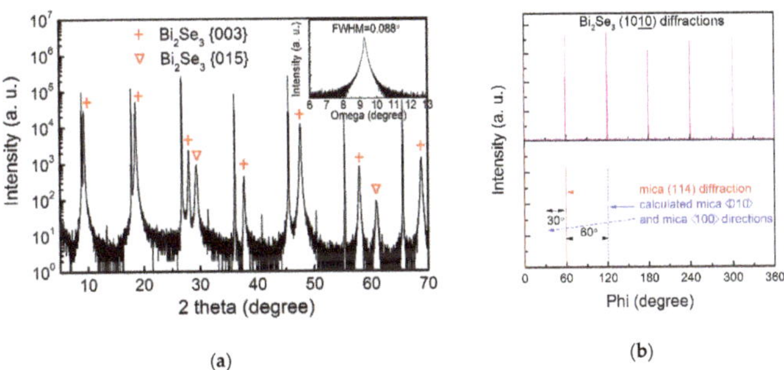

Figure 1. The XRD diffraction patterns of (**a**) 2θ-ω scan and (**b**) in-plane phi scan of Bi_2Se_3 on mica substrate. The inset in (**a**) gives the rocking curve.

As shown in the AFM top view image in Figure 2a, rhombohedral-shaped planes were observed, suggesting the highly c-oriented van der Waals (vdW) epitaxial growth. At the same time, fence-like vertical planes emerged, as shown clearly in the AFM perspective view in Figure 2b. These fence-like planes are parallel to the edge of the rhombohedral planes and are oriented at 60° and 120° relative to each other.

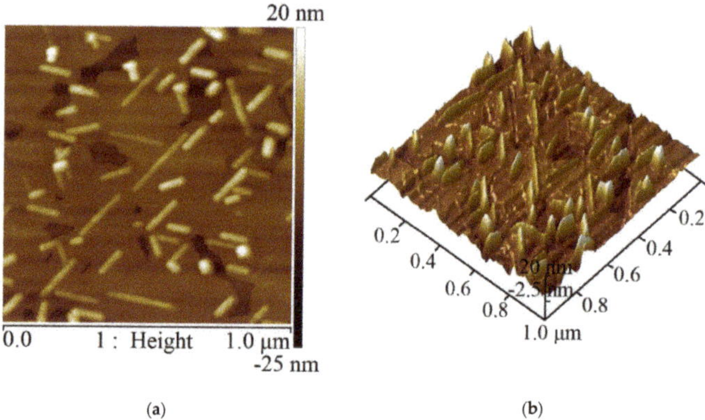

(a) (b)

Figure 2. The AFM images of (**a**) top view and (**b**) perspective view of Bi_2Se_3 on mica substrate.

High-resolution TEM measurements were conducted to reveal the Bi_2Se_3 microstructure. The hexagonal atomic arrangement, as shown in Figure 3a, confirms the c-axis-oriented growth direction of the film. The average distance between the planes is found to be 3.57 Å, as indicated by the red arrows, signifying the lattice constant as $a = 4.12$ Å. Figure 3c shows the atomic arrangement within the vertical planes. The average spacing between the planes indicated by the white rectangle was found to be 2.20 Å, in good agreement with the calculated value of 2.21 Å between planes along the Bi_2Se_3 $<5, 10, \bar{2}>$ direction. The distance of 2.06 nm between the five atoms along the longer side of the white rectangle coincided nicely with the lattice along Bi_2Se_3 $<500>$, in which $5a = 20.61$ Å. These results confirm that the fence-like vertical planes are those that grew along the Bi_2Se_3 (015) normal. The crystallographic relationship of Bi_2Se_3 and the mica (001) surface at these different growth orientations is illustrated by using the interface models, as shown in Figure 3b,d,e.

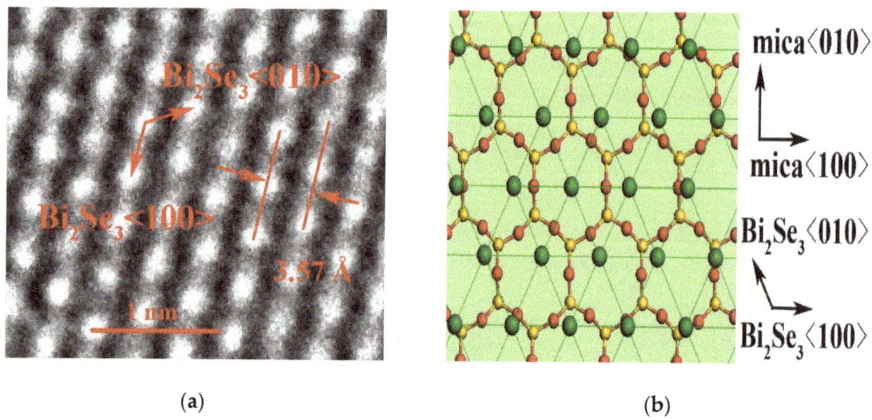

(a) (b)

Figure 3. *Cont.*

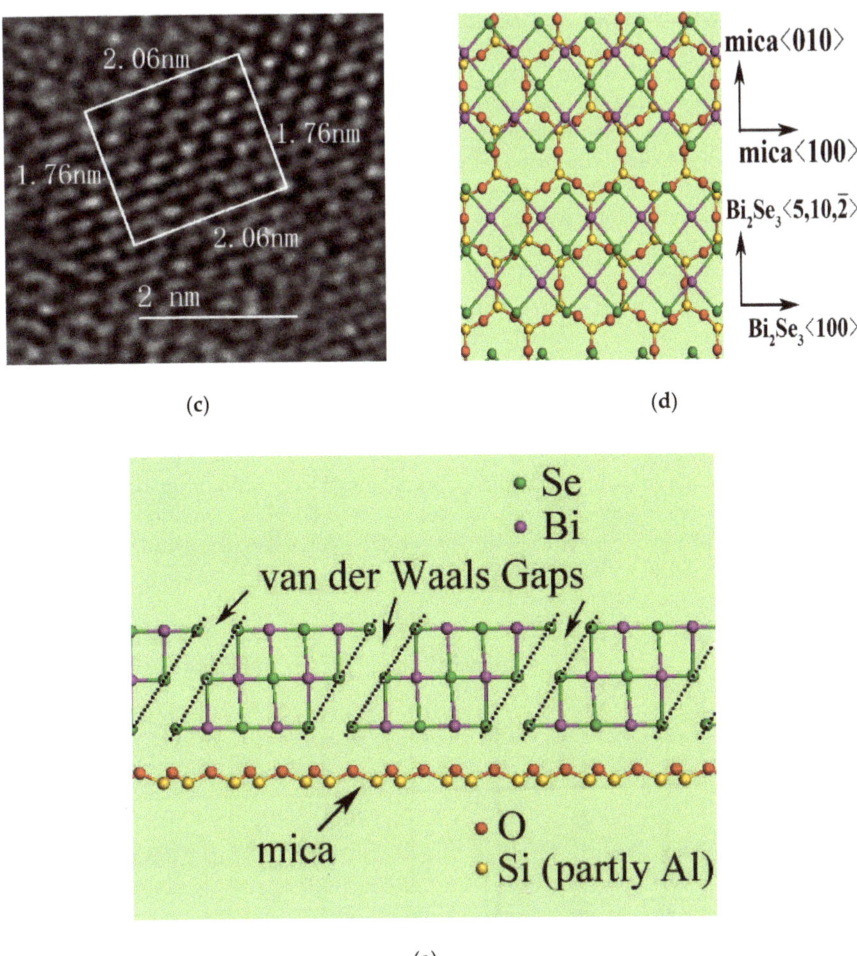

Figure 3. The TEM images and the interface models of Bi_2Se_3 on mica. (**a,b**) show the Bi_2Se_3 (001) // mica (001) planes; (**c**)–(**e**) give the Bi_2Se_3 (015) // mica (001) planes. In the interface models, O and Si (partly Al) atoms in mica are colored respectively by red and yellow, while Bi and Se atoms in Bi_2Se_3 are colored respectively by violet and green.

For van der Waals epitaxial growth with a large lattice misfit between the epilayer and substrate, the matching condition can be evaluated via the coincidence lattice mismatch [17], δ, which is defined as $(md_{epi}-nd_{sub})/nd_{sub}$, where d_{epi} and d_{sub} are the atomic distances of the epilayer and substrate, respectively. Lattice coincidence occurs when d_{epi}/d_{sub} = n/m, where n and m are positive integers and n/m is the smallest non-reducible integral ratio [24,25]. Then, the concept of coincidence site lattice unit (CSLU) can be established as CSLU = $n_x \times n_y$, where x and y denote the directions along the substrate lattice vectors *a* and *b*, respectively. It can be concluded that the smaller the CSLU, the better the lattice match between the epilayer and the substrate, as a small CSLU represents a larger number of coincidence sites per unit area [17]. For the case of planar growth, as shown in Figure 3b, the CSLU is 4 × 2 = 8. By contrast, for the vertical growth as illustrated in Figure 3d,e, the CSLUs are calculated to be 4 × 1 = 4 and 4 × 5 = 20 regardless of and counting the period of van der Waals gaps, respectively. It is apparent that the growth direction along the Bi_2Se_3 <5, 10, 2̄> direction will develop a larger CSLU size when involving more van der Waals gaps. Therefore, the growth with a smaller CSLU of 4×1 will be favored, leading to the preferential growth directions along Bi_2Se_3 <100> and the Bi_2Se_3 (015) normal and producing the fence-like planes.

The fence-like planes make the film surface uneven, e.g., a film 90 nm thick has a root-mean-square (rms) roughness of 17.8 nm on the scale of 5×5 μm^2 (see Figure S1 in the Supplementary Materials). This would bring harm to the fabrication and performance of electronic devices. To eliminate these fence-like vertical planes, a thin amorphous Bi_2Se_3 buffer layer was prepared at room temperature, prior to the Bi_2Se_3 growth at high substrate temperature. Figure 4 shows the XRD diffraction patterns of Bi_2Se_3 with an amorphous buffer layer on the mica substrate. It was clearly observed that the Bi_2Se_3 {015} diffractions, corresponding to the fence-like vertical planes, totally disappeared. As displayed in the AFM images in Figure 5, the surface morphology of the prepared Bi_2Se_3 film of ~90 nm with a buffer layer of ~5 nm is greatly improved, reflected by the reduction in the rms roughness to 4.9 nm over an area of 5×5 μm^2. Fence-like planes have totally disappeared, and triangular terraces are clearly observed, signifying that only the planar growth along Bi_2Se_3 <001> occurs. The buffer layer presenting as an amorphous state significantly weakens the interaction between the first stacking layer of Bi_2Se_3 and the mica surface, favoring the van der Waals epitaxial growth on top of it. However, it is also because of the much weaker interaction; the lateral growth alignment of van der Waals epitaxies will be weakened as well as reflected by the much broader peaks of the azimuthal phi scan with an average FWHM of 13.8 (see Figure S2 in the Supplementary Materials), which may contribute a large number of grain boundaries (GBs) in the film.

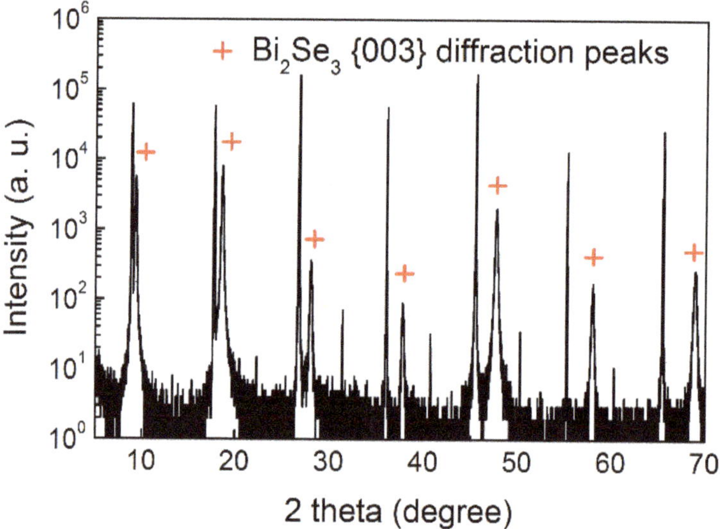

Figure 4. XRD diffraction pattern of Bi_2Se_3 on mica with buffer layer. The Bi_2Se_3 {003} family of planes are labeled by red crosses, while other peaks originate from the mica substrate.

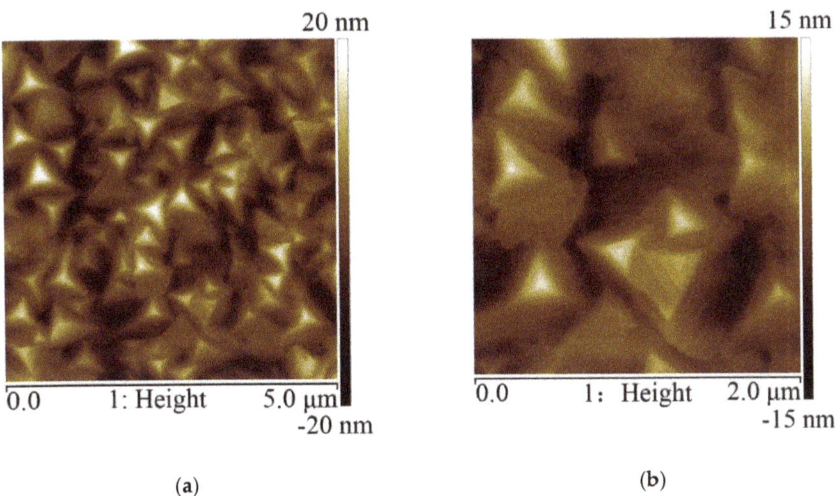

(**a**)　　　　　　　　　　　(**b**)

Figure 5. AFM images of Bi_2Se_3 on mica with an amorphous buffer layer. (**a**) is on a 5×5 μm^2 scale; (**b**) is on a 2×2 μm^2 scale.

The transport performances of the carrier density and the mobility dependence on temperature for the Bi_2Se_3 thin film were investigated by employing low-temperature Hall measurement. Figure 6a,b show the electrical properties of 90 nm thick Bi_2Se_3 films on mica without and with a buffer layer, respectively. When grown directly on mica, the Bi_2Se_3 Hall mobility increased from 697 cm^2/(Vs) at room temperature to 1761 cm^2/(Vs) at ~12 K, while the sheet electron density remained nearly constant at about 1.0×10^{14} cm^{-2}. By applying an amorphous Bi_2Se_3 buffer layer, the mobility at room temperature was slightly boosted to 726 cm^2/(Vs) with a reduced sheet carrier density of ~5.9×10^{13} cm^{-2}. The mobility increased at a near linear trend from 726 to 1469 cm^2/(Vs) as the temperature fell. However, the increase in the mobility of Bi_2Se_3 with the buffer layer lagged that without the buffer layer when the temperature decreased below ~115 K. The fence-like planes in the Bi_2Se_3 film grown directly on mica could contribute to the higher carrier density via introducing a larger surface area. However, the lower the mobility in the low-temperature range of Bi_2Se_3 with the buffer layer can be ascribed to the poor lateral alignment of crystals as discussed above. The weak interaction between the amorphous buffer layer and the Bi_2Se_3 crystals on top, although leading to removal of the fence-like planes, also weakened the lateral growth alignment, which produced a number of GBs mirrored by the misoriented sub-micrometer-sized domains in the AFM images in Figure 5 and the broad FWHM of the XRD in-plane phi scan in Figure S2. According to [26,27], the GBs play an important role in the mobility decline of 2D MoS_2 by means of interdomain scattering. In this study, the interdomain scattering arising from the misorientation of Bi_2Se_3 crystals on the buffer layer may contribute a lot to the slow rise in mobility with the decrease in temperature in the low-temperature range. By contrast, the lattice scattering may dominate the mobility variation in the Bi_2Se_3 crystals directly grown on mica, as they are perfectly aligned and connected laterally. The electrical transport property of 2D materials requires further investigation.

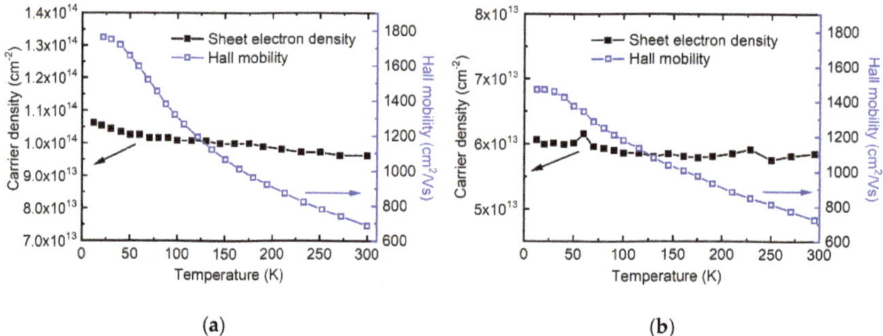

Figure 6. Sheet carrier density and mobility as a function of temperature for the Bi_2Se_3 film on mica substrate. (**a**) is without and (**b**) is with a buffer layer.

An optoelectronic device comprising the Bi_2Se_3/mica structure and organic absorber was fabricated, as shown in Figure 7a. The energy band alignment between the device components is illustrated in Figure 7b, in which the work function of the Bi_2Se_3 thin film is measured as 4.43 eV by using UPS (see Figure S3 in the Supplementary Materials). Figure 7c displays the *I–V* characteristics of the device, in which a short-circuit photocurrent of 2.75 mA/cm^2 and an open-circuit voltage of 697 mV are achieved. It is noteworthy that the values of photoelectric parameters are relatively lower compared to the conventional PTB7:PC$_{71}$BM-based solar cells, for which TCEs such as ITO coated on glasses are commonly used as the substrates instead of Bi_2Se_3 deposited on the mica. The absorption range of Bi_2Se_3 covers the broad visible light region due to the small bandgap of Bi_2Se_3, causing a reduced amount of light reaching the active layer (PTB7:PC$_{71}$BM) of the device to generate photocarriers. To address this issue, the thickness of Bi_2Se_3 should be further optimized or the opaque aluminum electrode should be replaced with transparent electrodes so that the light can be transmitted from both sides. The interface quality of the devices should also be investigated as Bi_2Se_3 coated on flexible mica will exhibit a different strain on the functional layers in the optoelectronic devices compared to the devices using the conventional rigid ITO-coated glass substrates. The presence of interfacial defects leads to carrier recombination, resulting in a poor fill factor of devices. Some interfacial passivation strategies [28,29] can be performed to relieve their negative impact on the device performance. Nevertheless, the demonstration of using Bi_2Se_3 in PTB7:PC$_{71}$BM-based optoelectronic devices in this work aims to pave the way to explore 2D transition metal dichalcogenides and exhibit their potential in the optoelectronic applications.

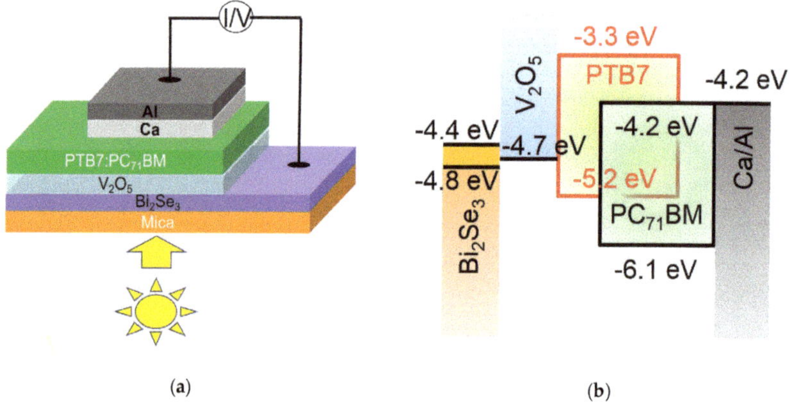

(**a**)

(**b**)

Figure 7. *Cont.*

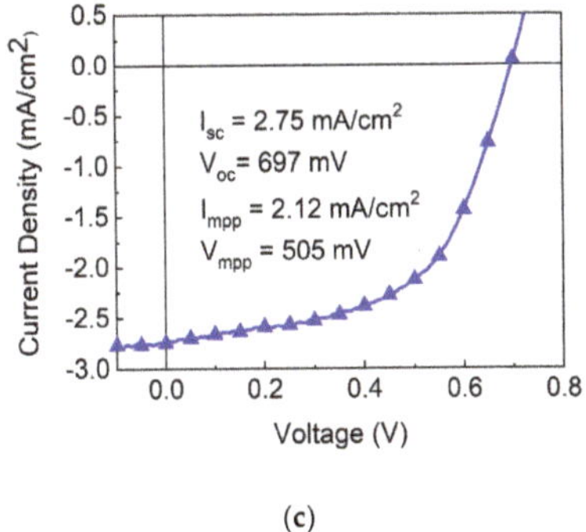

(**c**)

Figure 7. The Bi_2Se_3-based organic optoelectronic device. (**a**) Schematic diagram of the device structure; (**b**) energy band alignment; (**c**) *I–V* characteristics.

4. Conclusions

High-quality Bi_2Se_3 thin films have been prepared on a mica substrate via vdW epitaxial growth. The film exhibited a highly c-axis-oriented growth with an extraordinary crystallinity, reflected by a narrow XRD rocking curve FWHM of ~317 arc seconds. At the same time, fence-like planes with Bi_2Se_3 (015) parallel to the mica (001) surface emerged, which was explained by using the interface model associated with the CSLU concept developed especially for vdWE. To eliminate the fence-like planes, an amorphous Bi_2Se_3 buffer layer prepared at the substrate temperature of room temperature was applied, prior to the Bi_2Se_3 crystal growth. As a result, only the Bi_2Se_3 (001) growth direction with vdW gaps parallel to the substrate showed up, with a smoother surface than that without the amorphous buffer layer. However, this suppression in the fence-like planes sacrifices to a certain degree the crystal lateral growth alignment, mirrored by a broadening phi scan peak FWHM. A high Hall mobility of 1761 cm^2/(Vs) for Bi_2Se_3 on mica was obtained at ~12 K, and 697 cm^2/(Vs) at room temperature with a nearly constant sheet electron density of ~1.0×10^{14} cm^{-2}. By applying an amorphous Bi_2Se_3 buffer layer, the carrier density was reduced to about half that without the buffer layer. The mobility at room temperature was boosted to 726 cm^2/(Vs), and linearly increased to 1469 cm^2/(Vs) as the temperature dropped to 12 K. An optoelectronic device consisting of a Bi_2Se_3/mica TCE, organic absorber, and Ca/Al metal back electrode was designed and prepared, generating a short-circuit photocurrent of 2.75 mA/cm^2 and an open-circuit voltage of 697 mV under one sun irradiation. The results demonstrate the great potential of 2D Bi_2Se_3 along with mica for flexible optoelectronic applications.

Supplementary Materials: The following are available online at http://www.mdpi.com/2079-4991/10/9/1653/s1, Figure S1: AFM images of 90 nm-thick Bi_2Se_3 film with an rms roughness of 17.8 nm, Figure S2: In-plane phi scan of Bi_2Se_3 on mica with buffer layer, Figure S3: UPS spectrum of Bi_2Se_3.

Author Contributions: Conceptualization, S.W. and A.N.; methodology, S.W., A.N. and Q.H.; validation, Y.L.; formal analysis, Y.L.; investigation, Y.L., Q.Z., X.L. and H.L.; resources, Q.H.; data curation, S.W. and A.N.; writing—original draft preparation, S.W.; writing—review and editing, A.N.; supervision, S.W.; project administration, S.W. and A.N.; funding acquisition, S.W. All authors have read and agreed to the published version of the manuscript.

Funding: This research was funded by Wuhan University of Technology-Tibet University Joint Innovation Fund, grant number LZJ2020003, and the Reform and Development Funds for Local Region Universities from China Government in 2020, grant number ZCKJ 2020-11.

Conflicts of Interest: The authors declare no conflict of interest.

References

1. Novoselov, K.S.; Geim, A.K.; Morozov, S.V.; Jiang, D.; Zhang, Y.; Dubonos, S.V.; Grigorieva, I.V.; Firsov, A.A. Electric Field Effect in Atomically Thin Carbon Films. *Science* **2004**, *306*, 666–669. [CrossRef] [PubMed]
2. Wang, F.; Li, L.; Huang, W.; Li, L.; Jin, B.; Li, H.; Zhai, T. Submillimeter 2D Bi_2Se_3 Flakes toward High-Performance Infrared Photodetection at Optical Communication Wavelength. *Adv. Funct. Mater.* **2018**, *28*, 1802707. [CrossRef]
3. Zhang, X.; Wang, J.; Zhang, S.C. Topological insulators for high-performance terahertz to infrared applications. *Phys. Rev. B* **2010**, *82*, 245107. [CrossRef]
4. Sharma, A.; Bhattacharyya, B.; Srivastava, A.K.; Senguttuvan, T.D.; Husale, S. High performance broadband photodetector using fabricated nanowires of bismuth selenide. *Sci. Rep.* **2016**, *6*, 19138. [CrossRef] [PubMed]
5. Wang, Q.H.; Kalantar-Zadeh, K.; Kis, A.; Coleman, J.N.; Strano, M.S. Electronics and Optoelectronics of Two-Dimensional Transition Metal Dichalcogenides. *Nat. Nanotechnol.* **2012**, *7*, 699–712. [CrossRef]
6. Zhang, Q.; Zhang, Z.; Zhu, Z.; Schwingenschlögl, U.; Cui, Y. Exotic Topological Insulator States and Topological Phase Transitions in Sb_2Se_3–Bi_2Se_3 Heterostructures. *ACS Nano* **2012**, *6*, 2345–2352. [CrossRef]
7. Zhang, H.; Liu, C.X.; Qi, X.L.; Dai, X.; Fang, Z.; Zhang, S.C. Topological insulators in Bi_2Se_3, Bi_2Te_3 and Sb_2Te_3 with a single Dirac cone on the surface. *Nat. Phys.* **2009**, *5*, 438–442. [CrossRef]
8. Moore, J. Topological Insulators: The Next Generation. *Nat. Phys.* **2009**, *5*, 378. [CrossRef]
9. Zhang, Y.; He, K.; Chang, C.-Z.; Song, C.-L.; Wang, L.-L.; Chen, X.; Jia, J.-F.; Fang, Z.; Dai, X.; Shan, W.-Y.; et al. Crossover of the three-dimensional topological insulator Bi_2Se_3 to the two-dimensional limit. *Nat. Phys.* **2010**, *6*, 584–588. [CrossRef]
10. Pejova, B.; Grozdanov, I.; Tanuševski, A. Opitcal and thermal band gap energy of chemically deposited bismuth (III) selenide thin films. *Mater. Chem. Phys.* **2004**, *83*, 245–249. [CrossRef]
11. Kou, X.F.; He, L.; Xiu, F.X.; Lang, M.R.; Liao, Z.M.; Wang, Y.; Fedorov, A.V.; Yu, X.X.; Tang, J.S.; Huang, G.; et al. Epitaxial growth of high mobility Bi_2Se_3 thin films on CdS. *Appl. Phys. Lett.* **2011**, *98*, 242102. [CrossRef]
12. Peng, H.; Lai, K.; Kong, D.; Meister, S.; Chen, Y.; Qi, X.L.; Zhang, S.C.; Shen, Z.X.; Cui, Y. Aharonov-Bohm interference in topological insulator nanoribbons. *Nat. Mater.* **2010**, *9*, 225–229. [CrossRef] [PubMed]
13. Xiu, F.; He, L.; Wang, Y.; Cheng, L.; Chang, L.T.; Lang, M.; Huang, G.; Kou, X.; Zhou, Y.; Jiang, X.; et al. Manipulating surface states in topological insulator nanoribbons. *Nat. Nanotechol.* **2011**, *6*, 216–221. [CrossRef] [PubMed]
14. Steinberg, H.; Gardner, D.R.; Lee, Y.S.; Jarillo-Herrero, P. Surface state transport and ambipolar electric field effect in Bi_2Se_3 nanodevices. *Nano Lett.* **2010**, *10*, 5032–5036. [CrossRef] [PubMed]
15. Teweldebrhan, D.; Goyal, V.; Balandin, A.A. Exfoliation and characterization of bismuth telluride atomic quintuples and quasi-two-dimensional crystals. *Nano Lett.* **2010**, *10*, 1209–1218. [CrossRef]
16. Peng, H.L.; Dang, W.H.; Cao, J.; Chen, Y.L.; Wu, D.; Zheng, W.S.; Li, H.; Shen, Z.X.; Liu, Z.F. Topological insulator nanostructures for near-infrared transparent flexible electrodes. *Nat. Chem.* **2012**, *4*, 281–286. [CrossRef]
17. Wang, S.F.; Fong, W.K.; Wang, W.; Surya, C. Growth of highly textured SnS on mica using an SnSe buffer layer. *Thin Solid Films* **2014**, *564*, 206–212. [CrossRef]
18. Wang, S.F.; Wang, W.; Fong, W.K.; Yu, Y.; Surya, C. Tin Compensation for the SnS Based Optoelectronic Devices. *Sci. Rep.* **2017**, *7*, 39704. [CrossRef]
19. Brom, J.E.; Ke, Y.; Du, R.; Won, D.; Weng, X.; Andre, K.; Gagnon, J.C.; Mohney, S.E.; Li, Q.; Chen, K.; et al. Structural and electrical properties of epitaxial Bi_2Se_3 thin films grown by hybrid physical-chemical vapor deposition. *Appl. Phys. Lett.* **2012**, *100*, 162110. [CrossRef]
20. Tarakina, N.V.; Schreyeck, S.; Borzenko, T.; Grauer, S.; Schumacher, C.; Karczewski, G.; Gould, C.; Brunner, K.; Buhman, H.; Molenkamp, L.W. Microstructural characterisation of Bi_2Se_3 thin films. *J. Phys. Conf. Ser.* **2013**, *471*, 012043. [CrossRef]
21. Liu, Y.; Weinert, M.; Li, L. Spiral Growth without Dislocations: Molecular Beam Epitaxy of the Topological Insulator Bi_2Se_3 on Epitaxial Graphene/SiC(0001). *Phys. Rev. Lett.* **2012**, *108*, 115501. [CrossRef] [PubMed]
22. Zhuang, A.; Li, J.J.; Wang, Y.C.; Wen, X.; Lin, Y.; Xiang, B.; Wang, X.P.; Zeng, J. Screw-Dislocation-Driven Bidirectional Spiral Growth of Bi_2Se_3 Nanoplates. *Angew. Chem. Int. Ed.* **2014**, *53*, 6425–6429. [CrossRef] [PubMed]

23. Tabor, P.; Keenan, C.; Urazdhin, S.; Lederman, D. Molecular beam epitaxy and characterization of thin Bi_2Se_3 films on Al_2O_3 (110). *Appl. Phys. Lett.* **2011**, *99*, 013111. [CrossRef]

24. Sun, C.J.; Kung, P.; Saxler, A.; Ohsato, H.; Haritos, K.; Razeghi, M. A Crystallographic model of (00·1) Aluminum Nitride Epitaxial Thin Film Growth on (00·1) Sapphire Substrate. *J. Appl. Phys.* **1994**, *75*, 3964. [CrossRef]

25. Kato, T.; Kung, P.; Saxler, A.; Sun, C.J.; Ohsato, H.; Razeghi, M.; Okuda, T. Simultaneous Growth of Two Differently Oriented GaN epilayers on (11·0) Sapphire II. A Growth Model of (00·1) and (10·0) GaN. *J. Cryst. Growth* **1998**, *183*, 131. [CrossRef]

26. Ly, T.H.; Perello, D.J.; Zhao, J.; Deng, Q.; Kim, H.; Han, G.H.; Chae, S.H.; Jeong, H.Y.; Lee, Y.H. Misorientation-angle-dependent Electrical Transport across Molybdenum Disulfide Grain Boundaries. *Nat. Commun.* **2016**, *7*, 10426. [CrossRef]

27. Giannazzo, F.; Bosi, M.; Fabbri, F.; Schiliro, E.; Greco, G.; Roccaforte, F. Direct Probing of Grain Boundary Resistance in Chemical Vapor Deposition-Grown Monolayer MoS_2 by Conductive Atomic Microscopy. *Phys. Status Solidi RRL* **2019**, *14*, 1900393. [CrossRef]

28. Aidarkhanov, D.; Ren, Z.; Lim, C.-K.; Yelzhanova, Z.; Nigmetova, G.; Taltanova, G.; Baptayev, B.; Liu, F.; Cheung, S.H.; Balanay, M.; et al. Passivation engineering for hysteresis-free mixed perovskite solar cells. *Sol. Energy Mater. Sol. Cells.* **2020**, *215*, 110648. [CrossRef]

29. Walker, B.; Choi, H.; Kim, J.Y. Interfacial engineering for highly efficient organic solar cells. *Curr. Appl. Phys.* **2017**, *17*, 370–391. [CrossRef]

MDPI

St. Alban-Anlage 66

4052 Basel

Switzerland

Tel. +41 61 683 77 34

Fax +41 61 302 89 18

www.mdpi.com

Nanomaterials Editorial Office

E-mail: nanomaterials@mdpi.com

www.mdpi.com/journal/nanomaterials